Methods in Molecular Biology

Series Editor
John M. Walker
School of Life and Medical Sciences
University of Hertfordshire Hatfield,
Hertfordshire, UK

For further volumes:
http://www.springer.com/series/7651

For over 35 years, biological scientists have come to rely on the research protocols and methodologies in the critically acclaimed *Methods in Molecular Biology* series. The series was the first to introduce the step-by-step protocols approach that has become the standard in all biomedical protocol publishing. Each protocol is provided in readily-reproducible step-by-step fashion, opening with an introductory overview, a list of the materials and reagents needed to complete the experiment, and followed by a detailed procedure that is supported with a helpful notes section offering tips and tricks of the trade as well as troubleshooting advice. These hallmark features were introduced by series editor Dr. John Walker and constitute the key ingredient in each and every volume of the *Methods in Molecular Biology* series. Tested and trusted, comprehensive and reliable, all protocols from the series are indexed in PubMed.

Microbial Steroids

Methods and Protocols

Edited by

Carlos Barreiro

Área de Bioquímica y Biología Molecular, Departamento de Biología Molecular, Facultad de Veterinaria, Universidad de León, León, Spain; Department of Biotechnology, Curia Spain, León, Spain

José-Luis Barredo

Department of Biotechnology, Curia Spain, León, Spain

Humana Press

Editors
Carlos Barreiro
Área de Bioquímica y Biología Molecular, Departamento de Biología Molecular, Facultad de Veterinaria
Universidad de León
León, Spain

Department of Biotechnology
Curia Spain
León, Spain

José-Luis Barredo
Department of Biotechnology
Curia Spain
León, Spain

ISSN 1064-3745 ISSN 1940-6029 (electronic)
Methods in Molecular Biology
ISBN 978-1-0716-3384-7 ISBN 978-1-0716-3385-4 (eBook)
https://doi.org/10.1007/978-1-0716-3385-4

Cover Illustration Caption: 'Mycobacterium neoaurum' NRRL B-3805, an androstenedione (AD) producer bacterium used as workhorse of the European Union projects: MySterI (EIB.12.010) (PCIN-2013-024-C02-01) and Syntheroids (Project ID: PCI2018-093066). Author: Carlos Barreiro.

This Humana imprint is published by the registered company Springer Science+Business Media, LLC, part of Springer Nature.
The registered company address is: 1 New York Plaza, New York, NY 10004, U.S.A.

Preface

Sterols are essential components of cell membranes that regulate membrane fluidity and permeability in eukaryotic organisms for their stability, cell growth, and proliferation, and are precursors of bile acids and hormonal steroids. Animals generally synthesize cholesterol, whereas fungi produce ergosterol and plants show an array of sterols with β-sitosterol and campesterol as the most common ones.

Steroids occur widely in living systems, with over 250 sterols and related compounds reported in plants, insects, vertebrates, and lower eukaryotes such as yeasts. They comprise compounds of vital importance to life, including cholesterol, bile acids, sex hormones, vitamin D, corticoid hormones, cardiac aglycones, antibiotics, and insect molting hormones. Steroid-based drugs have a wide range of therapeutic purposes, such as anti-inflammatory, immunosuppressive, progestational, diuretic, anabolic and contraceptive, as well as other applications. Steroid drugs fall into two categories: anabolic and corticosteroids. Anabolic steroids are involved in the treatment of muscle loss or late puberty, while corticosteroids are used as anti-inflammatory drugs or allergic treatment. During the COVID-19 pandemic, the corticosteroid methylprednisolone was reported to have positive effects, and more recently, another corticosteroid, dexamethasone, has been found to reduce the death by roughly one-third of the patients on ventilators. As a result, about 300 approved steroid drugs exist to date, and the global market for steroid-containing APIs reaches 1 million tons annually. As a result, the global corticosteroids market was in the range of $5 billion USD in 2022 and is expected to reach $5.9–6.3 billion USD in the period 2026–2027 increasing up to $8.19 billion USD by 2030.

Scientific research on steroid chemistry began in the early twentieth century and was encouraged in the 1950s, with the discovery of the pharmacological effects of cortisol and progesterone. Their production is being done by chemical synthesis, but new pathways based on the replacement of some of the chemical steps by microbial bioconversions is allowing us, in many cases, to reach more competitive and robust processes.

One of the major raw materials for the steroid industry is the natural steroid sapogenin called diosgenin. The conversion of diosgenin to valuable steroids is done by a well-established chemical route, despite presenting several shortcomings such as higher costs, multistep syntheses, waste of land resources, and exhaustion of wild plant resources. Alternative starting materials for the steroid industry are the natural sterols, e.g., phytosterols, which are by-products of the oil refining process or the cellulose production processes.

Microbial bioconversion of phytosterols into steroid intermediates remains a focus of research in the field of steroids. Growing numbers of microbial biotransformations of steroids have been reported, with an emphasis mainly on steroid hydroxylation, Δ1-dehydrogenation, and sterol side-chain cleavage. Many of these biotransformation reactions, in combination with chemical synthesis, enabled the production of large quantities of commercial steroids, which are of special interest for the industry.

This book is intended to provide practical experimental laboratory procedures for a wide range of steroid bioconversions. Although not an exhaustive treatise, it provides a detailed "step-by-step" description of the most recent developments in these biotechnological processes. The detailed protocols are cross-referenced in the Notes section, providing special details, minor problems, troubleshooting, and safety comments that may not normally appear in journal articles easing the technical understanding.

The lead chapter of this volume is an overview on the current trends and perspectives of the microbial steroid bioconversions, followed by four chapters on microbial screening and synthetic biology applied to microorganisms able to catabolize sterols. The next five chapters show comprehensive experimental methods on strain characterization, including omics and biochemical analyses. Following that section, methods of fermentation and biocatalysis for steroids production are shown, including bioconversion of phytosterols to produce androstenedione (AD) and 24-norchol-4-ene-3,22-dione, 7- and 11-hydroxylation of steroids (e.g., from DHEA to 7α-OH-DHEA), and the use of a thermostable enzyme from a microorganism adapted to grow at temperatures above 50 °C. The last chapter presents an example of the medical use of glucocorticoids in cancer patients.

This book has been written by outstanding experts in their field and provides a reference source for laboratory and industrial professionals, as well as for students in a number of biological disciplines (biotechnology, microbiology, genetics, omics, molecular biology, etc.).

We are indebted to the authors who, in spite of their professional activities, agreed to participate in this book, to Dr. John Walker, Series Editor, for his encouragement and advice in reviewing the manuscripts, and to the staff of Springer, for their assistance in assembling this volume and their efforts in keeping this project on schedule. Last but not least, we warmly acknowledge our families, friends, and colleagues at the Universidad de León and Curia for their patience and kind support.

León, Spain

Carlos Barreiro
José-Luis Barredo

Contents

Contributors

JORGE ARANDA HERNANDEZ • *Department of Biochemical Engineering, University College London, London, UK*
JOSÉ-LUIS BARREDO • *Department of Biotechnology, Curia Spain, León, Spain*
CARLOS BARREIRO • *Instituto de Biotecnología de León, INBIOTEC, León, Spain; Área de Bioquímica y Biología Molecular, Departamento de Biología Molecular, Facultad de Veterinaria, Universidad de León, León, Spain*
JOSÉ BERENGUER • *Centro de Biología Molecular Severo Ochoa (UAM-CSIC). Facultad de Ciencias. Universidad Autónoma de Madrid, Madrid, Spain*
MARTA MARÍA BLANCO-NISTAL • *Complejo Asistencial Universitario de León, León, Spain*
SANDRA BOSCH • *Centro de Biología Molecular Severo Ochoa (UAM-CSIC). Facultad de Ciencias. Universidad Autónoma de Madrid, Madrid, Spain*
CARLA C. C. R. DE CARVALHO • *iBB-Institute for Bioengineering and Biosciences, Department of Bioengineering, Instituto Superior Técnico, Universidade de Lisboa, Lisbon, Portugal; Associate Laboratory i4HB – Institute for Health and Bioeconomy, Instituto Superior Técnico, Universidade de Lisboa, Lisbon, Portugal*
ALEJANDRO CHAMIZO-AMPUDIA • *Área de Bioquímica y Biología Molecular, Departamento de Biología Molecular, Facultad de Veterinaria, Universidad de León, León, Spain*
SOLEDAD CHAMORRO • *Water Research Center for Agriculture and Mining (CRHIAM), ANID Fondap Center, Victoria, Concepción, Chile; Engineering and Environmental Biotechnology Group, Environmental Science Faculty & Center EULA-Chile, University of Concepción, Concepción, Chile*
NATALIA O. DESHCHEREVSKAYA • *Institute of Biochemistry & Physiology of Microorganisms, Pushchino Scientific Center for Biological Research of the Russian Academy of Sciences, Federal Research Center, Pushchino, Russia*
MARINA V. DONOVA • *G.K. Skryabin Institute of Biochemistry and Physiology of Microorganisms, Federal Research Center "Pushchino Scientific Center for Biological Research of the Russian Academy of Sciences", Pushchino, Russia; G.K. Skryabin Institute of Biochemistry and Physiology of Microorganisms, Russian Academy of Sciences, Federal Research Center "Pushchino Center for Biological Research of the Russian Academy of Sciences", Pushchino, Russia; Federal Research Center "Pushchino Center for Biological Research of the Russian Academy of Sciences", G.K. Skryabin Institute of Biochemistry and Physiology of Microorganisms, Russian Academy of Sciences, Pushchino, Russia; Institute of Biochemistry & Physiology of Microorganisms, Pushchino Scientific Center for Biological Research of the Russian Academy of Sciences, Federal Research Center, Pushchino, Russia; Pharmins LTD, Pushchino, Russia*
DMITRY V. DOVBNYA • *Institute of Biochemistry & Physiology of Microorganisms, Pushchino Scientific Center for Biological Research of the Russian Academy of Sciences, Federal Research Center, Pushchino, Russia*
OLA EL ATAB • *Department of Biology, University of Fribourg, Fribourg, Switzerland*
CARMEN FELPETO-SANTERO • *Department of Microbial and Plant Biotechnology, Margarita Salas Centre for Biological Research-CSIC, Madrid, Spain*
PEDRO FERNANDES • *Associate Laboratory i4HB – Institute for Health and Bioeconomy, Instituto Superior Técnico, Universidade de Lisboa, Lisbon, Portugal; iBB-Institute for*

Bioengineering and Biosciences, Instituto Superior Técnico, Universidade de Lisboa, Lisbon, Portugal; DREAMS and Faculty of Engineering, Universidade Lusófona de Humanidades e Tecnologias, Lisbon, Portugal

JESÚS ANTONIO FERNÁNDEZ-FERNÁNDEZ • *Complejo Asistencial Universitario de León, León, Spain*

VICTORIA V. FOKINA • *G.K. Skryabin Institute of Biochemistry and Physiology of Microorganisms, Russian Academy of Sciences, Federal Research Center "Pushchino Center for Biological Research of the Russian Academy of Sciences", Pushchino, Russia*

BEATRIZ GALÁN • *Department of Microbial and Plant Biotechnology, Margarita Salas Centre for Biological Research-CSIC, Madrid, Spain*

JOSÉ LUIS GARCÍA • *Department of Microbial and Plant Biotechnology, Margarita Salas Centre for Biological Research-CSIC, Madrid, Spain*

LUIS GETINO • *Área de Bioquímica y Biología Molecular, Departamento de Biología Molecular, Facultad de Veterinaria, Universidad de León, León, Spain*

AURELIO HIDALGO • *Centro de Biología Molecular Severo Ochoa (UAM-CSIC). Facultad de Ciencias. Universidad Autónoma de Madrid, Madrid, Spain*

ANA M. IBÁÑEZ • *Instituto de Investigación de la Viña y el Vino, Escuela de Ingeniería Agraria, Universidad de León, León, Spain; Instituto Tecnológico Agrario de Castilla y León (ITACyL), Área de Investigación Agrícola, Valladolid, Spain*

TANYA V. IVASHINA • *Institute of Biochemistry & Physiology of Microorganisms, Pushchino Scientific Center for Biological Research of the Russian Academy of Sciences, Federal Research Center, Pushchino, Russia*

KJELL DOMAAS JOSEFSEN • *Department of Biotechnology and Nanomedicine, SINTEF Industry, Trondheim, Norway*

MIKHAIL V. KARPOV • *G.K. Skryabin Institute of Biochemistry and Physiology of Microorganisms, Russian Academy of Sciences, Federal Research Center "Pushchino Center for Biological Research of the Russian Academy of Sciences", Pushchino, Russia*

SERGEY M. KHOMUTOV • *Institute of Biochemistry & Physiology of Microorganisms, Pushchino Scientific Center for Biological Research of the Russian Academy of Sciences, Federal Research Center, Pushchino, Russia*

VYACHESLAV V. KOLLEROV • *Federal Research Center "Pushchino Center for Biological Research of the Russian Academy of Sciences", G.K. Skryabin Institute of Biochemistry and Physiology of Microorganisms, Russian Academy of Sciences, Pushchino, Russia*

SIMONE BALZER LE • *Department of Biotechnology and Nanomedicine, SINTEF Industry, Trondheim, Norway*

JOSÉ M. LUENGO • *Área de Bioquímica y Biología Molecular, Departamento de Biología Molecular, Facultad de Veterinaria, Universidad de León, León, Spain; Departamento de Biología Molecular (Área de Bioquímica y Biología Molecular), Facultad de Veterinaria, Universidad de León, León, Spain*

MARCO P. C. MARQUES • *Department of Biochemical Engineering, University College London, London, UK*

SONIA MARTÍNEZ-CÁMARA • *Department of Biotechnology, Curia Spain, León, Spain*

MARÍA-ANGÉLICA MONDACA • *Microbiology Department, University of Concepción, Concepción, Chile*

AURÉLIE MOSSION • *Nantes Université, Institut des Substances et Organismes de la Mer, ISOMer, Nantes, France*

ANNA NORDBORG • *Department of Biotechnology and Nanomedicine, SINTEF Industry, Trondheim, Norway*

ELIAS R. OLIVERA • *Área de Bioquímica y Biología Molecular, Departamento de Biología Molecular, Facultad de Veterinaria, Universidad de León, León, Spain; Departamento de Biología Molecular (Área de Bioquímica y Biología Molecular), Facultad de Veterinaria, Universidad de León, León, Spain*

SILJE MALENE OLSEN • *Department of Biotechnology and Nanomedicine, SINTEF Industry, Trondheim, Norway*

ISABELLE OURLIAC-GARNIER • *Nantes Université, Cibles et Médicaments des Infections et de l'Immunité, IICiMed, Nantes, France*

ANA-LUISA RIBEIRO • *Centro de Biología Molecular Severo Ochoa (UAM-CSIC). Facultad de Ciencias. Universidad Autónoma de Madrid, Madrid, Spain*

ANTONIO RODRÍGUEZ-GARCÍA • *Área de Microbiología, Departamento de Biología Molecular, Facultad de Ciencias Biológicas y Ambientales, Universidad de León, León, Spain; Instituto de Biotecnología de León, INBIOTEC, León, Spain*

MARTA RODRÍGUEZ-SÁIZ • *Department of Biotechnology, Curia Spain, León, Spain*

MERCEDES SÁNCHEZ • *Centro de Biología Molecular Severo Ochoa (UAM-CSIC). Facultad de Ciencias. Universidad Autónoma de Madrid, Madrid, Spain*

ROGER SCHNEITER • *Department of Biology, University of Fribourg, Fribourg, Switzerland*

ANDREI A. SHUTOV • *Federal Research Center "Pushchino Center for Biological Research of the Russian Academy of Sciences", G.K. Skryabin Institute of Biochemistry and Physiology of Microorganisms, Russian Academy of Sciences, Pushchino, Russia; Institute of Biochemistry & Physiology of Microorganisms, Pushchino Scientific Center for Biological Research of the Russian Academy of Sciences, Federal Research Center, Pushchino, Russia*

HÅVARD SLETTA • *Department of Biotechnology and Nanomedicine, SINTEF Industry, Trondheim, Norway*

ALBERTO SOLA-LANDA • *Instituto de Biotecnología de León, INBIOTEC, León, Spain; Fundación Cesefor, León, Spain*

SHIKUI SONG • *Laboratory of Protein Engineering and Biopharmaceutical Sciences, Key Laboratory of Industrial Fermentation and Cooperative Innovation Center of Industrial Fermentation, Hubei University of Technology, Wuhan, Hubei, China*

ZHENGDING SU • *Laboratory of Protein Engineering and Biopharmaceutical Sciences, Key Laboratory of Industrial Fermentation and Cooperative Innovation Center of Industrial Fermentation, Hubei University of Technology, Wuhan, Hubei, China*

ALEXEY V. SVIRIDOV • *G.K. Skryabin Institute of Biochemistry and Physiology of Microorganisms, Russian Academy of Sciences, Federal Research Center "Pushchino Center for Biological Research of the Russian Academy of Sciences", Pushchino, Russia*

MANUEL DE LA TORRE • *Department of Biotechnology, Curia Spain, León, Spain*

GLADYS VIDAL • *Water Research Center for Agriculture and Mining (CRHIAM), ANID Fondap Center, Victoria, Concepción, Chile; Engineering and Environmental Biotechnology Group, Environmental Science Faculty & Center EULA-Chile, University of Concepción, Concepción, Chile*

MARICEL VIDAL • *Microbiology Department, University of Concepción, Concepción, Chile*

GAETANE WIELGOSZ-COLLIN • *Nantes Université, Institut des Substances et Organismes de la Mer, ISOMer, Nantes, France*

CLAUDIA XAVIER • *Water Research Center for Agriculture and Mining (CRHIAM), ANID Fondap Center, Concepción, Chile; Engineering and Environmental Biotechnology Group, Environmental Science Faculty & Center EULA-Chile. University of Concepción, Concepción, Chile; Federal University of Technology – Paraná–UTFPR. Campus Curitiba, Curitiba, PR, Brazil*

Part I

General Overview

Chapter 1

Current Trends and Perspectives in Microbial Bioconversions of Steroids

Marina V. Donova

Abstract

The microbiological transformation of sterols is currently the technological basis for the industrial production of valuable steroid precursors, the so-called synthons, from which a wide range of steroid and indane isoprenoids are obtained by combined chemical and enzymatic routes. These compounds include value-added corticoids, neurosteroids, sex hormones, bile acids, and other terpenoid lipids required by the medicine, pharmaceutical, food, veterinary, and agricultural industries.

Progress in understanding the molecular mechanisms of microbial degradation of steroids, and the development and implementation of genetic technologies, opened a new era in steroid biotechnology. Metabolic engineering of microbial producers makes it possible not only to improve the biocatalytic properties of industrial strains by enhancing their target activity and/or suppressing undesirable activities in order to avoid the formation of by-products or degradation of the steroid core, but also to redirect metabolic fluxes in cells towards accumulation of new metabolites that may be useful for practical applications. Along with whole-cell catalysis, the interest of researchers is growing in enzymatic methods that make it possible to carry out selective structural modifications of steroids, such as the introduction of double bonds, the oxidation of steroidal alcohols, or the reduction of steroid carbonyl groups. A promising area of research is strain engineering based on the heterologous expression of foreign steroidogenesis systems (bacterial, fungal, or mammalian) that ensure selective formation of demanded hydroxylated steroids.

Here, current trends and progress in microbial steroid biotechnology over the past few years are briefly reviewed, with a particular focus on the application of metabolic engineering and synthetic biology techniques to improve existing and create new whole-cell microbial biocatalysts.

Key words Steroid, Microbial transformation, Bioconversion, Phytosterol, Side chain degradation, Hydroxylation, Dehydrogenation, Whole-cell biocatalysis

1 Introduction

Sustainable development strategies involve the steady replacement of chemical processes by appropriate microbial biotechnologies. Steroid field is one of the areas where the advantages of microbial biotechnologies over the multistep, complicated, and ecologically risky chemical syntheses are especially evident.

Carlos Barreiro and José-Luis Barredo (eds.), *Microbial Steroids: Methods and Protocols*, Methods in Molecular Biology, vol. 2704, https://doi.org/10.1007/978-1-0716-3385-4_1,

On some estimations, the global market of steroids is growing annually and prognosed to reach $17 billion in 2025 [1]. The market of only corticosteroid pharmaceuticals reached $5.05 billion in 2022 and is forecasted to grow further in perspective to 2030 [2]. The SARS-CoV pandemic evidenced the efficiency of corticosteroids as an adjunctive therapy for the treatment of the disease at the early phase of infection and similar applications tend to grow; the use of dexamethasone is a clear example [3]. There are growth forecasts for the androgens and anabolic steroid medicine market, which was valued at $56.45 billion in 2021 and is expected to exceed $131 billion by 2029 [4]. The growth is also prognosed for other segments of the steroid market.

Two major directions in steroid industry are recognized as (i) microbial conversion of the primary raw materials, mainly phytosterols to produce value-added intermediates, also called synthons, and (ii) the following syntheses of the active pharmaceutical ingredients (API) from the synthons using chemical or chemical–enzymatic methods. A growing steroid market needs to increase the production of raw materials such as phytosterols, which are also used in other industries, so the problem of available and cheap sterols may arise.

A relevant issue is the creation of new industrial microbial producers and development of new generation biotechnologies for the production of steroid and other isoprenoid compounds. The designing of novel microbial biocatalysts for industrial use in sterol bioconversion processes is based on a deep knowledge of the molecular mechanisms, features of functioning, and regulation of metabolic pathways of sterol degradation/bioconversion in actinobacteria. Metabolic engineering of microbial whole-cell catalysts makes it possible not only to improve the biocatalytic properties of industrial strains by suppressing undesirable activities resulting in the formation of by-products, but also to re-direct metabolic fluxes in the cell towards the formation of new products. Despite significant progress in recent years in studying the catabolism of natural sterols in actinobacteria, some gaps remain in understanding the genetic control of key steps in the oxidative degradation of the side chain and steroid core C/D rings in nonpathogenic species. Filling these gaps is necessary to create high-performance recombinant strains capable of selectively producing valuable steroid metabolites that are in demand in the production of new generation drugs.

An urgent issue also remains the expansion of the arsenal of methods for the targeted structural modification of steroids with microorganisms, such as oxyfunctionalization of inactive carbon atoms, regio- and stereospecific hydroxylation of the gonane core of steroids, and the side chain of sterols. These problems are almost impossible to solve by chemical methods, and the ways of their solutions are associated with the genetic engineering of microorganisms, including the heterologous expression of foreign steroidogenesis systems in microbial hosts. Controlled expression of such systems in appropriate microbial chassis opens new opportunities in selective production of the oxyfunctionalized steroids. The results are important not only for biotechnological applications, but also for biomedical research, since they open up prospects for identifying new targets for creating targeted drugs.

Large quantities of steroids, including sterols, bile acids, natural or synthetic steroid hormones, and other drugs and intermediates enter the environment through biomass decay or excretion by humans and animals, and as industrial waste from steroid manufacturing plants. Microbial bioremediation from recalcitrant, toxic, and endocrine-disrupting steroid pollutants is one of the most important fields of research. These aspects are detailed in several recent reviews [5–7] and are not discussed in this review.

This review highlights the major advances in microbial steroid bioconversions over the past few years and discusses future developments in this field.

2 Microbial Bioconversions of Phytosterol

Sterols are widespread in nature, and phylogenetically heterogeneous bacteria have evolved to utilize them as growth substrates. The study of 346 publicly available metagenomes showed global distribution of the steroid-degrading bacteria in natural, engineered, and host-associated environments with a prevalence in wastewater treatment plants, soil, plant rhizospheres, and the marine environment, including marine sponges [8]. The aerobic 9(10)-*seco* pathway has been identified in *Actinobacteria* and *Proteobacteria*, but the activity is more characteristic of actinobacteria.

Interestingly, the 9(10)-*seco* pathway has been identified not only in mesophilic but also in thermophilic actinobacteria. Recently, the genes related to the sterol and cholate degradation pathways have been identified in aerobic thermophilic actinobacterium of *Saccharopolyspora hirsuta* VKM Ac-666 [9]. The strain was able to utilize cholesterol and formed 26-hydroxylated derivatives such as 26-hydroxycholesterol and 3β-hydroxy-cholest-5-en-26-oic acid; those were not reported earlier for other organisms [10] (Fig. 1). However, in general, the presence of the 9(10)-*seco* pathway does not seem to be very common among thermophilic bacteria: the analysis of 52 publicly available genomes of thermophile microbes revealed the presence of key genes of the pathway in only 7 strains [9].

Until recently, the 9(10)-*seco* pathway (Fig. 2) was considered the only aerobic catabolic route for bacterial steroid degradation. In a latter study, the pathways via $\Delta^{1,4}$ or $\Delta^{4,6}$ -intermediates were revealed at the degradation of 7-hydroxy steroids such as cholate by members of the family *Sphingomonadaceae*, thus evidencing biochemical diversity of steroid core destruction [11].

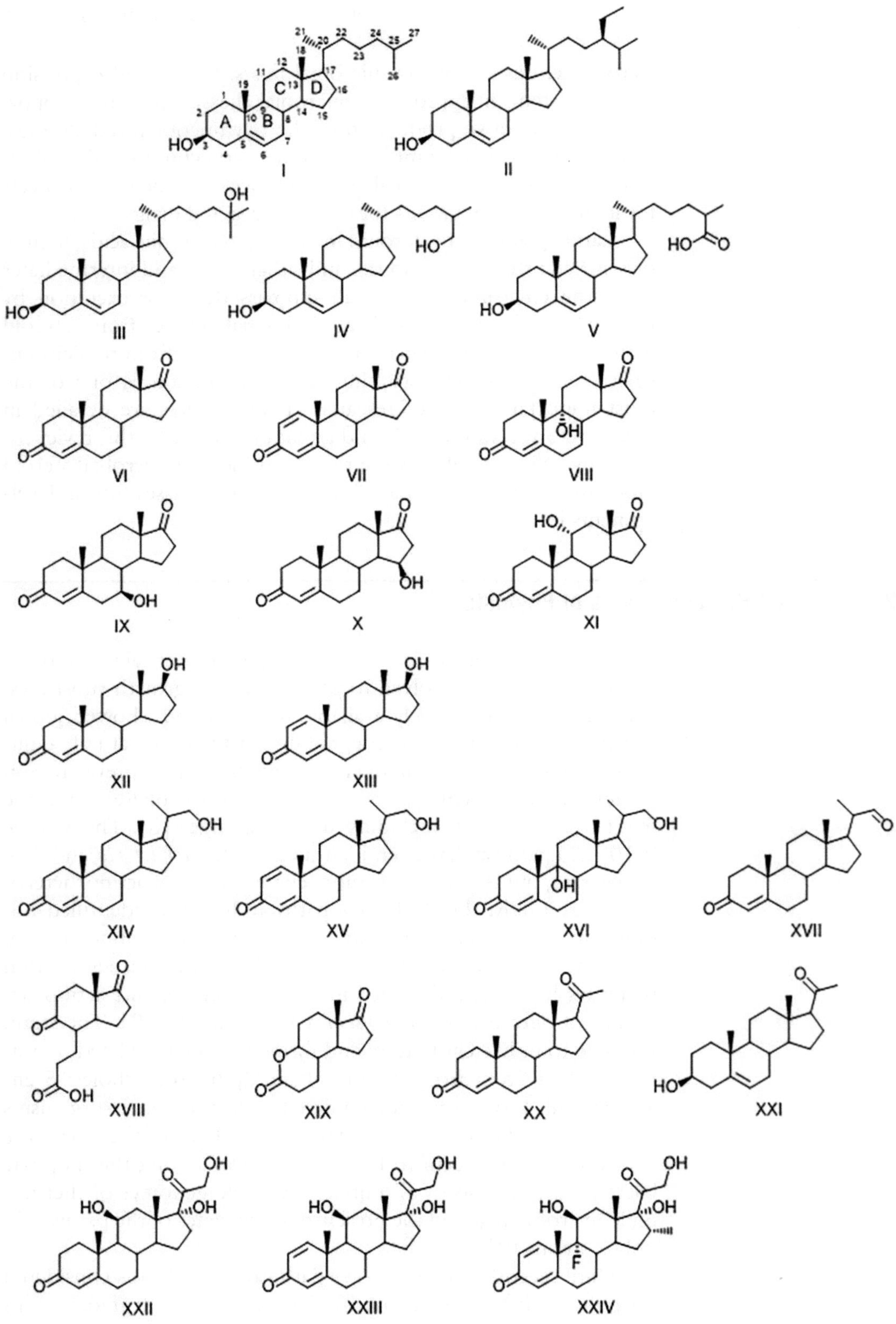

Fig. 1 Structures of some steroids mentioned in the review. **I**, cholesterol; **II**, β-sitosterol (major component of phytosterol); **III**, 25-hydroxy cholesterol; **IV,** 26-hydroxy cholesterol; **V**, 3β-hydroxy cholest-5-en-26-oic acid; **VI**, androst-4-ene-3,17-dione (AD); **VII**, androsta-1,4-diene-3,17-dione (ADD); **VIII,** 9α-hydroxy androst-4-

The ability to oxidize sterols was revealed for various phylogenetic groups of microorganisms, but all the currently known industrial strains for obtaining valuable steroids from sterols were obtained on the basis of representatives of the genus *Mycolicibacterium* (previously *Mycobacterium*). It should be noted that based on the genome sizes, analysis of complete genomic sequences, and results of polyphase taxonomy, many non-pathogenic mycobacteria

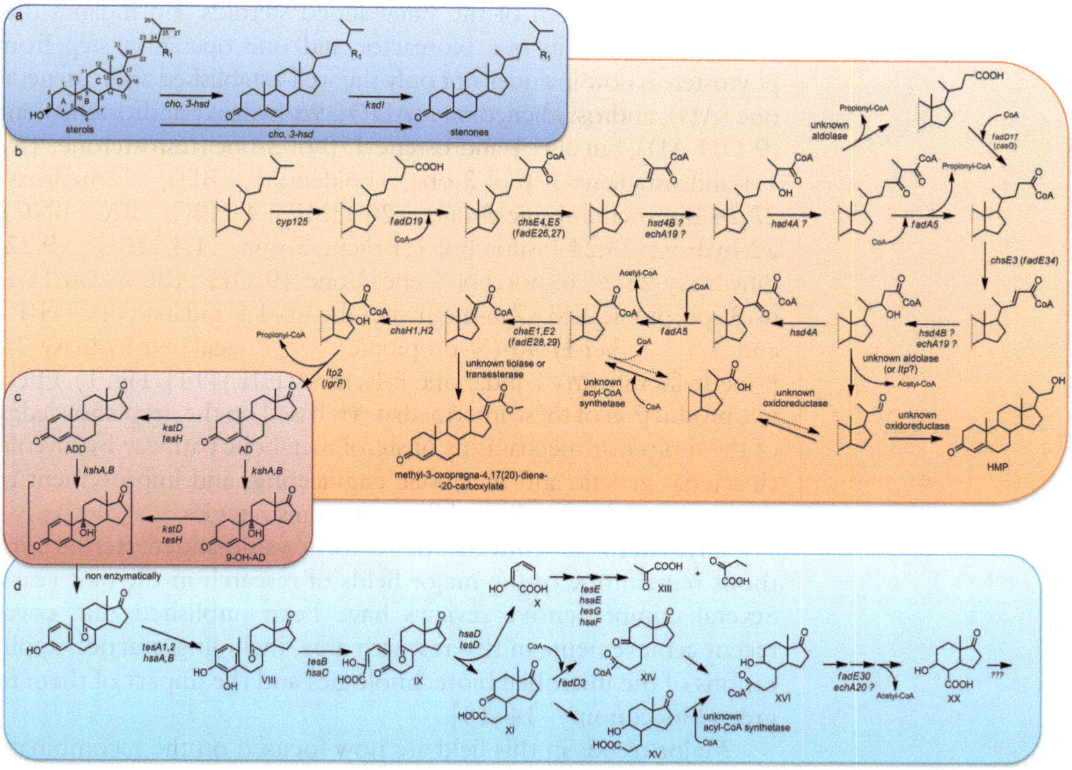

Fig. 2 The 9(10)-*seco* pathway of sterol degradation by actinobacteria. (Adopted from Ref. [46]). (**a**) Initial step of steroid core degradation–modification of 3β-hydroxy-5-ene to 3-keto-4-ene structure; (**b**) oxidative degradation of the side chain; (**c**) oxidation of ring A/B; (**d**) oxidation of ring C/D

Fig. 1 (continued) ene-3,17-dione (9-OH-AD); **IX**, 7β-hydroxy androst-4-ene-3,17-dione (7β-OH-AD); **X**, 15β-hydroxy androst-4-ene-3,17-dione (15β-OH-AD); **XI**, 11α-hydroxy-androst-4-ene-3,17-dione (11α-OH-AD); **XII**, 17β-hydroxy-androst-4-ene-3-one (testosterone); **XIII**, 17β-hydroxy-androst-1,4-diene-3,17-dione (boldenone, 1-dehydrotestosterone); **XIV**, 20-hydroxymethyl-pregna-4-ene-3-one (22-hydroxy-23,24-bisnorchol-4-ene-3-one, 20-HMP, 4-HBC, BA, BNC); **XV**, 20-hydroxymethyl-pregna-1,4-diene-3-one (22-hydroxy-23,24-bisnorchol-1,4-diene-3-one, 1,4-HBC, HMPD, 1,4- BNC); **XVI**, 9α,20-dihydroxy-methyl-pregna-4-ene-3-one (9,22-dihydroxy-23,24-bisnorchol-4-ene-3-one, 9-OH-HBC, 9-OH-HMP, 9-OH-BNC); **XVII**, pregn-4-ene-3-one-20-carbaldehyde (20-POA); **XVIII**, 3aα-H-4α-(3′-propionic acid)-7aβ--methylhexahydro-1,5-indanedione (HIP); **XIX**, 3aα-H-4α-(3′-propionic acid)-5α-hydroxy-7aβ--methylhexahydro-1-indanone-δ-lactone (HIL, sitolactone); **XX**, progesterone; **XXI**, pregnenolone; **XXII**, hydrocortisone (cortisol); **XXIII**, prednisolone; **XXIV**, dexamethasone

are re-classified into a separate genus, *Mycolicibacterium*, whose representatives are known among others for their ability to effectively oxidize sterols [12].

The capability of *Mycobacterium* sp. to degrade cholesterol and phytosterol as growth substrate was discovered in 1944, and since then it is intensively studied [13]. Nowadays, effective whole-cell biocatalysts have been created mainly based on the platform of *Mycolicibacterium smegmatis* mc^{155} or *Mycolicibacterium neoaurum* as described below. The spectrum of the value-added steroids and indanes that can be produced in one bioreactor and one operation step from phytosterols now includes not only the well-established androstenedione (AD), androstadienedione (ADD), 9α-hydroxy androstenedione (9-OH-AD), but also 4-androstene-17β-ol-3-one (testosterone, TS), 1,4-androstadiene-17β-ol-3-one (boldenone, BD), 22-hydroxy-23,24-bisnorchol-4-ene-3-one (20-HMP, 4-HBC, BA, BNC), 22-hydroxy-23,24-bisnorchol-1,4-diene-3-one (1,4-HBC), 9,22-dihydroxy-23,24-bisnorchol-4-ene-3-one (9-OH-HBC), 3aα-H-4-α-(3′-propionic acid)-7aβ-methylhexahydro-1,5-indanedione (HIP), and 3aα-H-4α-(3′-propionic acid)-5α-hydroxy-7a-β-methylhexahydro-1-indanone-δ-lactone (HIL) [14] (Fig. 1). Effective production of these intermediates is based on the deep knowledge of the molecular mechanisms of sterol metabolic pathway by mycolicibacteria, genetic and metabolic engineering, and improvement of the bioconversion media and fermentation strategies.

Phytosterol bioconversions to form value-added steroid synthons remain one of the major fields of research in the past years. Several comprehensive reviews have been published that cover recent achievements in this research area, including practical applications of the microbial biotechnologies and the impact of them to green bioeconomy [14–18].

Major trends in this field are now focused on the recombinant DNA technologies and synthetic biology for the creation of the engineered strains, mainly belonging to *Mycolicibacterium* genus, to provide effective production of C19, C22, and C26 steroids from phytosterol [19]. In addition, research on the re-directing of the steroid catabolic pathways towards accumulation of the valuable indane derivatives such as HIP or HIL is also in the focus.

2.1 Phytosterols

Phytosterols of various origins are useful as main raw materials for the production of value-added steroids by microorganisms. Commercial phytosterols are usually produced from vegetable oils, such as soya or rapeseed oil, or from wastes of paper and pulp mills (tall oil or tall oil pitch). New opportunities are associated with the use of the biowastes or wastes of the growing biogas industry as a novel and sustainable source of natural sterols: the feedstock sterols are not degraded during anaerobic digestion by methanogenic microbial community and accumulate in the digestates. These sterols may be transformed by appropriate microbial strains, e.g., *M. neoaurum,* to form the valuable steroids, such as ADD, without prior extraction and purification. The approach is of significance for a circular bioeconomy, considering huge volumes of the sterol-containing anaerobic digestate produced worldwide [20].

Another potentially sustainable source of sterols is microalgae. The calculations suggested that microalgae could yield 678–6035 kg per ha of phytosterols annually that is comparable with the current production of phytosterols from rapeseeds [21]. A promising approach may be de novo synthesis of sterols and steroids by yeasts [22]. There are green and safe production routes based on the recent achievements in metabolic engineering of the yeasts.

In addition to being used for the biotechnological production of synthons, sterols are also applied in food and cosmetics industry, as well as carriers in lipid nanoparticles in drug-delivery systems. There are data evidencing that phytosterols and phytosterol-enriched diets can control glucose and lipid metabolism and contribute to insulin resistance [23]. In general, growing industrial applications of phytosterols necessitate a search for new promising sources and effective technologies for their production.

With regard to the bioconversions of phytosterols to value-added synthons, the main areas of research are related to the improvement of well-established microbial biocatalysts, such as *M. smegmatis*, *M. neoaurum*, or *M. fortuitum*, in order to increase their productivity and the selectivity of the targeted reactions carried out by them.

2.2 Androstenedione (AD) and Androstadienedione (ADD)

Androstenedione (AD) remains a key synthon produced from phytosterol using mycolicibacteria strains. It is a key intermediate in the body's steroid metabolism and used as a precursor in the production of various steroid medicines such as testosterone, estradiol, estrone, cortisol, prednisolone, and their derivatives. The market size of AD is evaluated over $1 billion annually [24], and the production of AD and ADD exceeded 10,000 tons per year. As summarized in [25] most papers and patents in the field of microbial AD production from phytosterols are focused on the process improvement and genetic modification of the strains, mainly *Mycolicibacterium* sp. or *M. neoaurum*. With regard to the process improvement, the effects of the additions of solvents, cyclodextrins, surfactants, ionic liquids, oils, etc. as well as design and optimization of the nutrient media composition, temperature, and agitation regimens were studied. Strain improvement was mainly aimed at preventing the undesirable side reactions and overall increase of the productivity.

For example, a combined strategy of metabolic pathway regulation and two-step bioprocess has been developed for AD production with an engineered *M. neoaurum* [24]. To strengthen the metabolic flux of phytosterol to AD, the gene *choD* coding for cholesterol oxidase and *cyp125* (*smo*) encoding steroid

C27-monooxygenase were overexpressed in *M. neoaurum* VKM Ac-2015D, while the expression of *hsd4A* encoding 17-β-hydroxysteroid dehydrogenase was enhanced to prevent undesirable formation of 20-hydroxymethylpregnenone (20-HMP) as a major by-product (Fig. 2). The engineered strain was cultured at 30 °C, and the bioconversion was carried out by the resting cells at 37 °C, thus resulting in the production of 25.8 g/L of AD from 50 g/L phytosterol. The yield was estimated as the highest reported for AD production and demonstrates the efficiency of the approach applied [24].

Impressive results have been obtained by manipulating of energy metabolism by constructing a novel citrate-based ATP futile cycle (AFC) and pyruvate-based AFC in AD-producing *M. neoaurum* [26]. This approach made possible to strengthen ATP consumption, enhance propionyl-CoA metabolism, and promote NAD+/NADH ratio and cell viability, thus resulting in an increase of AD conversion yield from 60.6% to 97.3%. Repeated batch fermentation using untreated cane molasses ensured 181% higher productivity as compared with original strain.

To improve production of another key synthon, androstadienedione (ADD) from phytosterol, Shao et al. (2019) identified the key gene coding for steroid 27-monooxygenase in *M. neoaurum* JC 12. Along with the genes encoding cholesterol oxidase and 3-ketosteroid-Δ^1-dehydrogenase, this gene was overexpressed, while the genes coding for 9α-hydroxylase were disrupted. The recombinant strain obtained produced up to 20.1 g/L ADD [27].

2.3 9α-Hydroxy Androstenedione (9-OH-AD)

Other approaches useful to strengthen phytosterol bioconversion by mycolicibacteria are based on the cell envelope engineering, i.e., modification of mycobacterial cell wall to enhance cell wall permeability for steroids and nutrients. As an example, one can mention the identification and inactivation of the *embC* gene involved in the synthesis of lipoarabinomannan from lipomannan in *M. neoaurum* strain capable of producing 9α-hydroxy androstenedione (9-OH-AD) from phytosterol. Cells deficient in lipoarabinomannan provided more efficient phytosterol bioconversion: the increase was about 52.4% after 72 h conversion [28]. The efficiency of phytosterol bioconversion by *M. neoaurum* also increased with the deletion of the gene *kasB* related to the synthesis of mycolic acids. Cell wall permeability in the *kasB*-deficient strain was almost twice higher as compared with that of the wild-type strain, thus ensuring higher yield of 9-OH-AD while shortening conversion time. Similar enhancement was also demonstrated for the *kasB*-deficient strain producing 22-hydroxy-23,24-bisnorchol-4-ene-3-one (4-HBC, also known as 20-HMP) [29].

Significant progress is also observed in the field related to the development of synthetic biology toolkits for engineering mycobacterial strains capable of effective phytosterol conversion. For example, ten artificial promoters and ribosomal binding sites (RBS) have been constructed and examined. Application of a high-strength promoter (CP6) and a medium-strength RBS (R1) chosen ensured overexpression of the genes encoding 3-keto-steroid-9α-hydroxylase (*kshA* and *kshB*) in *M. neoaurum* ATCC 25795, thus providing a considerable increase of 9-OH-AD production from phytosterol [30].

2.4 9,21-Dihydroxy-20-methylpregna-4-en-3-one (9-OH-4-HP)

During the past few years, considerable progress was also achieved in the field of microbial production of the steroids with partly oxidized side chain, such as C22, C24, and C26(27) steroids. This progress is mainly due to the expansion of knowledge about the genetic control of sterol side chain degradation.

For instance, a novel steroid precursor, 9,21-dihydroxy-20-methylpregna-4-en-3-one (9-OH-4-HP) (Fig. 1), was obtained from phytosterols by modifying multiple genes and improving the intracellular environment in *M. neoaurum* [31]. This compound can be used for the synthesis of corticosteroids. To prevent steroid core degradation, 1(2)-dehydrogenation was blocked in *M. neoaurum* DSM 44074. In addition, the genes *hsd4A* coding for a β-hydroxyacyl-CoA dehydrogenase and *fadA5* encoding acyl-CoA thiolase (Fig. 2) were knocked out. To further improve the target activity, the catalase (*katE*) from *M. neoaurum* and an NADH oxidase (*nox*) from *Bacillus subtilis* were overexpressed in the strain. The resulting recombinant strain produced 3.58 g/L of 9-OH-4-HP from 5 g/L phytosterol [31]. Another new C22-steroid precursor with similar structure, 9-hydroxy-3-oxo-4,17-pregnadiene-20-carboxylic acid methyl ester, was produced by recombinant *M. neoaurum* DSM 44074 with deficiency in enoyl-CoA and modification of multiple genes (*3-ksd, chsE1-chsE2, hsd4a*) [32].

2.5 Hydroxy-cholesterol

Interest in microbiological methods for obtaining 25, or 26(27)-hydroxysteroids is caused by the expansion of their potential application in medicine. There is growing body of the data evidencing antiviral activity of 25-hydroxycholesterol (25-HC) against the enveloped viruses and some non-enveloped viruses such as human papillomavirus-16 (HPV-16), human rotavirus (HRoV), and human rhinovirus (HRhV). Significant antiviral activity against HPV-16, HRoV, and HRhV was also confirmed for 26-hydroxycholesterol [33]. In 2020, the inhibitory effect of 26 (27)-hydroxycholesterol was demonstrated against SARS-CoV-2 and its level was markedly decreased in COVID-19 patients [34]. Production of 26-hydroxycholesterol is possible using appropriate wild-type strains capable of cholesterol conversion as described above for the thermophilic *S. hirsuta* [9, 10].

2.6 Sitolactone (HIL) and HIP

Production of the important indane compounds, 3aα-H-4α-(3′-propionic acid)-5α-hydroxy-7aβ-methylhexahydro-1-indanone-δ-lactone (sitolactone, also known as HIL) and 3aα-H-4α-(3′-propionic acid)-7aβ-methylhexahydro-1,5-indanedione (well-known as HIP) from phytosterol, became possible due to the progress in the understanding of the molecular mechanisms of C/D ring degradation of the steroid core in mycobacteria [35]. These compounds are the precursors of the steroid pharmaceuticals with an α-methyl group or without the methyl group at the C10 position such as retroprogesterone, estradiol, and their derivatives. Knockouts of the genes *fadD3* and *fadE30* in *M. fortuitum* ATCC 6841 as well as inactivation of the genes *car1* and *car2* involved in the degradation of HIP allowed accumulation of HIP or HIL, respectively. The recombinant mycobacteria were shown to accumulate HIP or HIL in high yields (88% and 75%, respectively) from 20 g/L phytosterol [36], thus evidencing their biotechnological potential.

2.7 Testosterone and Boldenone

As mentioned above, androgens and anabolic steroids constitute a significant segment of the global steroid market. Production of testosterone, which is a main male hormone, is possible in one operation step from phytosterol using an engineered mycobacterial strain [37]. Heterologous expression of 17β-hydroxysteroid dehydrogenase from the fungus *Cochliobolus lunatus* in *M. smegmatis* mc2^{155} allowed the most effective conversion of 1.8 mM cholesterol to testosterone. In a recent study, manipulations of aeration conditions and glucose supplement during phytosterol bioconversion ("oxidative" regimen at the first stage and microaerophilic conditions at the second phase) ensured effective testosterone production (about 4.6 g/L from 10 g/L phytosterol) using *M. neoaurum* VKM Ac-2015D [38].

Another approach for production of steroid 17β-alcohols such as testosterone and 1-dehydrotestosterone (also called boldenone) is based on the cascade biotransformations using different microorganisms in one bioreactor without the intermediate isolation of AD or ADD. For instance, collaborative microbial community of *M. neoaurum* and the yeasts *Pichia pastoris* was applied to obtain boldenone which is an important veterinary hormone [39]. To enhance the productivity, 3-ketosteroid-Δ^1-dehydrogenase (KsdD) and 17β-hydroxysteroid dehydrogenase (17β-HSD) were overexpressed in *M. neoaurum* and *P. pastoris*, respectively. After process optimization exploiting semi-batch and glucose supplementation strategies, the productivity of boldenone increased significantly (from 10% to 76%). Cascade phytosterol biotransformation by *M. neoaurum* VKM Ac-1815D and *Nocardioides simplex* VKM Ac-2033D in one bioreactor ensured effective production of testosterone [40].

These few examples demonstrate the vast possibilities that have emerged in the field of bioconversion of phytosterols with actinobacteria, and in particular, mycolicibacteria, due to advances in omics technologies and metabolic engineering. The spectrum of steroid synthons that can be effectively produced from natural sterols using recombinant strains has expanded significantly in recent years and now includes not only AD, ADD, and 9-OH-AD, but also metabolites with a partially oxidized side chain, and products of deep oxidation of the steroid core.

3 Microbial Enzymes for Steroid Production

In recent years, there has been a growing interest of researchers in the production and study of enzymes that carry out key reactions of structural modification of steroids. The structures and molecular mechanisms of action of the enzymes involved cholesterol degradation in three major pathways: (i) aerobic β-oxidation (as in actinobacteria), (ii) the oxygen-independent degradation pathway in *Sterolibacterium denitrificans*, and (iii) the CYP11A1- and CYP17A1-catalyzed pathway in mammals, which have been recently reviewed [41].

3.1 Cholesterol Oxidase

Cholesterol oxidase (ChO) is a well-known enzyme that catalyzes oxidation of 3β-hydroxy-5-ene- to 3-keto-4-ene-steroids with the formation of hydrogen peroxide. The main field of application of microbial ChOs is cholesterol assay in biological fluids such as serum, food products, and feeds. In addition, the enzyme is in demand in agriculture and fine chemistry. Recent studies demonstrated in vitro anti-cancer activities of microbial ChOs against different cancer lines and in vivo apoptosis [42, 43]. In connection with these discoveries, one can predict an increase in interest in the search for and production of new microbial ChOs for medical use. Predominantly the strains of *Streptomyces* genera exhibit high level of extracellular ChOs production. In latter papers, highly active extracellular ChOs were also identified in *Nocardioides simplex* VKM Ac-2033D and *Chromobacterium* sp. DS1, and recombinant ChOs have been obtained and characterized [44, 45].

3.2 3-Ketosteroid-Δ^1-dehydrogenase (KstD)

Steroid 1(2)-dehydrogenation is generally recognized as one of the most important reactions of steroid bioconversions. Despite the effective methods developed in recent years for the chemical introduction of the 1(2)-double bond into the steroid backbone, microbiological and enzymatic approaches have not lost their relevance.

The strain of *N. simplex* VKM Ac-2033D (syn. *Pimelobacter simplex*) is well known for its high 3-ketosteroid Δ^1-dehydrogenase (3-KstD) activity towards various 3-oxosteroids. Its superior activity is due to the *kstDs* redundancy in the genome, with the highest expression level in response to cortisone 21-acetate (over 1200-fold increase) of the gene *KR76_27125* orthologous to *kstD2*. The product of the gene is considered as a promising candidate for further practical application [46, 47]. The results above are in accordance with those obtained for 3-KstDs (syn. KsdDs) from a closely relative strain, *Arthrobacter simplex* [48]. Higher catalytic activity has been shown for two of five 3-KstDs (KsdDs), namely, KstD1 and KstD2. Both enzymes have been purified and characterized, and exactly the KstD2 was estimated as the most prominent candidate for industrial application.

The highest specific activity and broad substrate range has been demonstrated for KstD from *Propionibacterium* sp. (PrKstD) [49]. When using resting recombinant *E. coli* cells harboring this enzyme, the conversion of hydrocortisone reached 92.5% at high substrate concentration (80 g/L).

Three *kstDs* identified in *M. neoaurum* DSM 1381 differed on their expression level in response to sterols and functionality [50]. The genes *kstD1* and *kstD2* were heterologously expressed in *E. coli* and *B. subtilis* hosts, and recombinant *kstD2* from *B. subtilis* showed higher activity towards AD and BNC (20-HMP) that is in good correlation with the abovementioned data obtained for KstD2 from non-mycobacterial species such as *N. simplex* and *A. simplex*.

Different aspects and progress in biotechnological application of microbial 3-KstDs were highlighted in a recent comprehensive review [51]. In addition to being used in the pharmaceutical industry in the production of valuable 1-dehydrosteroids, such as prednisolone, 6-methylprednisolone, triamcinolone, exemestane, dexamethasone, and others, 3-KstD is considered as a target for the therapeutic effect of drugs to combat *Mycobacterium tuberculosis* which is a well-known dangerous pathogen.

An interesting potential application of microbial 3-KstD in the prevention and treatment of cardiovascular diseases was proposed [52]. The idea is based on the assumption that it is possible to prevent or reverse the atherosclerotic process by introducing microbial enzymes into human cells that promote cholesterol degradation since 3-KstD and other cholesterol-degrading enzymes are absent in the human body. The first step towards testing this hypothesis was the construction and heterologous expression of the humanized actinobacterial 3-KstD in human cells of Hep3B and U-937 [52].

3.3 Laccases

Laccases are multicopper oxidases found in plants, fungi, and bacteria. Despite the wide use of laccases in various fields, including biocatalysis [53], their use for steroid bioconversion has not been studied enough. However, there are some reports illustrating the possibility of using laccases for biotransformation of steroids, for example, 3β-hydroxy-5-ene steroids (such as DHEA and pregnenolone) [54]. In a recent study, oxidation of 20-HMP (also known as BNC, or HBC) with a laccase-mediator system was reported to ensure the selective production of a pregn-4-en-3-one-20-carbaldehyde (20-POA) [55] (Fig. 1). This compound is a valuable precursor in the synthesis of progesterone, which is one of the most sought after the steroid hormones [56], as well as the intermediate in the new route for the synthesis of ursodeoxycholic acid (UDCA), which is in demand for prevention or therapy of different diseases in gastroenterology [57].

4 Heterologous Expression of Steroidogenesis Systems and Generation of Recombinant Strains for Oxyfunctionalization of Steroids

Physiological activity of steroid compounds depends on the degree of oxidation of the gonane core and the presence of various functional substituents. The regio- and stereospecific introduction of hydroxyl groups into the steroid core is a necessary step in the preparation of therapeutic steroids. Obtaining diastereomerically pure steroid alcohols is difficult (if not impossible) to carry out by chemical means. An extremely important role in the reaction of oxyfunctionalization of inactive carbon atoms belongs to cytochrome P450 monooxygenases (P450/CYP).

4.1 Bacterial Steroid Hydroxylases

New frontiers in the field of heterologous CYP expression are associated with bacterial steroid hydroxylases [58]. For example, the genes encoding P450 cytochromes CYP106A1 and CYP 106A2 from different *Bacillus megaterium* strains have been expressed in *M. smegmatis* with deletions in the key genes of steroid core degradation (*kstD, kshB*). The recombinant strain expressing CYP106A2 selectively hydroxylated AD at position 15β, thus evidencing that host proteins of *M. smegmatis* are capable of supplying electrons to heterologous cytochromes to support their hydroxylation activity [59].

Cytochrome P450 BM3 (CYP102A1) from *Priestia megaterium* (previously, *Bacillus megaterium*) has a number of unique functional features that make it promising for directed evolution and other synthetic applications. It is one of the most active hydroxylases due to its self-sufficient electron transfer [60]. Protein engineering allowed obtaining mutants capable of effective hydroxylating of the steroid core at position 7β [61]. This outstanding achievement opened new prospects for generation new whole-cell biocatalysts for steroid biotechnology. Zhao and co-workers constructed an engineered strain capable of producing 7β-OH-AD from phytosterol by introducing the mP450-BM3 into the AD-producing strain of *M. neoaurum*. To improve the target hydroxylating activity, the genes encoding NAD kinase and glucose-6-phosphate dehydrogenase were overexpressed in *M. neoaurum* to provide sufficient cofactor (NADPH) regeneration [62]. The approach is a promising way for one-pot production of steroid 7β-OH-alcohols from phytosterol.

4.2 Mammalian Steroidogenesis System

Expression of mammalian steroidogenesis system in mycobacterial hosts allowed the production of valuable C21 steroids, such as pregnenolone and progesterone, in one biotechnological step from phytosterol [63]. Progesterone is one of the key steroid hormones. It plays a crucial role in female fertility, maintaining pregnancy, prevention of preterm labor, lowering negative effects of menopause, and various gynecological pathologies. In recent years, new biological effects of progesterone were discovered such as neuro- and immunoprotective activities [64].

In a recent study, heterologous mutant cytochrome P450 CYP11A1 (mCYP11A1) and adrenodoxin reductase were connected by a flexible linker (L), and a chimera steroidogenic system mCYP11A1-L-ADR was co-expressed with adrenodoxin reductase homolog in *M. neoaurum*. A novel inorganic–biological hybrid system consisting of the engineered strain and a light-driven InP nanoparticles regenerating NADPH produced 235 + 50 mg/L progesterone from cholesterol (5 g/L) [65].

4.3 Fungal CYPs

In fungi, steroid hydroxylation reactions are usually carried out by two-component systems consisting of a CYP and NAD(P)H-cytochrome P450 reductase (CPR) [66]. Unlike bacterial ones, fungal CYPs are often membrane-bound, which creates additional difficulties in their expression in bacterial hosts. Nevertheless, methods have been developed to express fungal hydroxylases in mycolicibacteria and *Corynebacterium glutamicum*.

Expression of fungal hydroxylases in mycolicibacterial hosts opens up the prospects for obtaining target hydroxysteroids in a single biotechnological operation from phytosterol as it was demonstrated [67]. Production of 11α-hydroxy androstenedione from phytosterol became possible using recombinant *M. smegmatis* harboring the 11α-hydroxylase system from the fungus *Rhizopus oryzae*.

Unlike mycobacteria, *Corynebacterium glutamicum* does not have its own steroid degradation systems, which makes them suitable for more selective production of the target hydroxysteroid products. These bacteria have been served the chassis for expression of the genes encoding the CYP responsible for the 11β-hydroxylase activity of the fungus *Cochliobolus lunatus* and its corresponding CPR. The recombinant strain created was able to synthesize 11-β-hydroxysteroids from different steroid substrates [68]. The results give hope that efficient cellular biocatalysts for 11-β-hydroxylation, one of the most important reactions in the synthesis of corticosteroids, can be created in the near future.

A promising heterologous host for steroid bioproduction is yeasts, including *Saccharomyces cerevisiae*, *Yarrowia lipolytica*, and *Pichia pastoris*. To date, engineered yeasts have been described capable of producing boldenone, hydrocortisone, cholesterol, phytosterol, and their derivatives [69].

5 Conclusions

Steroid bioconversion is one of the most actively developing areas of applied microbiology and biotechnology. Significant advances have been made in the past few years in the development of engineered strains with improved or new biocatalytic capabilities. Effective "microbial cell factories" based on mycolicibacteria are capable of bioconverting phytosterols into not only the highly demanded C19 steroids (AD, ADD, 9-OH-AD) but also the C22 steroids (BA, BNC, or 20-HMP) and their hydroxylated derivatives.

Significant progress has been made in the field of obtaining and applying steroid-modifying enzymes and heterologous expression of foreign systems of steroidogenesis in mycolicibacteria and other hosts, which allows expanding the range of valuable steroids that can be obtained from phytosterols by one- or two-stage biotechnological operations. It can be expected that many developments will go beyond the laboratory and will be scaled up to industrial level in the near future.

Acknowledgments

The Russian Science Foundation is gratefully acknowledged for the support (grant no. 21-64-00024). The author is grateful to D. Dovbnya, A. Shutov, and A. Byakov for their assistance in preparing the manuscript and the figures.

References

1. QY Research (2018) Global steroids market: corticosteroids segment to reach value of US$ 8.6 Bn by 2025 end – QY Research, Inc. Available via PRNewswire. www.prnewswire.com/in/news-releases/global-steroids-market-corticosteroids-segment-to-reach-value-of-us-86-bn-by-2025-end—qy-research-inc-693708281.html
2. The Business Research Company (2023) Available via: www.databridgemarketresearch.com/reports/
3. Ahmed MH, Hassan A (2020) Dexamethasone for the treatment of coronavirus disease (COVID-19): a review. SN Compr Clin Med 2:2637–2646. https://doi.org/10.1007/s42399-020-00610-8
4. Global androgens and anabolic steroids market – industry trends and forecast to 2029. Available via: databridgemarketresearch.com/reports/global-androgens-and-anabolic-steroids-market
5. Olivera ER, Luengo JM (2019) Steroids as environmental compounds recalcitrant to degradation: genetic mechanisms of bacterial biodegradation pathways. Genes 10:512. https://doi.org/10.3390/genes10070512
6. Feller FM, Holert J, Yücel O, Philipp B (2021) Degradation of bile acids by soil and water bacteria. Microorganisms 9:1759. https://doi.org/10.3390/microorganisms9081759
7. Mendelski MN, Dölling R, Feller FM, Hofmann D, Ramos Fangmeier L, Ludwig KC, Yücel O, Mährlein A, Paul RJ, Philipp B (2019) Steroids originating from bacterial bile acid degradation affect *Caenorhabditis elegans* and indicate potential risks for the fauna of manured soils. Sci Rep 9:11120. https://doi.org/10.1038/s41598-019-47476-y
8. Holert J, Cardenas E, Bergstrand LH, Zaikova E, Hahn AS, Hallam SJ, Mohn WW (2018) Metagenomes reveal global distribution of bacterial steroid catabolism in natural, engineered, and host environments. mBio 9:e02345-17. https://doi.org/10.1128/mBio.02345-17
9. Lobastova T, Fokina V, Tarlachkov S, Shutov A, Bragin E, Kazantsev A, Donova M (2021) Steroid metabolism in thermophilic

actinobacteria *Saccharopolyspora hirsuta* VKM Ac-666T. Microorganisms 9(12):2554–2572. https://doi.org/10.3390/microorganisms9122554
10. Lobastova T, Fokina V, Pozdnyakova-Filatova I, Tarlachkov S, Shutov A, Donova M (2022) Insight into different stages of steroid degradation in thermophilic *Saccharopolyspora hirsuta* VKM Ac-666T strain. Int J Mol Sci 23(24): 16174. https://doi.org/10.3390/ijms232416174
11. Feller FM, Wöhlbrand L, Holert J, Schnaars V, Elsner L, Mohn WW, Rabus R, Philipp B (2021) Proteome, bioinformatic, and functional analyses reveal a distinct and conserved metabolic pathway for bile salt degradation in the *Sphingomonadaceae*. Appl Environ Microbiol 87:e00987–e00921. https://doi.org/10.1128/AEM.00987-21
12. Gupta RS, Lo B, Son J (2018) Phylogenomics and comparative genomic studies robustly support division of the genus *Mycobacterium* into an emended genus *Mycobacterium* and four novel genera. Front Microbiol 9:67. https://doi.org/10.3389/fmicb.2018.00067
13. Kieslich K (1985) Microbial side-chain degradation of sterols. J Basic Microbiol 25(7): 461–474. https://doi.org/10.1002/jobm.3620250713
14. Zhao A, Zhang X, Li Y, Wang Z, Lv Y, Liu J, Alam MA, Xiong W, Xu J (2021) *Mycolicibacterium* cell factory for the production of steroid-based drug intermediates. Biotechnol Adv. https://doi.org/10.1016/j.biotechadv.2021.107860
15. Fernández-Cabezón L, Galán B, García JL (2018) New insights on steroid biotechnology. Front Microbiol 9. https://doi.org/10.3389/fmicb.2018.00958
16. Galan B, Garcia-Fernandez J, Felpeto-Santero C, Fernandez-Cabezon L, Garcia JL (2019) Bacterial metabolism of steroids. In: Aerobic utilization of hydrocarbons, oils, and lipids, pp 315–336. https://doi.org/10.1007/978-3-319-50418-6_43
17. Giorgi V, Pilar Menandez P, Garcia Carnelli C (2019) Microbial transformation of cholesterol: reactions and practical aspects—an update. World J Microbiol Biotechnol 35: 131. https://doi.org/10.1007/s11274-019-2708-8
18. Feng J, Wu Q, Zhu D, Ma Y (2022) Biotransformation enables innovations toward green synthesis of steroidal pharmaceuticals. ChemSusChem 15(9):e202102399. https://doi.org/10.1002/cssc.202102399
19. Galan B, Uhia I, Garcia-Fernandez E, Martinez I, Bahillo E, de la Fuente JL, Barredo JL, Fernandez-Cabezon L, Garcia JL (2017) *Mycobacterium smegmatis* is a suitable cell factory for the production of steroidic synthons. Microb Biotechnol 10:138–150. https://doi.org/10.1111/1751-7915.12429
20. Weckerle T, Ewald H, Guth P, Knorr K-H, Philipp B, Holert J (2022) Biogas digestate as a sustainable phytosterol source for biotechnological cascade valorization. Microb Biotechnol:1–13. https://doi.org/10.1111/1751-7915.14174
21. Randhir A, Laird DW, Maker G, Trengove R, Moheimani NR (2020) Microalgae: a potential sustainable commercial source of sterols. Algal Res 46:101772. https://doi.org/10.1016/j.algal.2019.101772
22. Gu Y, Jiao X, Ye L et al (2021) Metabolic engineering strategies for *de novo* biosynthesis of sterols and steroids in yeast. Bioresour Bioprocess 8:110. https://doi.org/10.1186/s40643-021-00460-9
23. Prasad M, Jayaraman S, Eladl MA, El-Sherbiny M, Abdelrahman MAE, Veeraraghavan VP, Vengadassalapathy S, Umapathy VR, Jaffer Hussain SF, Krishnamoorthy K et al (2022) A comprehensive review on therapeutic perspectives of phytosterols in insulin resistance: a mechanistic approach. Molecules 27:1595. https://doi.org/10.3390/molecules27051595
24. Chang H, Zhang H, Zhu L, Zhang W, You S, Qi W, Qian J, Su R, He Z (2020) A combined strategy of metabolic pathway regulation and two-step bioprocess for improved 4-androstene-3,17-dione production with an engineered *Mycobacterium neoaurum*. Biochem Eng J 164:107789. https://doi.org/10.1016/j.bej.2020.107789A
25. Nunes VO, Vanzellotti NC, Fraga JL, Pessoa FLP, Ferreira TF, Amaral PFF (2020) Biotransformation of phytosterols into androstenedione—a technological prospecting study. Molecules 27:3164. https://doi.org/10.3390/molecules27103164
26. Zhou X, Zhang Y, Shen Y, Zhang X, Zan Z, Xia M, Luo J, Wang M (2020) Efficient repeated batch production of androstenedione using untreated cane molasses by *Mycobacterium neoaurum* driven by ATP futile cycle. Bioresour Technol 309:123307. https://doi.org/10.1016/j.biortech.2020.123307
27. Shao M, Zhang X, Rao Z, Xu M, Yang T, Xu Z, Yang S (2019) Identification of steroid C27 monooxygenase isoenzymes involved in sterol catabolism and stepwise pathway engineering of *Mycobacterium neoaurum* for improved androst-1,4-diene-3,17-dione production. J Ind Microbiol Biotechnol 46:635–647. https://doi.org/10.1007/s10295-018-02135-5

28. Xiong L-B, Liu H-H, Song X-W, Meng X-G, Liu X-Z, Ji Y-Q, Wang F-Q, Wei D-Z (2020) Improving the biotransformation of phytosterols to 9α-hydroxy-4-androstene-3,17-dione by deleting embC associated with the assembly of cell envelope in *Mycobacterium neoaurum*. J Biotechnol 323:341–346. https://doi.org/10.1016/j.jbiotec.2020.09.019
29. Xiong LB, Liu HH, Zhao M et al (2020) Enhancing the bioconversion of phytosterols to steroidal intermediates by the deficiency of kasB in the cell wall synthesis of *Mycobacterium neoaurum*. Microb Cell Factories 19:80. https://doi.org/10.1186/s12934-020-01335-y
30. Sun H, Yang J, Song H (2020) Engineering mycobacteria artificial promoters and ribosomal binding sites for enhanced sterol production. Biochem Eng J:107739. https://doi.org/10.1016/j.bej.2020.107739
31. Yuan CY, Ma ZG, Zhang JX et al (2021) Production of 9,21-dihydroxy-20-methyl-pregna-4-en-3-one from phytosterols in *mycobacterium neoaurum* by modifying multiple genes and improving the intracellular environment. Microb Cell Factories 20:229. https://doi.org/10.1186/s12934-021-01717-w
32. Yuan C, Song S, He J, Zhang J, Liu X, Pena EL, Sun J, Shi J, Su Z, Zhang B (2022) Bioconversion of phytosterols to 9-hydroxy-3-oxo-4,17-pregnadiene-20-carboxylic acid methyl ester by enoyl-CoA deficiency and modifying multiple genes in *Mycolicibacterium neoaurum*. Appl Environ Microbiol 88:22. https://doi.org/10.1128/aem.01303-22
33. Civra A, Cagno V, Donalisio M et al (2014) Inhibition of pathogenic non-enveloped viruses by 25-hydroxycholesterol and 27-hydroxycholesterol. Sci Rep 4:7487. https://doi.org/10.1038/srep07487
34. Marcello A, Civra A, Bonotto RM, Alves LN, Rajasekharan S, Giacobone C, Caccia C, Cavalli R, Adami M, Brambilla P, Lembo D, Poli G, Leoni V (2020) The cholesterol metabolite 27-hydroxycholesterol inhibits SARS-CoV-2 and is markedly decreased in COVID-19 patients. Redox Biol 36:101682. https://doi.org/10.1016/j.redox.2020.101682
35. Crowe AM, Casabon I, Brown KL, Liu J, Lian J, Rogalski JC, Hurst TE, Snieckus V, Foster LJ, Eltis LD (2017) Catabolism of the last two steroid rings in *Mycobacterium tuberculosis* and other bacteria. mBio 8:e00321-17. https://doi.org/10.1128/mBio.00321-17
36. Liu N, Feng J, Zhang R, Chen X, Li X, Yao P, Wu Q, Ma Y, Zhu D (2019) Efficient microbial synthesis of key steroidal intermediates from bio-renewable phytosterols by genetically modified *Mycobacterium fortuitum* strains. Green Chem 21:4076. https://doi.org/10.1039/C9GC01267B
37. Fernández-Cabezón L, Galán B, García JL (2017) Engineering *Mycobacterium smegmatis* for testosterone production. Microb Biotechnol 10:151–161. https://doi.org/10.1111/1751-7915.12433
38. Tekucheva DN, Nikolayeva VM, Karpov MV, Timakova TA, Shutov AV, Donova MV (2022) Bioproduction of testosterone from phytosterol by *Mycolicibacterium neoaurum* strains: "one-pot", two modes. Bioresour Bioprocess 9:116. https://doi.org/10.1186/s40643-022-00602-7
39. Tang R, Shen Y, Xia M, Tua L, Luo J, Geng Y, Gao T, Zhou H, Zhao Y, Wang M (2019) A highly efficient step-wise biotransformation strategy for direct conversion of phytosterol to boldenone. Bioresour Technol 283:242–250. https://doi.org/10.1016/j.biortech.2019.03.058
40. Tekucheva DN, Fokina VV, Nikolaeva VM, Shutov AA, Donova MV (2022) Cascade biotransformation of phytosterol to testosterone by *Mycolicibacterium neoaurum* VKM Ac-1815D and *Nocardioides simplex* VKM Ac-2033D Strains. Microbiology 91:303–312. https://doi.org/10.1134/S0026261722300099
41. Kong K, Zhang M, Zhang H et al (2022) Structures and molecular mechanisms of action of the cholesterol C17 side-chain-degrading enzymes. Syst Microbiol Biomanuf. https://doi.org/10.1007/s43393-022-00083-x
42. El-Naggar NEA, Soliman HM, El-Shweihy NM (2018) Extracellular cholesterol oxidase production by *Streptomyces aegyptia*, *in vitro* anticancer activities against rhabdomyosarcoma, breast cancer cell-lines and in vivo apoptosis. Sci Rep 8:2706. https://doi.org/10.1038/s41598-018-20786-3
43. Alapati K, Handanahal SS (2021) Cytotoxic activity of cholesterol oxidase produced by *Streptomyces* sp. AKHSS against cancerous cell lines: mechanism of action in HeLa cells. World J Microbiol Biotechnol 37:141. https://doi.org/10.1007/s11274-021-03076-5
44. Fokina VV, Karpov MV, Kollerov VV, Bragin EY, Epiktetov DO, Sviridov AV, Kazantsev AV, Shutov AA, Donova MV (2022) Recombinant extracellular cholesterol oxidase from *Nocardioides simplex*. Biochem Moscow 87:903–915. https://doi.org/10.1134/S0006297922090048
45. Fazaeli A, Golestani A, Lakzaei M, Rasi Varaei SS, Aminian M (2019) Expression optimization, purification, and functional

characterization of cholesterol oxidase from *Chromobacterium* sp. DS1. PLoS One 14(2): e0212217. https://doi.org/10.1371/journal.pone.0212217

46. Shtratnikova VY, Schelkunov MI, Fokina VV, Bragin EY, Shutov AA, Donova MV (2021) Different genome-wide transcriptome responses of *Nocardioides simplex* VKM Ac-2033D to phytosterol and cortisone 21-acetate. BMC Biotechnol 21:7. https://doi.org/10.1186/s12896-021-00668-9
47. Shtratnikova VY, Schelkunov MY, Fokina VY, Bragin EY, Lobastova TG, Shutov AA, Kazantsev AV, Donova MV (2020) Genome-wide transcriptome profiling provides insight on cholesterol and lithocholate degradation mechanisms in *Nocardioides simplex* VKM Ac-2033D. Genes (Basel). https://doi.org/10.3390/genes11101229
48. Mao S, Sun J, Wang L, Gao X, Liu X, Lu F, Qin H-M (2022) Mining and characterization of 3-ketosteroid-Δ1-dehydrogenases from *Arthrobacter simplex* genome and applications for steroid dehydrogenation. Biochem Eng J 181:108383. https://doi.org/10.1016/j.bej.2022.108383
49. Wang Y, Zhang R, Feng J, Wu Q, Zhu D, Ma Y (2022) A new 3-ketosteroid-D1–dehydrogenase with high activity and broad substrate scope for efficient transformation of hydrocortisone at high substrate concentration. Microorganisms 10:508. https://doi.org/10.3390/microorganisms10030508
50. Zhang R, Liu X (2018) Wang Y (2018) identification, function, and application of 3-ketosteroid Δ^1-dehydrogenase isozymes in *Mycobacterium neoaurum* DSM 1381 for the production of steroidic synthons. Microb Cell Factories 17:77. https://doi.org/10.1186/s12934-018-0916-9
51. Rohman A, Dijkstra BW (2021) Application of microbial 3-ketosteroid Δ^1-dehydrogenases in biotechnology. Biotechnol Adv 49:107751. https://doi.org/10.1016/j.biotechadv.2021.107751
52. D'Arcy BM, Swingle MR, Schambeau L et al (2019) Development of a synthetic 3-ketosteroid Δ^1-dehydrogenase for the generation of a novel catabolic pathway enabling cholesterol degradation in human cells. Sci Rep 9:5969. https://doi.org/10.1038/s41598-019-42046-8
53. Bassanini I, Ferrandi EE, Riva S, Monti D (2020) Biocatalysis with laccases: an updated overview. Catalysts 11(1):26. https://doi.org/10.3390/catal11010026
54. Khomutov SM, Shutov AA, Chernikh AM, Myasoedova NM, Golovleva LA, Donova MV (2016) Laccase-mediated oxyfunctionalization of 3β-hydroxy-Δ5-steroids. J Mol Catal B Enzym 123:47–52. https://doi.org/10.1016/j.molcatb.2015.11.004
55. Khomutov S, Shutov A, Averin A, Dovbnya D, Donova M (2022) Laccase-mediated oxidation of steroid alcohols in the presence of methylated β-cyclodextrin: from inhibition to selective synthesis. J Chem Technol Biotechnol 97(5):1162–1170. https://doi.org/10.1002/jctb.7000
56. Peng H, Wang Y, Jiang K, Chen X, Zhang W, Zhang Y et al (2021) A dual role reductase from phytosterols catabolism enables the efficient production of valuable steroid precursors. Angew Chem Int Ed 60:5414–5420. https://doi.org/10.1002/anie.202015462
57. Wang J, Gu XZ, He LM, Li CC, Qiu WW (2020) Synthesis of ursodeoxycholic acid from plant-source (20S)-21-hydroxy-20-methylpregn-4-en-3-one. Steroids 157: 108600. https://doi.org/10.1016/j.steroids.2020.108600
58. Szaleniec M, Wojtkiewicz AM, Bernhardt R, Borowski T, Donova M (2018) Bacterial steroid hydroxylases: enzyme classes, their functions and comparison of their catalytic mechanisms. Appl Microbiol Biotechnol 102(19):8153–8171. https://doi.org/10.1007/s00253-018-9239-3
59. Karpov MV, Nikolaeva VM, Fokina VV, Shutov AA, Kazantsev AV, Strizhov NI, Donova MV (2022) Creation and functional analysis of *Mycolicibacterium smegmatis* recombinant strains carrying the bacillary cytochromes CYP106A1 and CYP106A2 genes. Appl Biochem Microbiol 58(9):947–957. https://doi.org/10.1134/S0003683822090058
60. Munro AW, Leys DG, McLean KJ, Marshall KR, Ost TW, Daff S, Miles CS, Chapman SK, Lysek DA, Moser CC (2002) P450 BM3: the very model of a modern flavocytochrome. Trends Biochem Sci 27:250–257. https://doi.org/10.1016/s0968-0004(02)02086-8
61. Li A, Acevedo-Rocha CG, D'Amore L, Chen J, Peng Y, Garcia-Borras M, Gao C, Zhu J, Rickerby H, Osuna S (2020) Regio- and stereoselective steroid hydroxylation at C7 by cytochrome P450 monooxygenase mutants. Angew Chem Int Ed 59:12499–12505. https://doi.org/10.1002/anie.202003139
62. Zhao YQ, Liu YJ, Ji WT, Liu K, Gao B, Tao XY, Zhao M, Wang FQ, Wei DZ (2022) One-pot biosynthesis of 7β-hydroxyandrost-4-ene-3,17-dione from phytosterols by cofactor regeneration system in engineered *Mycolicibacterium neoaurum*. Microb Cell Factories 21:

59. https://doi.org/10.1186/s12934-022-01786-5

63. Strizhov N, Fokina V, Sukhodolskaya G, Dovbnya D, Karpov M, Shutov A, Novikova L, Donova M (2014) Progesterone biosynthesis by combined action of adrenal steroidogenic and mycobacterial enzymes in fast growing mycobacteria. New Biotechnol 31:67. https://doi.org/10.1016/j.nbt.2014.05.1766

64. Kolatorova L, Vitku J, Hill M, Parizek A (2022) Progesterone: a steroid with wide range of effects in physiology as well as human medicine. Int J Mol Sci 23(14):7989. https://doi.org/10.3390/ijms23147989

65. Liu K, Wang FQ, Liu K, Zhao Y, Gao B, Tao X, Wei D (2022) Light-driven progesterone production by InP–(*Mneoaurum*) biohybrid system. Bioresour Bioprocess 9:93. https://doi.org/10.1186/s40643-022-00575-7

66. Kristan K, Rižner TL (2012) Steroid-transforming enzymes in fungi. J Steroid Biochem Mol Biol 129(1–2):79–91. https://doi.org/10.1016/j.jsbmb.2011.08.012

67. Felpeto-Santero C, Galán B, García JL (2021) Production of 11α-hydroxysteroids from sterols in a single fermentation step by *Mycolicibacterium smegmatis*. Microb Biotechnol 14:2514–2524. https://doi.org/10.1111/1751-7915.13735

68. Felpeto-Santero C, Galán B, Luengo JM, Fernández-Cañon JM, del Cerro C, Medrano FJ, García JL (2019) Identification and expression of the 11β-steroid hydroxylase from *Cochliobolus lunatus* in *Corynebacterium glutamicum*. Microbial Biotechnol 12(5):856–868. https://doi.org/10.1111/1751-7915.13428

69. Xu S, Li Y (2020) Yeast as a promising heterologous host for steroid bioproduction. J Ind Microbiol Biotechnol 47(9–10):829–843. https://doi.org/10.1007/s10295-020-02291-7

Part II

Microbial Screening and Genetic Manipulation

Chapter 2

Isolation of Environmental Bacteria Able to Degrade Sterols and/or Bile Acids: Determination of Cholesterol Oxidase and Several Hydroxysteroid Dehydrogenase Activities in *Rhodococcus*, *Gordonia*, and *Pseudomonas putida*

Alejandro Chamizo-Ampudia, Luis Getino, José M. Luengo, and Elias R. Olivera

Abstract

Interest about the isolation and characterization of steroid-catabolizing bacteria has increased over time due to the massive release of these recalcitrant compounds and their deleterious effects or their biotransformation derivatives as endocrine disruptors for wildlife, as well as their potential use in biotechnological approaches for the synthesis of pharmacological compounds. Thus, in this chapter, an isolation protocol to select environmental bacteria able to degrade sterols, bile acids, and androgens is shown. Moreover, procedures for the determination of cholesterol oxidase or different hydroxysteroid dehydrogenase activities in *Pseudomonas putida* DOC21, *Rhodococcus* sp. HE24.12, *Gordonia* sp. HE24.4J and *Gordonia* sp. HE24.3 are also detailed.

Key words *Pseudomonas*, *Rhodococcus*, *Gordonia*, Catabolism, Cholic acid, Cholesterol, Androstanolone, Androstenedione, Testosterone, Cholesterol oxidase, 3α-Hydroxysteroid dehydrogenase, 17-β-Hydroxysteroid dehydrogenase, Enzyme activity

1 Introduction

Steroidogenesis has been proposed as one of the processes defining eukaryogenesis [1]. However, although the synthesis of steroids is restricted to eukaryotic cells, fulfilling critical functions in them, only bacteria can degrade these environmentally recalcitrant compounds [2]. However, steroids are highly recalcitrant to microbial mineralization because of their extremely low solubility in water due to the inherent hydrophobic nature of their common sterane system of rings and the low presence of polar functional groups in their structure. Nevertheless, the entry of these compounds into environment, through the excreta of animals or biomass

Carlos Barreiro and José-Luis Barredo (eds.), *Microbial Steroids: Methods and Protocols*, Methods in Molecular Biology, vol. 2704, https://doi.org/10.1007/978-1-0716-3385-4_2,

decomposition, has been occurring since more than 635 million years ago [3], allowing bacteria to develop metabolic mechanisms using these compounds as carbon and energy sources for growth. In that sense, steroids constitute an attractive nutrient for bacteria, particularly in niches with low content of available carbon and energy sources or even in situations of high competition for nutrients in which steroids could also be toxic for some microorganisms. Thus, the metabolic potential of soil bacteria has evidenced the important role of carbon cycle in allowing the mineralization of steroid compounds [4, 5].

On the other hand, during the Anthropocene, massive livestock farming and release of pharmaceutical industry and urban waste waters have contributed to the massive spread of this kind of compounds, along with other steroids of synthetic origin, not observed until now in nature (e.g., ethinylestradiol, dexamethasone, and others) [6–10].

In addition, it must also be considered that microbial activities acting over steroids found in the environment could result into the synthesis of steroid hormones contributing to their environmental dissemination [11]. For example, it has been described that some microorganisms from river sediments can release androgens from phytosterols present in pulp and paper mill effluents [12, 13].

According to that, concerns about the ecotoxicological impact of the presence of this kind of compounds, mainly those affecting endocrine system of wildlife and even human, have arisen [14–21]. This has originated an increasing need for new approaches for the bioremediation of steroid compounds which has triggered the isolation of bacterial strains able to metabolize these compounds as well as the identification and analysis of the metabolic answers that different bacteria have adopted for their use as carbon and energy source [4, 5, 22].

Additionally, the interest on these microorganisms is also increased by the fact that they can be used in pharmaceutical industry to obtain synthons useful to produce synthetic steroids with clinical relevance [23–26].

In bacteria, different pathways have been identified for the degradation of different kinds of steroids, according to their chemical structure or the environmental conditions in which they are metabolized [27]. Thus, aerobic metabolism of sterols, bile acids, and androgens is performed through the so-called 9,10-*seco* pathway. In this pathway, cholesterol and bile acids suffer from the oxidation of the A and B rings from the sterane nucleus at the time that the C-17 side chain is removed leading to the formation of a common intermediate, androst-4-en-3,17-dione (or a hydroxylated derivative if using a hydroxylated bile acid) [4].

In that sense, the first step in the degradation of sterols and bile acids through this 9,10-*seco* pathway involves oxidation of the 3-hydroxyl substituent to an oxo derivative.

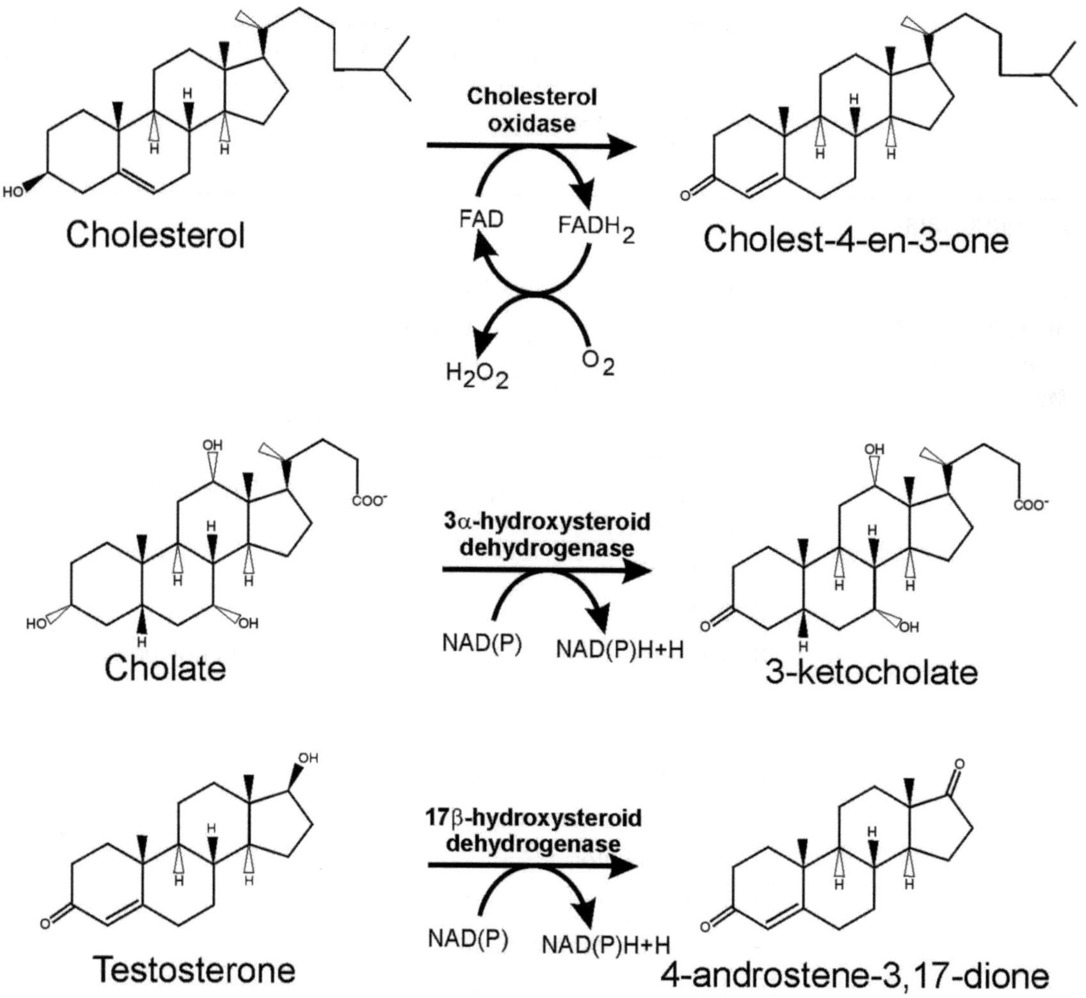

Fig. 1 Reactions catalyzed by cholesterol oxidase, 3α-hydroxysteroid dehydrogenase, and 17β-hydroxysteroid dehydrogenase

For sterols, this initial reaction involves, in most of the cases, the participation of cholesterol oxidases catalyzing the oxidation of the 3β-hydroxy group at the time that isomerizes the C5–C6 unsaturation present in these molecules to yield 3-keto-4-ene molecules. Taking as a model compound cholesterol, cholesterol oxidases use molecular oxygen to render 4-cholesten-3-one (Fig. 1). Most of the bacteria catabolizing cholesterol produce extracellular cholesterol oxidases, either released to extracellular media or linked to cellular surface [28].

Bile acids also have a hydroxyl group in the C3 position, although in α-configuration, that should be oxidized to an oxo-group (Fig. 1). These molecules are not substrate for cholesterol oxidases but 3α-hydroxysteroid dehydrogenases, NAD(P)-dependent enzymes belonging to the short-chain dehydrogenase/reductase superfamily [29–31].

Testosterone catabolism converges at the level of androst-4-en-3,17-dione directly through the NAD-dependent oxidation of the 17β-hydroxy substituent leading to the formation of a 17-oxo-derivative (Fig. 1) [32].

2 Materials

2.1 Preparation of Steroids as Carbon Sources for Culture Media

1. Cholesterol (Sigma-Aldrich).
2. Testosterone (Sigma-Aldrich).
3. Cholic acid (Sigma-Aldrich).
4. Androsterone (3α-hydroxy-5α-androstan-17-one) (Acros Organics).
5. Randomly methylated β-cyclodextrins, Trappsol® (TRMB-T, CTD Inc.).
6. Chloroform.
7. 2-Propanol.
8. Tyloxapol [4-(1,1,3,3-tetramethylbutyl)phenol polymer with formaldehyde and oxirane].
9. Sodium hydroxide.
10. Potassium dihydrogen phosphate.
11. Potassium chloride.
12. Sodium chloride.
13. Dipotassium hydrogen phosphate.
14. Laboratory Water Bath.
15. Ultrasonic Bath 360 W.
16. Autoclave equipment.

2.2 Isolation of Environmental Strains Able to Degrade Cholesterol and/or Cholic Acid

1. *Pseudomonas* chemically defined minimal medium (MM) [33]: 13.6 g/L of KPO_4H_2, 2.0 g/L of $(NH_4)_2SO_4$, 0.25 g/L of $MgSO_4.7H_2O$, and 0.005 g/L of $FeSO_4.7H_2O$, pH 7.0 [33]; add carbon source at the final concentration required in each experiment (usually 2–5 mM when bile acids, testosterone, and cholesterol are used) prepared as indicated in Subheadings 3.1.1 and 3.1.2. In other cases (4-hydroxyphenylacetic acid or other non-steroid compounds), 10 mM final concentration is used. For solid MM, 2% (w/v) of agar must be added.
2. Stock solutions of cholesterol, cholic acid, and/or testosterone (Subheading 3.1).
3. Erlenmeyer flasks (500 mL).
4. Rotary shaker.
5. Thermal cycler.
6. Polymerase chain reaction reagents.

7. Autoclave equipment.
8. Specific oligonucleotide primers for the amplification of housekeeping genes to check the isogenicity of isolated strains and to ascribe them to a phylogenetic group: 16S rDNA (primer 6F, 5′- GGAGAGTTAGATCTTGGCTCAG -3′, and primer 1510R, 5′- GTGCTGCAGGGTTACCTTGTTAC GACT-3′) and *rpoB* (primer LAPS5, 5′-TGGCCGAGAACCA GTTCCGCGT-3′, and primer LPAS27, 5′-CGGCTTCGTC CAGCTTGTTCAG-3′).

2.3 Identification of Strains Showing Extracellular Cholesterol Oxidase Activities

1. Tryptic soy agar medium (TSA): 17 g/L of pancreatic digest of casein, 3 g/L of papaic digest of soybean meal, 5 g/L of sodium chloride, 2.5 g/L of glucose, 2.5 g/L of dipotassium hydrogen phosphate, and 15 g/L of agar.
2. Solid *Pseudomonas* chemically defined minimal medium (MM with 2% agar) (Subheading 2.2) [33], containing 2 or 5 mM cholesterol as carbon source.
3. Cholesterol stock solution (Subheading 3.1).
4. Petri dishes.
5. Rotary shaker.
6. Autoclave.

2.4 Measurement of Extracellular Cholesterol Oxidase Activity in Culture Supernatants

1. *Pseudomonas* MM (Subheading 2.2), containing 2 mM cholesterol as a sole carbon source.
2. Cholesterol (Sigma-Aldrich).
3. Glucose.
4. Erlenmeyer flasks (500 mL).
5. Rotary shaker.
6. Centrifuge (e.g., Eppendorf 5810R).
7. 0.45 μm pore size filter (e.g., SLHA0335B, Millipore).
8. Ammonium sulfate.
9. Refrigerator 4 °C.
10. Milli-Q® grade sterile water.
11. Desalting PD-10 (G25) columns (Cytiva).
12. Bradford protein assay kit (Thermo Fisher).
13. Tris–HCl 50 mM, pH 7.5.
14. 2-Propanol.
15. Ethanol.
16. 4-Cholesten-3-one (Sigma-Aldrich).
17. Tris–HCl 125 mM, pH 7.5.
18. Autoclave.

2.5 Measurement of 3α-Hydroxysteroid Dehydrogenase Activity in Cell Extracts from Cells Grown with Cholic Acid as Sole Carbon Source

1. *Pseudomonas* chemically defined minimal medium (MM) (Subheading 2.2), containing 2 mM cholic acid or testosterone as a sole carbon source.
2. Sodium cholate (Sigma-Aldrich).
3. Testosterone (Sigma-Aldrich).
4. Glucose.
5. Erlenmeyer flasks (500 mL).
6. Rotary shaker.
7. Centrifuge (e.g., Eppendorf 5810R).
8. Phosphate buffered saline (PBS): 137 mM NaCl, 2.7 mM KCI, 8 mM Na_2HPO_4, and 2 mM KH_2PO_4, pH 7.4.
9. Phenylmethylsulfonyl fluoride (PMSF) (Thermo Fisher).
10. Ultrasonic homogenizer (e.g., Sonifier 450, Branson).
11. Phosphate buffer 200 mM, pH 7.4.
12. Bradford protein assay kit (Thermo Fisher).
13. Milli-Q® water.
14. $NADP^+$ 100 mM (Roche).
15. Androsterone (Acros Organics).
16. Methanol (UHPLC quality).
17. Spectrophotometer (e.g., DU 640, Beckman Coulter Ltd.).
18. Freezer −20 °C.
19. HPLC equipment with a tims-TOF detector.
20. Zorbax StableBond-C18 (4.6 mm × 250 mm) chromatographic column (Agilent).
21. Autoclave.

2.6 Measurement of 17β-Hydroxysteroid Dehydrogenase Activity in Cell Extracts

1. *Pseudomonas* MM (Subheading 2.2) containing 2 mM cholic acid or testosterone as a sole carbon source.
2. Sodium cholate (Sigma-Aldrich).
3. Testosterone (Sigma-Aldrich).
4. Glucose.
5. Erlenmeyer flasks (500 mL).
6. Rotary shaker.
7. Centrifuge (e.g., Eppendorf 5810R).
8. Phosphate buffered saline solution (PBS) pH 7.4 (Subheading 2.5).
9. Phenylmethylsulfonyl fluoride, PMSF (Thermo Fisher).
10. Ultrasonic homogenizer (e.g., Sonifier 450, Branson).
11. Phosphate buffer 200 mM pH 7.4.
12. Bradford protein assay kit (Thermo Fisher).

13. Milli-Q® grade water.
14. 100 mM $NADP^+$ solution in Milli-Q® water (Roche).
15. Methanol (UHPLC quality).
16. Spectrophotometer (e.g., DU 640, Beckman Coulter Ltd.).
17. Freezer −20 °C.
18. HPLC equipment with a tims-TOF detector.
19. Zorbax StableBond-C18 (4.6 mm × 250 mm) chromatographic column (Agilent).
20. Autoclave.

2.7 UHPLC Analyses of Selected Steroids

1. Chloroform.
2. Vortex mixer.
3. Centrifuge (e.g., Eppendorf 5810R).
4. Estrone (Sigma-Aldrich).
5. Thermoblock.
6. Methanol (UHPLC quality).
7. Acetonitrile (UHPLC quality).
8. Milli-Q® grade water.
9. Sonication bath (360 W).
10. 0.22 μm pore size filters (Millipore).
11. HPLC vials of 2 mL (Thermo Fisher).
12. UHPLC equipment with a tims-TOF detector.
13. Zorbax StableBond-C18 (4.6 mm × 250 mm) column (Agilent).

3 Methods

3.1 Preparation of Steroids as Carbon Sources for Culture Media

In general, steroids are poorly soluble in polar compounds, so to use them in culture media, their highly homogeneous dispersion should be obtained. This could be reached by the use of methylated β-cyclodextrins [34] or other surfactants (e.g., tyloxapol, a non-ionic surfactant) [35]. Thus, methylated cyclodextrins possess a hydrophobic inner cavity where steroids are included, whereas their surface is hydrophilic, allowing the dispersion in the culture media (*see* **Note 1**).

3.1.1 Preparation of Cholesterol, Bile Acids, or Testosterone Emulsified with Methyl-β-cyclodextrins

1. Dissolve 1 g of methyl-β-cyclodextrins in 11 mL of PBS.
2. On the other hand, dissolve 45 mg of cholesterol, BAs, or testosterone in 400 μL of 2-propanol/chloroform (2:1) in a glass tube.

3. Later, heat the cyclodextrin solution to 80 °C and add the steroid solution drop by drop while shaking the mixture (*see* **Note 2**).
4. In this new solution of methyl-β-cyclodextrins with cholesterol, the concentration of cholesterol is 10.2 mM.
5. Once dissolved, sterilize in autoclave for 15 min at 121 °C.

3.1.2 Preparation of Cholesterol or Testosterone Emulsified in Tyloxapol

1. Dissolve 10% (v/v) tyloxapol in Milli-Q® water.
2. Once the tyloxapol is dissolved, add the cholesterol, BAs, or testosterone to a final concentration of 5 mM.
3. Once these compounds have been added, sonicate in a bath at 80 °C for approximately 30 min (or until complete solution).
4. Once dissolved, autoclave for 15 min at 121 °C.

3.2 Isolation of Environmental Strains Able to Degrade Cholesterol and/or Cholate

This is the protocol of routine used in our laboratory to select the strains shown in this chapter [36].

1. Make suspensions of soil samples (1 g of soil in 200 mL of sterile distilled water) and shake them vigorously for 1 h at 25 °C.
2. Inoculate 500 mL Erlenmeyer flasks containing 100 mL of MM supplied with CA or cholesterol (4 mM) as a carbon source with 500 μL of the suspensions obtained in **step 1**.
3. Incubate the flasks on a rotary shaker (250 rpm) for 48 h at 30 °C.
4. After 24 and 48 h of incubation, spread serial dilutions from these cultures into MM plates containing CA 5 mM as the sole carbon source.
5. Incubate the plates for 48–72 h at 30 °C, select isolated colonies, and streak them on new MM plates containing CA (5 mM, MM-CA). Incubate the plates 48 h at 30 °C.
6. Pick the colonies from these new plates with sterile toothpicks and suspend the cells in 200 μL of sterile Milli-Q^R water. Plate serial dilutions of these cultures on MM-CA and incubate the plates for 48 h at 30 °C.
7. Streak the new colonies in MM-CA and test the isogenicity of cultures through colony PCR amplification and sequencing of housekeeping genes (16S rDNA using primers 6F and 1510R and *rpoB* with primers LAPS5 and LPAS27). Sequences of these oligonucleotides are indicated in Subheading 2.2.

3.3 Isolation of Environmental Strains with Cholesterol Oxidase Activity

3β-Hydroxysteroid oxidase or cholesterol oxidase is an extracellular enzyme that produces the oxidation of cholesterol. This oxidation occurs at carbon 3, oxidizing a hydroxylated carbon by a keto group carbon. This oxidation is the previous step for its

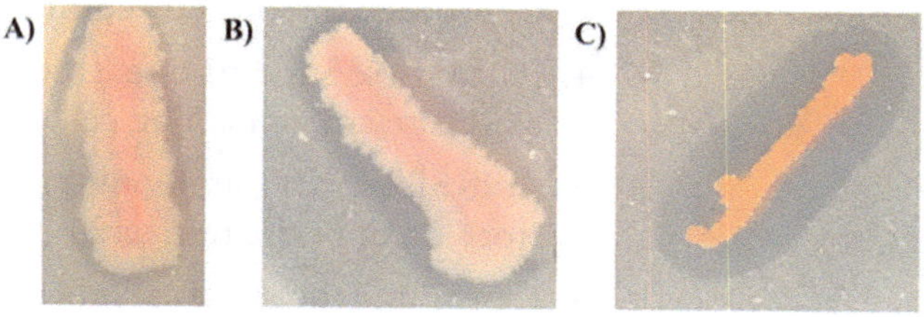

Fig. 2 Cholesterol plate used to see the cholesterol degradation halo. (**a**) *Gordonia* sp. HE-24.4J. (**b**) *Gordonia* sp. HE-24.3. (**c**) *Rhodococcus* sp. HE-24.12. Growth conditions are as indicated in Subheading 3.3. Incubation was performed at 30 °C for 3 days

metabolization in the degradation pathway. In many microorganisms this activity has always been identified extracellularly.

1. Make suspensions of TSA-plated bacterial isolated colonies (pick the colonies and resuspend them in 1 mL of PBS) and shake them vigorously for 1 h at 25 °C.
2. Sterilize the medium salts, agar, and cholesterol separately (*see* **Note 3**). Once sterilized, mix them so that the *Pseudomonas* medium is 1×, containing 1.5% (w/v) agar, and 2 or 5 mM cholesterol (from the stock solutions prepared in Subheading 3.1).
3. Dispose the medium in 90 mm Petri dishes and observe that they should show a cloudy aspect.
4. Seed the plates from suspensions obtained from **step 1**, or dilutions from them, to get isolated colonies.
5. Incubate the plates for 48–72 h at 30 °C.
6. Observe growth over time and the formation of a clarified halo around the colonies due to cholesterol consumption (Fig. 2).

3.4 In Vitro Determination of Cholesterol Oxidase Activity

3β-Hydroxysteroid oxidase or cholesterol oxidase belongs to the FAD (flavin adenine dinucleotide)-dependent oxidoreductase family of proteins. This enzyme carries out the cholesterol oxidation reaction by utilizing molecular oxygen as an electron acceptor leading to products as 4-cholesten-3-one and hydrogen peroxide.

The procedure here described is a slight modification from the one previously described [37].

1. Prepare two different cultures in sterile 500 mL flask: one culture in a medium containing 2 mM cholesterol and the other culture supplied with 5 mM glucose (negative control), as sole carbon sources.
2. Culture the bacteria in 500 mL flasks containing 100 mL of each media, on a rotary shaker (250 rpm) at 30 °C the two different media until reach approximately the half of the exponential growth phase. Culture growth is determined as a function of their increase of absorbance at 600 nm.

3. Centrifuge the bacteria cultures and recover the supernatants.
4. Filter these supernatants through 0.45 μm size pore filters to eliminate any remaining cell.
5. Add 50% (w/v) ammonium sulfate to the supernatants and store them in the refrigerator at 4 °C overnight (*see* **Note 4**).
6. Centrifuge the supernatants to 10,000 × *g* during 30 min.
7. Carefully discard the supernatants an resuspend the pellets containing the extracellular proteins in 2 mL sterile MQ® grade water.
8. The collected extracellular proteins should be desalted through a PD-10 being eluted with 50 mM Tris–HCl pH 7.5 (*see* **Note 5**).
9. The concentration of the obtained proteins is measured with the Bradford protein assay kit to normalize samples.
10. Solve cholesterol and 4-cholesten-3-one in 2-propanol to a final concentration of 12 mM (*see* **Note 6**).
11. Prepare the reaction mix containing 400 μL of 125 mM Tris–HCl buffer pH 7.5 and 100 μL of extracellular proteins. Incubate in water bath at 37 °C during 3 min.
12. Start the reaction by adding 25 μL of the 12 mM cholesterol solution and incubate 30 min at 30 °C.
13. The negative control of the reaction contains the same components, although 25 μL of 2-propanol is added instead of cholesterol solution.
14. Add 2.5 mL absolute ethanol to stop the reactions.
15. The cholesterol oxidase transforms cholesterol in 4-cholesten-3-one. This reaction product is determined by measuring its absorbance at 240 nm (*see* **Note** 7) (Fig. 3).
16. Determination of a reference calibration line with different known concentrations of 4-cholesten-3-one is mandatory. For that, samples resembling the reaction mix, in which the volume of extracellular protein is replaced by 50 mM Tris–HCl pH 7.5 and the volume of cholesterol is changed by solutions of 4-cholesten-3-one at known concentrations, are spectrophotometrically analyzed at 240 nm (Fig. 4).
17. Extrapolate the values obtained in **step 15** over the reference line from **step 16**.
18. Store the collected samples at −20 °C for ulterior extraction of steroids in the reaction mixture and later analyses using UHPLC (Subheading 3.7).

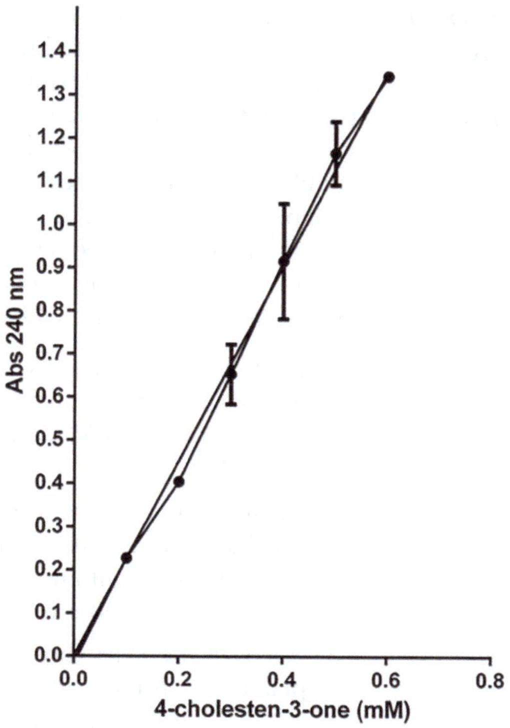

Fig. 3 Absorbance determination at 240 nm of different concentrations of 4-cholesten-3-one, showing the linearity at indicated concentrations (mM). The graph shown represents the measurement of triplicates of each condition. The reaction was performed as described in Subheading 3.4. $R^2 = 0.9976$

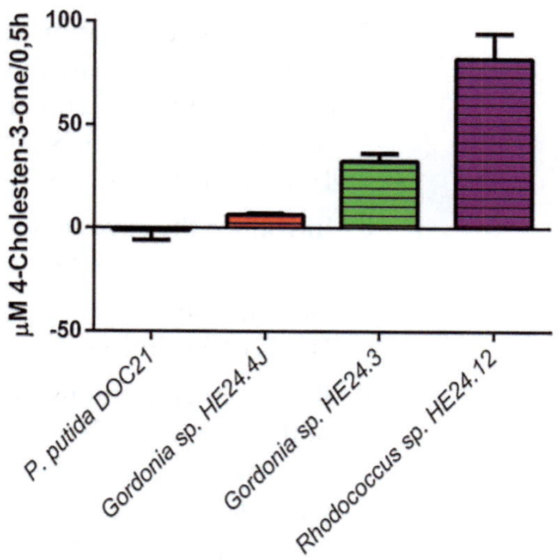

Fig. 4 Cholesterol oxidase activity measured twofold by the appearance of 4-cholesten-3-one in the medium. The reaction was performed as described in Subheading 3.4 and incubated 30 min

3.5 Determination of 3α-Hydroxysteroid Dehydrogenase Activity in Cell Extracts

3α-Hydroxysteroid dehydrogenase is an enzyme generally belonging to the aldo/keto reductase family of proteins using NAD^+ and/or $NADP^+$ as electron acceptors in the oxidative process of the α-hydroxyl substituent at carbon 3 present in steroids as bile acids or androsterone. Moreover, it has been proposed in some of the characterized enzymes that the reverse reaction, reducing a keto group at C-3 of some steroids by means of the electron donor NADH + H^+, is also possible. Here, a procedure for the determination of the oxidative process generating NADH + H^+ and androstanedione from NAD^+ and androsterone is described [38].

1. Prepare three different cultures in sterile 500 mL flask containing 100 mL of MM, using as carbon source for each one 2 mM testosterone, 2 mM cholic acid, and 5 mM glucose (negative control).
2. Incubate the flask on a rotary shaker (250 rpm) at 30 °C until the bacterial cultures reach half of the exponential growth phase by determining their growth as an increase of their absorbance at 600 nm.
3. Recover the cells through centrifugation and resuspend cell pellets in 1 mL of 100 mM phosphate buffer pH 7.4.
4. Obtain cell-free extracts through sonication using 6 pulses of 10 s with 15 s of incubation on ice between them (*see* **Note 8**).
5. Determine protein concentration of cell lysate using the Bradford protein assay kit. Results from the different samples will allow the normalization between them for the assays.
6. Prepare a stock solution of 30 mM androsterone dissolved in methanol (*see* **Note 6**).
7. For the determination of 3α-hydroxysteroid dehydrogenase activity, a mix prepared in a spectrophotometer quartz cuvette contains 500 μL of 200 mM phosphate buffer pH 7.4, 50 μL of cell-free extract (*see* **Note 9**), 430 μL of Milli-Q® water, and 10 μL of 100 mM $NADP^+$.
8. Introduce the cuvette on the spectrophotometer and allow the absorbance stabilization of the mix at 340 nm.
9. Once the absorbance is stabilized, start the reaction by adding 10 μL of a solution of 30 mM androsterone.
10. Record in continuous in the spectrophotometer the behavior of the reaction at 340 nm each 20 s until a final time of 90 min (Fig. 5a).
11. Store the samples at −20 °C for ulterior extraction of products and analysis using UHPLC (Subheading 3.7) (*see* **Note 10**).

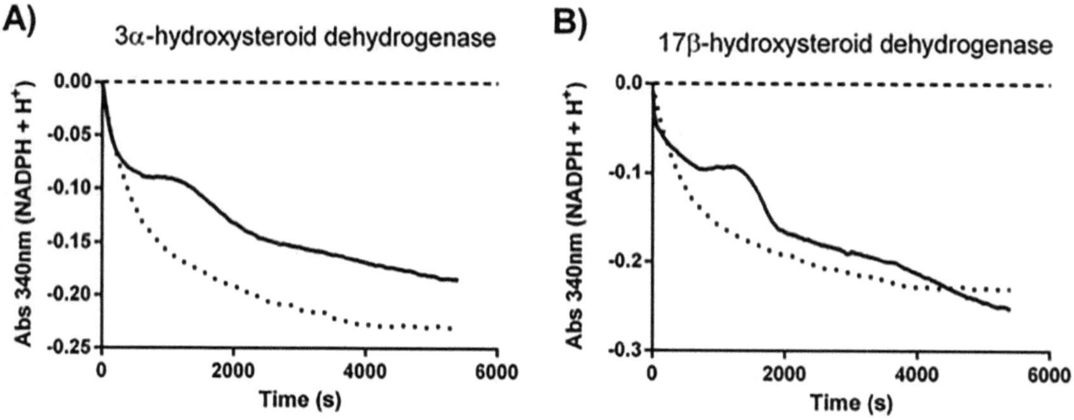

Fig. 5 3α-Hydroxysteroid dehydrogenase (**a**) and 17β-hydroxysteroid dehydrogenase (**b**) kinetics determined as NADPH + H^+ production from androsterone according to Subheading 3.5 (**a**) or testosterone as related in Subheading 3.6 (**b**). Solid lines represent the reactions with androsterone or testosterone, whereas dotted lines represent controls without steroid addition

3.6 17β-Hydroxysteroid Dehydrogenase Activity in Cell Extracts

Bacterial 17β-hydroxysteroid dehydrogenases belong to the aldo/keto reductase family of proteins. These enzymes carry out the oxidative reaction of the β-hydroxy group at carbon 17 in steroids as, e.g., testosterone using $NAD(P)^+$ as electron acceptor. It has also been proposed that some of these enzymes could reduce the keto group at carbon 17 of some steroids, by means of the electron donor NAD(P)H + H^+, rendering a β-hydroxy group in this position. The procedure here described corresponds to the in vitro assay of the NAD(P)-depending oxidation of testosterone generating androst-4-en-3,17-dione and NAD(P)H + H^+ [39].

1. Prepare three different cultures in sterile 500 mL flask. Each medium has a different carbon source: testosterone 2 mM, cholic acid 2 mM, or glucose 5 mM (negative control).
2. Grow each bacterium in the three different media until half of the exponential growth phase. Incubate the flasks (500 mL) on a rotary shaker (250 rpm) at 30 °C.
3. Determine the growth of each culture as a function of their absorbance at 600 nm.
4. Centrifuge all bacteria culture and recover the cell pellet. Resuspend cell pellet in 1 mL of 100 mM phosphate buffer pH 7.4.
5. Obtain cell-free extracts through sonication using the equipment at 70% of power, with 6 pulses of 10 s with 15 s of incubation on ice between them (*see* **Note 8**).
6. Determine the protein amount from cell-free extracts using the Bradford protein assay kit to normalize the concentration of proteins between samples.

7. Prepare a solution containing the substrate (30 mM testosterone) solved in methanol (*see* **Note 6**).
8. The reaction mixture for determination of 17β-hydroxysteroid dehydrogenase activity contains 500 μL of 200 mM phosphate buffer pH 7.4, 50 μL of cell-free extract (*see* **Note 9**), 430 μL of Milli-Q® water, and 10 μL of 100 mM $NADP^+$, prepared on a spectrophotometer quartz cuvette. Maintain the cuvette on the spectrophotometer until absorbance stabilizes at 340 nm.
9. Start the reaction through the addition of 10 μL of the solution 30 mM testosterone made in **step** 7.
10. Observe the evolution of absorbance at 340 nm on the spectrophotometer in continuous, recording absorbance each 20 s for 90 min (Fig. 5b).
11. Store the samples in freeze (−20 °C) for later extraction of steroids and load in UHPLC (Subheading 3.7) (*see* **Note 10**).

3.7 Analysis of Substrates and Products of In Vitro Reactions by UHPLC

Chromatographic separation of steroids is performed, based on their inherent hydrophobic character, using organic phases and chromatographic resins like C18. On the other hand, detection should be performed with a mass detector having a higher sensitivity for the concentration of these compounds. The separation between some substrates and products from the previously described protocols is performed on a timsTOF fleX at molecular mass level of 1 m/z unit. This protocol is an adaptation of the method previously described [40].

1. Take 500 μL of the sample and add an internal standard (1 mM estrone).
2. Add 1000 μL of chloroform to the mix of the sample and the internal standard (*see* **Note 11**).
3. Shake for 15 min with vortex mixer.
4. Centrifuge at 13000 × *g* for 5 min (*see* **Note 12**).
5. Take the organic phase and allow it to evaporate in a thermoblock maintained at 55–60 °C (*see* **Note 13**).
6. Resuspend the remaining solids in 40–100 μL of methanol (*see* **Note 14**).
7. Centrifuge at 13000 × *g* for 5 min and collect the supernatants.
8. Filter and dispose the samples in HPLC vials (*see* **Note 15**).
9. Prepare mobile phases for UHPLC of the samples: 64% methanol, 30% acetonitrile, and 6% water (*see* **Note 16**).
10. Run an isocratic UHPLC method using a column Zorbax StableBond-C18 (150 mm × 2.1 mm), a timsTOF detector, and a sample injection volume of 2 μL. Keep the temperature of column at 20 °C and an ionization energy of 7 eV (*see* **Note 17**).

4 Notes

1. Consider that, for some specific experiments, bile acids (except lithocholic acid) could be solubilized by adjusting pH to about 8.0–8.5 by the addition of NaOH or KOH, due to the increase of polar groups (hydroxy groups) in the molecule.
2. Solubility of cholesterol, lithocholic acid, or testosterone is very low in water, but their bioavailability in culture media could be increased using methylated β-cyclodextrins. To prepare the steroid stock solution, one drop of the steroid solubilized in non-polar solvents is added to the cyclodextrin solution maintained at 80 °C. Before the addition of another drop, check the inexistence of separated phases. If necessary, eliminate them using an ultrasonic bath. Add the next drop and repeat the process until the volume of the steroid solution is completed. Take care when adding chloroform to the cyclodextrin solution at 80 °C, as it evaporates very quickly.
3. Salts for the preparation of the culture media, agar, and carbon sources (especially the steroids) should be sterilized separately in the autoclave, avoiding the loss of some salts and/or the steroid decomposition catalyzed by metals present in the media.
4. The following link was used for the calculation of the percentages of ammonium sulfate: https://www.encorbio.com/protocols/AM-SO4.html.
5. The importance of washing of the column with 50 mM Tris–HCl several times beforehand to pH 7.5 could be critical for the appropriate developing of the technique. Once the PD-10 column has been equilibrated with several volumes of 3.5 mL, protein samples will be added, which contain the ammonium sulfate salts; once the samples have entered the gel, 3.5 mL of 50 mM Tris–HCl pH 7.5 is added simultaneously to the eluted volumes collected. In this way we will eliminate the ammonium sulfate that does not allow a correct folding of the protein, returning to its native form.
6. Steroids can be dissolved, at the indicated concentration, using 2-propanol or methanol as solvents, avoiding the use of methyl-β-cyclodextrins.
7. Measure at 240 nm with a quartz cuvette since quartz does not absorb in the far UV wavelengths.
8. The action of ultrasonic homogenizer could increase the temperature of the samples. Thus, between pulses, samples should be maintained on ice, to avoid a thermal denaturation of the proteins.

9. Cell extracts should not be centrifuged since both soluble proteins and proteins interacting with cell membranes could also be of interest.
10. The samples are frozen at −20 °C to stop the reaction. Different replicates of the samples should be frozen and after that the extractions for UHPLC will be performed (Subheading 3.7).
11. Strone is added as an internal steroid control since this steroid does not participate on the reactions. Therefore, once the samples are thawed, a concentration of 1 mM estrone is added. This estrone, when analyzed, will be used to normalize the obtained values after the extraction of the steroids between samples.
12. Samples are centrifuged to facilitate the separation of phases, where the upper one is the aqueous phase, while the lower phase corresponds to chloroform (organic phase).
13. Chloroform evaporation should be carried out in an extraction hood.
14. Solids from samples are resuspended in one of the solvents used in the chromatographic method to avoid the appearance of artifactual peaks in the chromatographic resolution.
15. The HPLC filters must have the minimal dead volume possible, due to the small sample volume available (40–100 μL). Once in the HPLC vials, samples should be incubated on the ultrasonic bath to avoid the presence of bubbles.
16. Mobile phase solutions should also be filtered and possible bubbles eliminated in the ultrasonic bath before use in the UHPLC.
17. Ionization energy is the key point for detection by a TOF detector. Thus, to resolve the desired compounds, the indicated value must be used.

Acknowledgments

Development of the methods here presented has been supported by the Ministerio de Economía y Competitividad (Madrid, España, grants BFU2009-11545-C03-01, BIO2012-39695-C02-02, and BIO2015-66960-C3-3R), by a CENIT Project RTC-2014-2249-1 (CDTI, Ministerio de Economía y Competitividad, Madrid, España), and by a grant from the Junta de Castilla y León (Consejería de Educación, Valladolid, España) LE246A11-2. The authors also want to thank the support to their actual research by the Horizon Europe Framework Programme (call: HORIZON-CL4-2021-RESILIENCE-01-11) through the ESTELLA project ("DESign of bio-based Thermoset polymer with rEcycLing capabiLity by dynAmic bonds for bio-composite manufacturing") (Project no. 101058371), the Ministerio de Ciencia e Innovación (grant TED2021-132593B-I00 belonging to the 2021 convocatory "Proyectos Estratégicos Orientados a la Transición Ecológica y a la Transición Digital" and RTI2018-095584-B-C43 from Proyectos de I+D+i RETOS INVESTIGACION), and the Junta de Castilla y León, grant LE250P20.

References

1. Kodner RB, Pearson A, Summons RE et al (2008) Sterols in red and green algae: quantification, phylogeny, and relevance for the interpretation of geologic steranes. Geobiology 6: 411–420
2. Bergstrand LH, Cardenas E, Holert J et al (2016) Delineation of steroid-degrading microorganisms through comparative genomic analysis. MBio 7:1–14
3. Bobrovskiy I, Hope JM, Nettersheim BJ et al (2021) Algal origin of sponge sterane biomarkers negates the oldest evidence for animals in the rock record. Nat Ecol Evol 5:165–168
4. Olivera ER, Luengo JM (2019) Steroids as environmental compounds recalcitrant to degradation: Genetic mechanisms of bacterial biodegradation pathways. Genes (Basel) 10:512
5. Feller FM, Holert J, Yücel O et al (2021) Degradation of bile acids by soil and water bacteria. Microorganisms (Basel) 9:1759
6. Lange IG, Daxenberger A, Schiffer B et al (2002) Sex hormones originating from different livestock production systems: fate and potential disrupting activity in the environment. Anal Chim Acta 473:27–37
7. Froehner S, Martins RF, Errera MR (2009) Assessment of fecal sterols in Barigui River sediments in Curitiba, Brazil. Environ Monit Assess 157:591–600
8. Chang H-S, Choo K-H, Lee B et al (2009) The methods of identification, analysis, and removal of endocrine disrupting compounds (EDCs) in water. J Hazard Mater 172:1–12
9. Santos LHMLM, Araújo AN, Fachini A et al (2010) Ecotoxicological aspects related to the presence of pharmaceuticals in the aquatic environment. J Hazard Mater 175:45–95
10. Matić Bujagić I, Grujić S, Laušević M et al (2019) Emerging contaminants in sediment core from the Iron Gate I Reservoir on the Danube River. Sci Total Environ 662:77–87
11. Mendelski MN, Dölling R, Feller FM et al (2019) Steroids originating from bacterial bile acid degradation affect *Caenorhabditis elegans* and indicate potential risks for the fauna of manured soils. Sci Rep 9:11120
12. Jenkins RL (2003) Androstenedione and progesterone in the sediment of a river receiving paper mill effluent. Toxicol Sci 73:53–59
13. Orrego R, Guchardi J, Hernandez V et al (2009) Pulp and paper mill effluent treatments have differential endocrine-disrupting effects on rainbow trout. Environ Toxicol Chem 28: 181
14. Howell WM, Black DA, Bortone SA (1980) Abnormal expression of secondary sex characters in a population of mosquitofish, *Gambusia affinis holbrooki*: evidence for environmentally-induced masculinization. Copeia:676
15. Kolodziej EP, Gray JL, Sedlak DL (2003) Quantification of steroid hormones with pheromonal properties in municipal wastewater effluent. Environ Toxicol Chem 22:2622
16. Orlando EF, Kolok AS, Binzcik GA et al (2004) Endocrine-disrupting effects of cattle feedlot effluent on an aquatic sentinel species, the fathead minnow. Environ Health Perspect 112:353–358
17. Oetken M, Bachmann J, Schulte-Oehlmann U et al (2004) Evidence for endocrine disruption in invertebrates. Int Rev Cytol 236:1–44
18. Sun Y, Huang H, Sun Y et al (2013) Ecological risk of estrogenic endocrine disrupting chemicals in sewage plant effluent and reclaimed water. Environ Pollut 180:339–344
19. Morthorst JE, Brande-Lavridsen N, Korsgaard B et al (2014) 17β-Estradiol causes abnormal development in embryos of the viviparous eelpout. Environ Sci Technol 48:14668–14676
20. Lambert MR, Giller GSJ, Barber LB et al (2015) Suburbanization, estrogen contamination, and sex ratio in wild amphibian populations. Proc Natl Acad Sci U S A 112:11881–11886
21. Regnault C, Usal M, Veyrenc S et al (2018) Unexpected metabolic disorders induced by endocrine disruptors in *Xenopus tropicalis* provide new lead for understanding amphibian

decline. Proc Natl Acad Sci U S A 115:E4416–E4425

22. Chiang Y, Wei ST, Wang P et al (2020) Microbial degradation of steroid sex hormones: implications for environmental and ecological studies. Microb Biotechnol 13:926–949
23. García JL, Uhía I, Galán B (2012) Catabolism and biotechnological applications of cholesterol degrading bacteria. Microb Biotechnol 5:679–699
24. Donova MV, Egorova OV (2012) Microbial steroid transformations: current state and prospects. Appl Microbiol Biotechnol 94:1423–1447
25. Fernández-Cabezón L, Galán B, García JL (2018) New insights on steroid biotechnology. Front Microbiol 9:958
26. Feng J, Wu Q, Zhu D et al (2022) Biotransformation enables innovations toward green synthesis of steroidal pharmaceuticals. ChemSusChem 15:e202102399
27. Olivera ER, de la Torre M, Barrientos Á et al (2018) Steroid catabolism in bacteria: Genetic and functional analyses of *stdH* and *stdJ* in *Pseudomonas putida* DOC21. Can J Biotechnol 2:88–99
28. Doukyu N (2009) Characteristics and biotechnological applications of microbial cholesterol oxidases. Appl Microbiol Biotechnol 83:825–837
29. Oppermann UCT, Maser E (1996) Characterization of a 3alpha-hydroxysteroid dehydrogenase/carbonyl reductase from the Gram-negative bacterium *Comamonas testosteroni*. Eur J Biochem 241:744–749
30. Maser E, Xiong G, Grimm C et al (2001) 3-α-Hydroxysteroid dehydrogenase/carbonyl reductase from *Comamonas testosteroni*: biological significance, three-dimensional structure and gene regulation. Chem Biol Interact 130–132:707–722
31. Birkenmaier A, Holert J, Erdbrink H et al (2007) Biochemical and genetic investigation of initial reactions in aerobic degradation of the bile acid cholate in *Pseudomonas* sp. strain Chol1. J Bacteriol 189:7165–7173
32. Genti-Raimondi S, Tolmasky ME, Patrito LC et al (1991) Molecular cloning and expression of the β-hydroxysteroid dehydrogenase gene from *Pseudomonas testosteroni*. Gene 105:43–49
33. Martínez-Blanco H, Reglero A, Rodriguez-Aparicio LB et al (1990) Purification and biochemical characterization of phenylacetyl-CoA ligase from *Pseudomonas putida*. A specific enzyme for the catabolism of phenylacetic acid. J Biol Chem 265:7084–7090
34. Boczar D, Michalska K (2022) Cyclodextrin inclusion complexes with antibiotics and antibacterial agents as drug-delivery systems—A pharmaceutical perspective. Pharmaceutics 14:1389
35. Regev O, Zana R (1999) Aggregation behavior of tyloxapol, a nonionic surfactant oligomer, in aqueous solution. J Colloid Interface Sci 210:8–17
36. Merino E, Barrientos A, Rodríguez J et al (2013) Isolation of cholesterol- and deoxycholate-degrading bacteria from soil samples: Evidence of a common pathway. Appl Microbiol Biotechnol 97:891–904
37. Yehia HM, Hassanein WA, Ibraheim SM (2015) Purification and characterisation of the extracellular cholesterol oxidase enzyme from *Enterococcus hirae*. BMC Microbiol 15:178
38. Steckelbroeck S, Jin Y, Gopishetty S et al (2004) Human cytosolic 3alpha-hydroxysteroid dehydrogenases of the aldo-keto reductase superfamily display significant 3beta-hydroxysteroid dehydrogenase activity: implications for steroid hormone metabolism and action. J Biol Chem 279:10784–10795
39. Schultz RM, Groman EV, Engel LL (1977) 3(17)beta-Hydroxysteroid dehydrogenase of *Pseudomonas testosteroni*. A convenient purification and demonstration of multiple molecular forms. J Biol Chem 252:3775–3783
40. Al-Khadhra RS (2020) The determination of common anabolic steroid and stimulants in nutritional supplements by HPLC-DAD and LC-MS. J Chromatogr Sci 58:355–361

Chapter 3

Selection of Biodegrading Phytosterol Strains

María-Angélica Mondaca, Maricel Vidal, Soledad Chamorro, and Gladys Vidal

Abstract

The phytosterol-biotransforming strains can be selected from *Mycobacterium* sp. using a high concentration of β-sitosterol. The selection is made by culturing the strains in a medium enriched with 14 g/L of β-sitosterol as the unique source of carbon. During 2 months, the bacterial cultures are transferred successively. The extraction of the biotransformation products is made with methanol and ethyl acetate. The qualitative and quantitative analyses are made by means of thin-layer chromatography, gas–liquid chromatography (GLC), and GLC–mass spectrometry. Under these conditions, it is observed that after seven transfers, the strains *Mycobacterium* sp. MB-3683 and *Mycobacterium fortuitum* B-11045 increase their biotransformation capacity from 20% to 64% and from 34% to 55%, respectively. The products in the highest proportion identified for each trial are androstenedione and androstadienedione. The results suggest that the high substrate concentration could be a selective mechanism to obtain strains more efficient in the biotransformation of β-sitosterol into steroidal bases.

Key words Steroids, Biotransformation, Natural products, β-sitosterol *Mycobacterium* sp.

1 Introduction

Growing industrial and human activity has led to an increase in the presence of steroids in the environment [1, 2]. The recalcitrant nature of these compounds, which is due to their structural stability, results in contamination of soil and water having a potential threat to biological processes like reproduction [1–3]. The use of high concentrations of phytosterol is an efficient way to select bacterial strains that can biotransform phytosterol. Several studies of steroid degradation pathways [4] suggest a likely convergence of steroids in common intermediates like 4-androsten-3,17-dione (AD) and 1,4-androstadien-3,17-dione (ADD) (Fig. 1) [5]. In this sense, steroid drugs are synthetized by chemical or microbial routes, both of which involve conversion of steroid precursors into drug intermediates and final conversion of intermediates. Microbial

Carlos Barreiro and José-Luis Barredo (eds.), *Microbial Steroids: Methods and Protocols*, Methods in Molecular Biology, vol. 2704,
https://doi.org/10.1007/978-1-0716-3385-4_3,

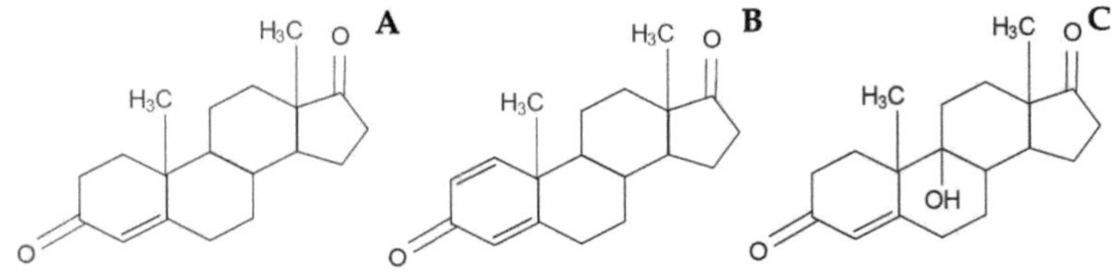

Fig. 1 Biotransformation products obtained during the selection of *Mycobacterium* sp. strains: (**a**) 4-Androstene-3,17-dione (AD). (**b**) 1,4-Androstadien-3,17-dione (ADD). (**c**) 9-Hydroxy-(4-androstene-3,17-dione) (9-OH-AD). The wild-type strain only produces AD and ADD. 9-OH-AD is also produced by the strains derived from *M. fortuitum* B-11045, beginning with the fourth assay

transformation cleaves complex side chains of steroid precursor steroids in a single step and incorporates desirable modifications in steroid nuclei [6, 7].

A better understanding of the genetic and biochemical catabolic routes of these compounds would enable us to use them as decontaminating biological agents and thereby mitigate the problems caused by steroid accumulation.

The goal of the present work is to select strains capable of transforming steroids using high concentrations of β-sitosterol.

2 Materials

2.1 Substrate

1. Fraction isolated from tall oil (Kraft mill effluent) rich in phytosterols. This fraction contains about 95% of sterols (85% corresponds to β-sitosterol and 15% corresponds to campesterol and stigmasterol).

2.2 Bacterial Strains

1. *Mycobacterium* sp. MB-3683 strains (Bombay) were obtained from *Mycobacterium* sp. NRRL B-3683 (supplied by Dr. Raltan Sood; University of Bombay) (*see* **Note 1**).
2. *Mycobacterium fortuitum* B-11045 strains according to Wovcha et al. [8].

2.3 Bacterial Growth Media

All solutions are prepared with purified and sterilized water.

1. Middlebrook agar 7H10: Formulation per liter of purified water: 0.025 g of $MgSO_4 \times H_2O$, 5.0 g of bovine albumin (fraction V), 0.04 g of ferric ammonium citrate, 0.004 g of catalase, 0.4 g of sodium citrate, 0.001 g of pyridoxine, 0.5 g of $(NH_4)_2\ HSO_4$, 0.001 g of $ZnSO_4 \times 7H_2O$, 0.5 g of monosodium glutamate, 0.001 g of $CuSO_4 \times 7H_2O$, 0.0005 g of biotin, 1.5 g of $Na_2HPO_4 \times H_2O$, 1.5 g of $KH_2PO_4 \times H_2O$, 0.0005 g of $CaCl_2 \times H_2O$, 0.00025 g of malachite green, 17.0 g of agar, 5.0 g of glycerol, 0.85 g of NaCl, 2.0 g of glucose, and 0.05 mL of oleic acid.

2. Inoculum medium: 10 g of yeast extract, 10 mL of glycerol, 1.5 g of $(NH_4)_2$ HPO_4, 1.3 g of $K_2HPO_4 \times 3H_2O$, 0.2 g of $MgSO_4$, 0.01 g of $FeSO_4 \times 7H_2O$, and 0.02 g of $ZnSO_4 \times 7H_2O$ per liter of distilled water.
3. Enrichment medium: 14 g of β-sitosterol, 60 g of Amberlite XAD-2 (Sigma-Chemical Company), 1.5 g of $(NH_4)_2$ HPO_4, 1.3 g of $K_2HPO_4 \times 3H_2O$, 0.2 g of $MgSO_4$, 0.01 g of $FeSO_4 \times 7H_2O$, and 0.02 g of $ZnSO_4 \times 7H_2O$ per liter of distilled water.
4. β-Sitosterol purity >98% (Sigma).
5. Amberlite XAD-2 resin surface area 330 m^2 g^{-1}, pore diameter 9 nm, and bead size 20–60 mesh (Sigma).

2.4 Equipment and Solvents

1. Thin-layer chromatography (TLC) is performed on a sheet of glass (20 × 20 cm), which is coated with a thin layer of cellulose and is conducted with methanol [8].
2. Gas–liquid chromatography (GLC) is conducted with a Varian Star 3400 Cx chromatograph and a Hewlett–Packard (HP 5890 30 m length × 0.25 μm) GLC–mass spectrometer with an FID detector HP 5972 mass detector and HP-5 ms column and helium as the carrier gas.
3. GLC–mass spectrometry is conducted with helium as the carrier gas.
4. Standard of AD purity >98% (Sigma).
5. Standard of ADD purity >97% (Sigma).
6. Incubator.
7. Orbital shaker.
8. 500 mL Erlenmeyer flasks.
9. Methanol purity ≥99.8%.
10. Ethyl acetate purity ≥99.8%.
11. Rotary evaporator.
12. Sephadex column.
13. 1H-nuclear magnetic resonance (NMR) Bruker AC 250-P spectrometer at 250 MHz.
14. CDCl3 solvent.
15. Nicolet 550 magno-IR spectrophotometer.

3 Methods

3.1 Selection of Strains Tolerant to High Concentrations of β-Sitosterol

1. Store *Mycobacterium* sp. strains in a freezer at −70 °C.
2. Seed 30 mL of inoculum medium with 300 μL of the frozen vial and incubate at 30 °C for 7 days.

Table 1
***Mycobacterium* sp. strains selected with 14 g/L of β-sitosterol in the culture medium**

	Biotransformation products (%)		
Strains	**AD**	**ADD**	**9-OH-AD**
MB-3683(1)	51.2	48.0	0
MB-3683(2)	70.4	29.3	0
MB-3683(3)	68.5	30.1	0
MB-3683(4)	81.8	17.5	0
MB-3683(5)	61.2	38.5	0
MB-3683(6)	60.0	39.5	0
MB-3683(7)	72.3	26.7	0
MB-3683(27)	71.0	28.9	0
MB-3683(29)	95.8	4.1	0
MB-3683(30)	30.7	67.8	0
B-11045(31)	39.5	41.4	18.5
B-11045(32)	50.0	48.7	1.3
B-11045(33)	19.5	60.0	18.3
B-11045(34)	31.3	64.5	4.2
B-11045(35)	9.8	80.0	10.1
B-11045(39)	10.0	79.5	9.8
B-11045(40)	5.5	80.7	12.0
B-11045(41)	31.2	57.4	11.1
B-11045(42)	29.5	60.1	10.4

AD 4-androstene-3,17-dione, *ADD* 1,4-androstadien-3,17-dione, *9-OH-AD* 9-hydroxy-4-androstene-3,17-dione

3. Add 5 mL of inoculum culture (10%) to 50 mL of enrichment medium.
4. Incubate the bacterial cultures at 30 °C and 180 rpm for 7 days.
5. Transfer 5 mL of the bacterial culture to 50 mL of enrichment medium every 7 days over a period of 2 months.
6. Seed the tolerant (growing) strains in Middlebrook agar 7H10 with 14 g/L, 28 g/L, and 56 g/L of β-sitosterol (Table 1) (*see* **Note 2**).

Table 2
Biotransformation percentages for the selected strains of *Mycobacterium* sp. MB-3683 and *M. fortuitum* B-11045 cultured at a concentration of 14 g/L of β-sitosterol

Strains	Successive transfer	Biotransformation percentages (%)
Mycobacterium sp. MB-3683	1	22.39
	2	40.28
	3	39.40
	4	46.27
	5	59.54
	6	55.79
	7	64.28
M. fortuitum B-11045	1	34.06
	2	33.10
	3	40.77
	4	45.65
	5	43.35
	6	50.01
	7	55.35

Biotransformation percentages were determined with reference to the wild strain

3.2 β-Sitosterol Biotransformation

1. Inoculate 30 mL (10% of the final volume) of the strains previously selected in 500 mL Erlenmeyer flasks with a final volume of 300 mL of enrichment medium (Table 2) (*see* **Notes 3** and **4**).
2. Incubate the flasks at 30 °C and 180 rpm for 7 days.

3.3 Extraction of Biotransformation Products

1. Sterilize the biotransformation cultures at 121 °C for 20 min.
2. Extract the biotransformation products after sterilizing the cultures from the Amberlite with 150 mL methanol and 150 mL ethyl acetate.
3. Concentrate the total extracts concentrated in a rotary evaporator at reduced pressure and then take to a final volume of 10 mL (*see* **Notes 4** and **5**).
4. Concentrate 300 mL of total extracts in a rotary evaporator at reduced pressure at 25 °C for about 3–4 h to obtain a final volume of 10 mL (Table 2) (*see* **Notes 6** and **7**).

3.4 Analysis of Biotransformation Products

3.4.1 Analysis by TLC

1. Take each sample obtained from the biotransformation processes to be qualitatively analyzed initially by means of thin-layer chromatography (TLC) [9].
2. Place each sample on a silica gel 60 F 254 plate and leave dry for 15 min at room temperature.
3. Place the silica gel plate inside the TLC box with a mixture of hexane/ethyl acetate (30/70) as a mobile phase.

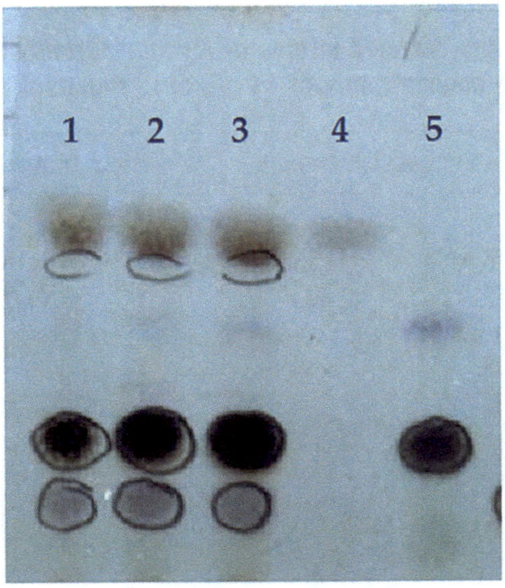

Fig. 2 Thin-layer chromatography of the *Mycobacterium* sp. strain MB-3683. Note the higher AD production by this strain. 1. Replica 1; 2. Replica 2; 3. Replica 3; 4. Standard β-sitosterol; 5. Standard AD

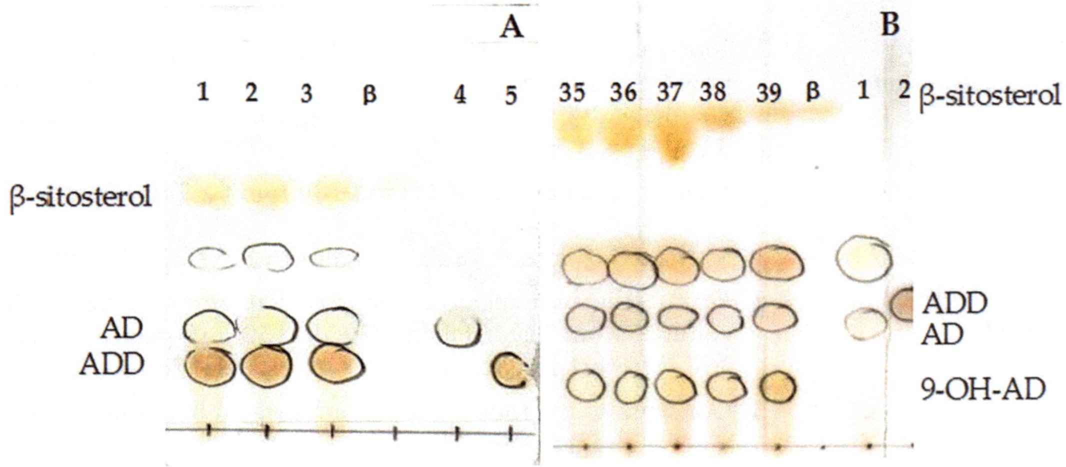

Fig. 3 Thin-layer chromatography of the *Mycobacterium fortuitum* strain B-11045. (**a**) Assay with the wild-type strain. 1. Replica 1; 2. Replica 2; 3. Replica 3; β. Standard β-sitosterol; 4. Standard AD; 5. Standard ADD. (**b**) *M. fortuitum* strains producing 9-OH-AD derived from B-11045 after the selection process. The band of 9-OH-AD is under the band of AD. 35. Strain 35; 36. Strain 36; 37. Strain 37; 38. Strain 38; 39. Strain 39; β. Standard β-sitosterol; 1. Standard AD; 2. Standard ADD

4. Dry the silica plate and check under UV light. The selected strains of *Mycobacterium* sp. MB-3683 produce AD and ADD (Fig. 2). The selected strains derived from *M. fortuitum* B-11045 (from the fifth test) produce 9-hydroxy-androst-4-en-3,17-dione (9-OH-AD) (Fig. 3).

3.4.2 Steroid Quantification by GLC–Mass Spectrometry Analysis

1. Perform the gas–liquid chromatography (GLC) with helium as a carrier gas, flow rate 2 mL/min, injector temperature 250 °C, detector temperature 280 °C, and oven temperature 350 °C.
2. Prepare each sample as a mixture (1:1) with an internal standard, which in this case corresponds to cholesterol 1 mg/mL.
3. Inject 1 μL of sample.
4. For qualitative biotransformation take as a reference the area of standard solution of cholesterol, androstenedione, campesterol, and β-sitosterol (concentration of 1 mg/mL of each of the components).
5. Determine the response factor of each of the components with reference to cholesterol.
6. Purify the compounds by chromatography on a Sephadex column and determine the structure of the pure compounds by 1H-nuclear magnetic resonance (NMR) and by infrared (IR) spectrophotometry.

4 Notes

1. The strains of *Mycobacterium* sp. are first trained to potentially transform steroid compounds. Mycobacteria have been used in this field, as well as successfully used to obtain hormones.
2. 14 g/L of β-sitosterol is used with the selected strain (30 strains from *Mycobacterium* sp. MB-3683 and 12 from *M. fortuitum* B-11045). Of these strains, 30 grew with concentrations of 28 g/L of β-sitosterol and 12 with 56 g/L of β-sitosterol (Table 1).
3. The experiment on the use of high steroidal substrate concentrations as a selective mechanism for biotransforming strains was successful. In the first transfer, the total biotransformation percentages are less than 35% for the *Mycobacterium* sp. MB-3683 and *M. fortuitum* B-11045 strains. For the MB-3683 strains, the main transformation product is AD, and in the case of the B-11045 strains, the main product is ADD.
4. The 42 strains isolated in the previous stage are subjected to trials in biotransformation cultures, which are carried out in triplicate in order to determine an average value for their β-sitosterol biotransformation capacity. After seven successive transfers, the *Mycobacterium* sp. MB-3683 strains and the *M. fortuitum* B-11045 strains increase their biotransformation capacity from 22% to 64% and from 34% to 55%, respectively (Table 2).
5. From the fourth trial onward, the biotransformation percentages increase to 50%, and the products for the strains derived

from MB-3683 are AD and ADD, although some strains begin to produce only AD. In the case of the strains derived from B-11045, although the main product is also ADD, an additional compound, a hydroxylated AD derivative, is also identified (Fig. 1).

6. The production of 9-hydroxyandrostenedione by strains from *M. fortuitum* B-11045 could be the result of selective pressure, due to the high concentrations of β-sitosterol. The high concentration of AD may have been responsible for the high level of enzymatic activity of 9-hydroxylase to convert AD into 9-hydroxyandrostenedione. Thus, these strains may use an alternative pathway in the degradation of phytosterols. In contrast, the strains from *Mycobacterium* sp. MB-3683 may use the traditional route, but the larger amount of substrate results in a higher yield in the end.

7. According to the results, the use of high substrate concentrations is an efficient methodology to select strains with phytosterol-biotransforming capacity. In addition, this approach has the advantage that it can be directed toward obtaining specific characteristics, such as a particular product or a high biotransformation capacity.

Acknowledgments

This work was supported by grant ANID/FONDAP/15130015.

References

1. Chamorro S, Hernández V, Matamoros V, Dominguez C, Becerra J, Vidal J, Piña B, Bayona J (2013) Chemical characterization of organic microcontaminant sources and biological effects in riverine sediments impacted by urban sewage and pulp mill discharges. Chemosphere 90:611–619
2. Pokhrel D, Viraraghavan T (2004) Treatment of pulp and paper mill wastewater—a review. Sci Total Environ 333(1–3):37–58
3. Vidal M, Becerra J, Mondaca MA, Silva M (2001) Selection of *Mycobacterium* sp. strain with capacity to biotransform high concentrations of β-sitosterol. Appl Microbiol Biotechnol 57:385–389
4. Kumar V, Dhall P, Kumar R, Singh YP, Kumar A (2012) Bioremediation of agro-based pulp mill effluent by microbial consortium comprising autochthonous bacteria. Sci World J 127014:1–7
5. Horinouchi N, Sakai E, Kawano T, Matsumoto S, Sasaki M, Hibi M, Shima J, Shimizu S, Ogawa J (2012) Construction of microbial platform for an energy-requiring bioprocess: practical 2′-deoxyribonucleoside production involving a C-C coupling reaction with high energy substrates. Microb Cell Factories 11(1):82–89
6. Rao S, Thakkar K, Pawar K (2013) Microbial transformation of steroids: current trends in cortical side chain cleavage. Quest 1:16–20
7. Seidel L, Horhold C (1992) Selection and characterization of new microorganisms for the manufacture of 9-OH-AD from sterols. J Basic Microbiol 32:49–55
8. Wovcha MG, Antosz FJ, Knight JC, Kominek LA, Pyke TR (1978) Bioconversion of sitosterol to useful steroidal intermediates by mutants of *Mycobacterium fortuitum*. Biochim Biophys Acta 531:308–321
9. Shah K, Mehdi I, Khan A, Vora V (1980) Microbial transformation of sterols. I. Microbial transformation of phytosterols to androsta-1,4-diene-3,17-dione by *Arthrobacter simplex*. Eur J Appl Microbiol Biotechnol 10:167–169

Chapter 4

Identification and Characterization of Some Genes, Enzymes, and Metabolic Intermediates Belonging to the Bile Acid Aerobic Catabolic Pathway from *Pseudomonas*

José M. Luengo and Elias R. Olivera

Abstract

The study of the catabolic potential of microbial species isolated from different habitats has allowed the identification and characterization of bacteria able to assimilate bile acids and/or other steroids (e.g., testosterone and 4-androsten-3,17-dione) under aerobic conditions through the 9,10-*seco* pathway. From soil samples, we have isolated several strains belonging to genus *Pseudomonas* that grow efficiently in chemically defined media containing some cyclopentane–perhydrophenanthrene derivatives as carbon sources. Genetic and biochemical studies performed with one of these bacteria (*P. putida* DOC21) allowed the identification of the genes and enzymes belonging to the route involved in bile acids and androgens, the 9,10-*seco* pathway in this bacterium. In this manuscript, we describe the most relevant methods used in our lab for the identification of the chromosomal location and nucleotide sequence of the catabolic genes (or gene clusters) encoding the enzymes of this pathway, and the tools useful to establish the role of some of the enzymes that participate in this route.

Key words *Pseudomonas*, Catabolism, Cholic acid, Chenodeoxycholic acid, Lithocholic acid, Deoxycholic acid, 4-Androstene-3,17-dione (AD), 1,4-Androstadiene-3,17-dione (ADD), Tn*5* mutagenesis, Genome editing, Acyl-CoA synthetase, 3-Ketosteroid dehydrogenase

1 Introduction

Bile acids (BAs) are the main end products of cholesterol metabolism in animals and thus are ubiquitously found in the environment as a result of excreta and animal decay. Those compounds, which in animals are produced exclusively in the liver (primary BAs, PBA), although biotransformed in other BA species by gut-associated microbiota (secondary BAs), represent about 90% of the actively metabolized cholesterol in the body; the remaining 10% of the cholesterol is used for steroid hormone biosynthesis. In humans, the final products of liver synthesis, PBAs, are cholic acid (CA) and chenodeoxycholic acid (CDCA). Both CA and CDCA are

Carlos Barreiro and José-Luis Barredo (eds.), *Microbial Steroids: Methods and Protocols*, Methods in Molecular Biology, vol. 2704, https://doi.org/10.1007/978-1-0716-3385-4_4,

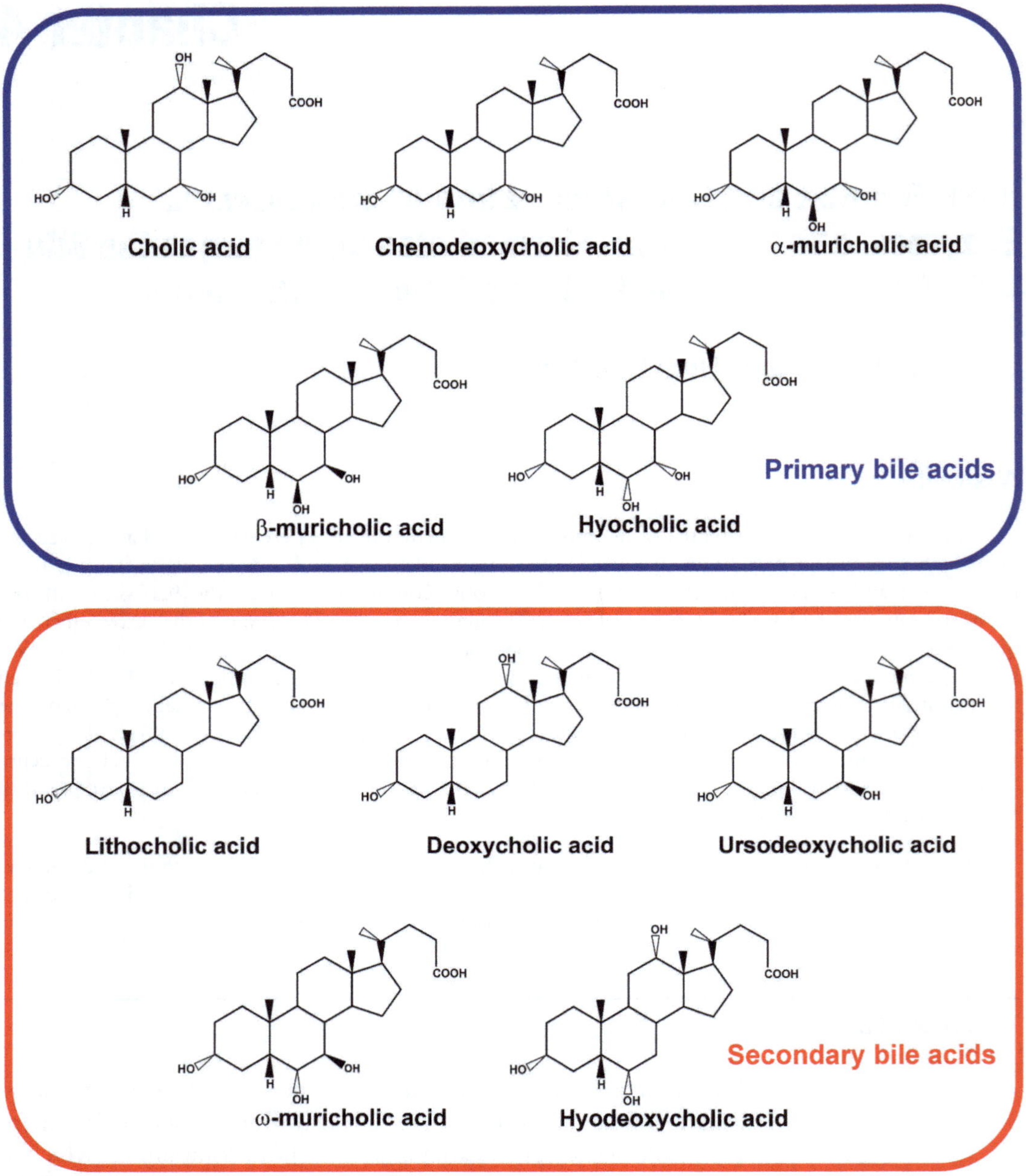

Fig. 1 Structure of some representative primary and secondary bile acids

synthesized from cholesterol through complex processes that involve hydroxylation, saturation of the double bond at C5–C6, epimerization of the 3-hydroxyl group (from β configuration in cholesterol to α configuration in bile acids), and oxidative cleavage of a 3-carbon unit from the lateral chain at C17 [1, 2] (Fig. 1). In the liver, the final metabolic process consists in the condensation of PBA with glycine or taurine, to generate bile salts (BS) [3]. These BS, which are stored in the gallbladder, are liberated into the duodenum because of cholecystokinin production triggered by food ingesta. They act as biological detergents and emulsifiers, allowing the absorption of fats and lipid-soluble vitamins (vitamins A, D, E, and K) along the small intestine tract [4]. Conjugated CA and CDCA are mainly reabsorbed from ileum (>95%) entering the portal vein for going back to the liver, to be reused. This process is known as enterohepatic circulation [5]. However, 3–5% of BS reach the colon (400–800 mg/day) and interact with the intestinal microbiota [6]. Then, BS are deconjugated by bacterial hydrolases, releasing free PBAs which are converted by gut bacterial enzymes in secondary SBAs (deoxycholic acid (DCA), lithocholic acid (LCA), and ursodeoxycholic acid (UDCA)—sometimes considered a tertiary BA-) (Fig. 1). 95% of these SBAs are reabsorbed through passive diffusion in the upper small intestine and in the colon, whereas 5% of BAs are not reabsorbed from the small intestine, which implies that 600 mg of bile acids/day are secreted into the feces. Although DCA and LCA are the most abundant SBA detected in human feces, more than 50 different secondary bile acids have been found, and composition of BAs (both PBAs and SBAs) seems to be species dependent (i.e., in rodents α- and β-muricholic acids are present in a considerable amount between PBAs, whereas significant amounts of ω-muricholic and hyodeoxycholic acids are found as SBAs in these animals) (Fig. 1) [7].

As a consequence, due to the decomposition of biomass and excreta, BAs are available to environmental microorganisms, and some microbes have developed metabolic systems which allow the specific assimilation of these compounds.

We have isolated from soil samples, and using the techniques indicated in this chapter, two different gram-negative bacteria are able to catabolize BAs and testosterone, although they are unable to metabolize sterols. Both strains were identified as *Pseudomonas putida* in base to homology and phylogenetic analysis using partial sequences of their 16S rDNA and *rpoB* genes, morphological aspect, and metabolic profiling. These two strains were referred as *P. putida* DOC19 and *P. putida* DOC21 [8].

P. putida strain DOC21 was used as a model to identify and characterize the genes encoding the catabolic enzymes required for the degradation of BAs. Identification of the clusters coding for enzymes involved was initially performed through random Tn*5* insertional mutagenesis following the protocol shown below. Homology of the encoded proteins as well as synteny comparison with other steroid degradative strains which catabolize CA and testosterone (*Comamonas testosteroni*) [9], cholesterol (*Mycobacterium*, *Rhodococcus*, and *Gordonia*) [10–14], or CA (*Comamonas*, *Rhodococcus*, and *Pseudomonas*) [9, 15–19] allowed the identification of the genes involved in BA catabolism as part of the so-called 9,10-*seco* pathway. The essentiality of some of the ORFs for BA catabolism in *P. putida* DOC21 was established by knockout of certain genes (following the strategies described in this chapter) and some of the metabolic intermediates accumulated by the metabolic blockade were identified [8, 17, 19, 20].

The obtained data allow us to conclude that *P. putida* DOC21 uses a 9,10-*seco* pathway for the assimilation of BA and testosterone, androstanolone, or 4-androstene-3,17-dione (AD) (Fig. 2), the same pathway used by actinobacteria to catabolize sterols [18, 19].

According to the proposed 9,10-*seco* catabolic pathway for CA mineralization, the first step corresponds to the oxidation of the hydroxyl function present in C-3 (α-configuration) to a keto group by StdD (*see* Chapter 2 by Chamizo-Ampudia et al. in this book). Later, or at the same time, the carboxy group from the side chain in C-17 is activated to a coenzyme A thioester throughout a reaction catalyzed by a bile acid–CoA ligase (StdA1) [17]. Once activated, a β-oxidation-like process for lateral chain elimination starts with a dehydrogenation in C-2/C-3 of the acyl-CoA side chain in the second step, and hydration of the α,β-double bond occurs, introducing a hydroxy group at C-3, as suggested by the characterization of the enoyl-CoA hydratase Shy1 in *Pseudomonas stutzeri* Chol1 [21]. However, the next enzymatic step differs from a canonical β-oxidation since an aldolytic cleavage of the C–C bond (catalyzed by Sal1), generating a steroid aldehyde with the concomitant release of an acetyl-CoA molecule, happens [21, 22]. This aldehyde is oxidized to the corresponding carboxylic acid by a specific aldehyde dehydrogenase (Sad in *Pseudomonas* sp. Chol1), and later, a second acyl-CoA synthetase (StdA2) thioesterifies this acid to a CoA derivative [17]. Degradation of the remaining side chain occurs through a similar mechanism, yielding a molecule of propionyl-CoA and a steroid derivative with a keto function in C-17 that continues the pathway.

At the same time that the elimination of the C-17 side chain occurs, the oxidation of the rings from the steroid nucleus starts with the introduction of double bonds in C-4 (catalyzed by StdI) and in C-1 (reaction driven by StdH) [20]. In sum, degradation of LCA (lacking the hydroxy groups in 7 and 12) renders at this stage androsta-1,4-diene-3,17-dione (ADD), which is a catabolite of convergence in the paradigmatic 9,10-*seco* pathway when testosterone, sterol, and AD degradation occurs. However, when CA is provided as only carbon source, the intermediate catabolite is 7α,12α-dihydroxy-androsta-1,4-diene-3,17-dione, the dehydroxylated analog of ADD. Hydroxylated ADD derivatives generated from CA and other bile acids and ADD are a convergence point for the degradative pathway of steroids in *Pseudomonas* species. All these metabolites suffer an α-hydroxylation in C-9, catalyzed by a two-component hydroxylase (StdJK) [20], originating in an unstable intermediate that suffers a spontaneous opening of the B ring.

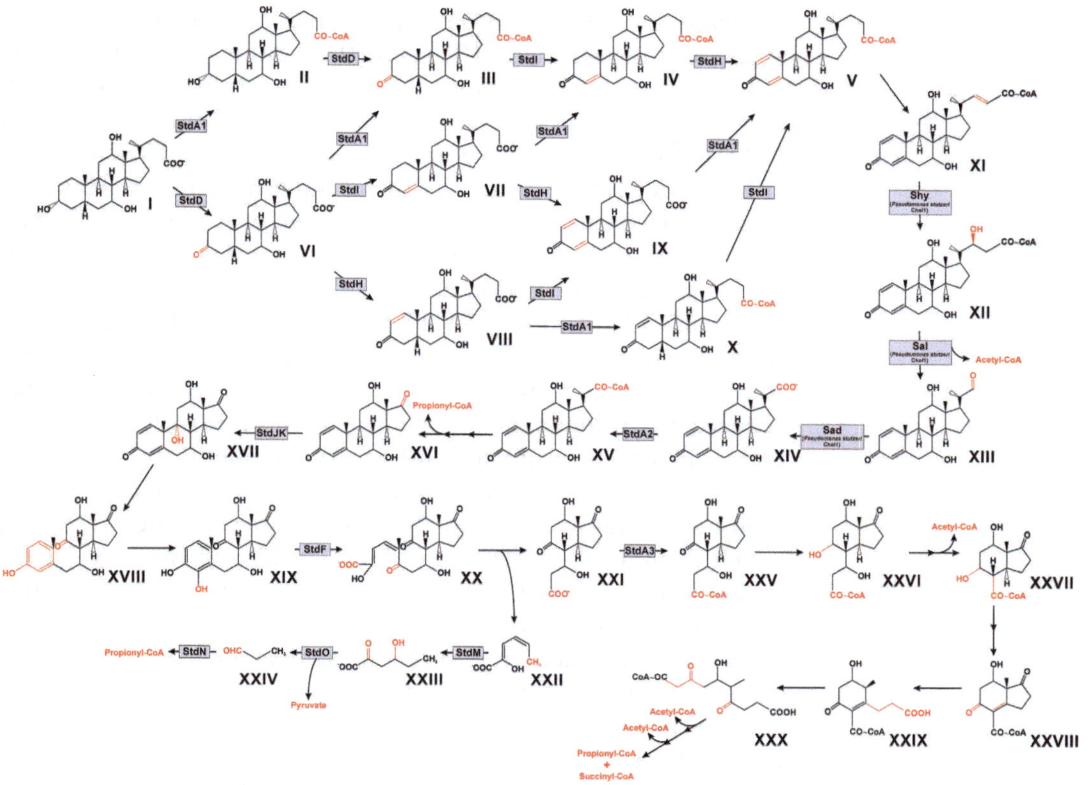

Fig. 2 Catabolic pathway proposed for aerobic assimilation of bile acids and other steroid compounds in *P. putida* DOC21, exemplified with the molecule of cholic acid. (I) Cholic acid; (II) cholyl-CoA; (III) 3-keto-5β-cholyl-CoA; (IV) Δ^4-3-keto-cholenoyl-CoA; (V) $\Delta^{1,4}$-3-keto-choladienoyl-CoA; (VI) 3-keto-5β-cholanoate; (VII) Δ^4-3-keto-cholenoate; (VIII) Δ^1-3-keto-5β-cholenoate; (IX) $\Delta^{1,4}$-3-keto-choladienoate; (X) Δ^1-3-keto-5β-cholenoyl-CoA; (XI) (22*E*)7α,12α-dihydroxy-3-oxochola-1,4,22*E*-triene-24-oyl-CoA; (XII) 7α,12α,22-trihydroxy-3-oxochola-1,4-diene-24-oyl-CoA; (XIII) 7α,12α-dihydroxy-3-oxopregna-1,4-diene-20*S*-carbaldehyde; (XIV) 7α,12α-dihydroxy-3-oxopregna-1,4-diene-20-carboxylate; (XV) 7α,12α-dihydroxy-3-oxopregna-1,4-diene-20-carboxylyl-CoA; (XVI) 7α,12α-dihydroxy-androsta-1,4-diene-3,17-dione; (XVII) 7α,9α,12-α-trihydroxy-androsta-1,4-diene-3,17-dione; (XVIII) 3,7,12-trihydroxy-9,10-secoandrosta-1,3,5(10)-triene-9,17-dione; (XIX) 3,4,7,12-tetrahydroxy-9,10-secoandrosta-1,3,5(10)-triene-9,17-dione; (XX) 4,5-9,10-diseco-3,7,12-trihydroxy-5,9,17-trioxoandrosta-1(10),2-dien-4-oate; (XXI) 3aα-*H*-4α(3′(*R*)-hydroxy-3-′-propanoate)-7α-hydroxy-7aβ-methylhexahydro-1,5-indanedione; (XXII) 2-hydroxyhexa-2,4-dienoate; (XXIII) 4-hydroxy-2-oxohexanoate; (XXIV) propionaldehyde; (XXV) 3aα-*H*-4α(3′(*R*)-hydroxy-3′-propanoyl-CoA)-7-α-hydroxy-7aβ-methylhexahydro-1,5-indanedione; (XXVI) 3aα-*H*-(3′(*R*)-hydroxy-3′propanoyl-CoA)-5,7-dihydroxy-7aβ-methylhexahydro-1-indanone; (XXVII) 3aα-*H*-4α(3′-carboxyl-CoA)-5,7-dihydroxy-7aβ-methylhexahydro-1-indenone; (XXVIII) (7a*S*)-7a-methyl-7-hydroxy-1,5-dioxo-2,3,5,6,7,7a-hexahydro-1*H*-indene-4-carboxyl-CoA; (XXIX) 2-(2-carboxyethyl)-4-hydroxy-3-methyl-6-oxocyclohex-1-ene-1-carboxyl-CoA; 3,6-dioxo-5-hydroxy-6-methyl-decanedioyl-CoA; (XXX) 3,6-dioxo-5-hydroxy-6-methyl-decanedioyl-CoA

Simultaneously, A ring is aromatized, leading to 3,7α,12-α-trihydroxy-9,10-secoandrosta-1,3,5(10)-triene-9,17-dione when CA is the BA that is being catabolized. The next step in CA degradation involves a hydroxylation at C-4 (A ring), a reaction catalyzed by a monooxygenase, yielding 3,4,7,12-tetrahydroxy-9,10-secoandrosta-1,3,5(10)-triene-9,17-dione. Now, the 3,4-dihydroxylated aromatized A ring is a substrate of StdF [8], an estradiol dioxygenase homologous to TesB from *C. testosteroni* [23] and to HsaC from *Rhodococcus jostii* [10] and *Mycobacterium tuberculosis* [24]. The study of the catabolism of testosterone and cholesterol revealed that the *meta*-cleavage of A ring is performed by TesB and HsaC. When catabolism of CA is approached in *Pseudomonas*, this reaction is catalyzed by StdF, giving 4,5-9,10-diseco-3,7,12-trihydroxy-5,9,17-trioxoandrosta-1(10)2-diene-4-oic acid, which is further degraded by a hydrolytic cleavage, leading to the products 2-hydroxyhexa-2,4-dienoic acid (HHD) and a fully reduced methylindanone derived from C and D rings [3aα-*H*-4α(3′(*R*)-hydroxy-3′propanoate)-7β-hydroxy-7aβ-methylhexahydro-1,5-indane-dione, 3′7-diOH-HIP]. The catabolism of HHD is then carried out in *P. putida* DOC21 by three enzymes: (i) StdM, which catalyzes the hydration of HHD yielding 4-hydroxy-2-oxo-hexanoic acid; (ii) StdO that catalyzes the aldolytic cleavage of this compound giving pyruvate and propionaldehyde; and (iii) StdN, a CoA-dependent acylating propionaldehyde dehydrogenase that transforms propionaldehyde in propionyl-CoA.

On the other hand, the metabolism of 3′7-diOH-HIP derived from CA, as well as other related compounds (originated from other bile acids or testosterone), continues through its C-3-side chain CoA activation, which is catalyzed by StdA3 in *P. putida* DOC21 [17]. Homologous CoA ligases recognizing HIP and hydroxylated derivatives have been identified in *R. jostii* RHA1 [25] and *C. testosteroni* TA441 [26, 27]. It has been proposed that after CoA activation, the 5-keto substituent is reduced, leading to 5-hydroxy-HIP and hydroxy derivatives, which suffer a β-oxidation-like process of the CoA-activated side chain, releasing an acetyl-CoA moiety and 3aα-*H*-4α(3′-carboxyl-CoA)-5,7-dihydroxy-7aβ-methylhexahydro-1-indanone. Then, (7aS)-7a-methyl-7-hydroxy-1,5-dioxo-2,3,5,6,7,7a-hexahydro-1H-indene-carboxyl-CoA is produced through the concomitant oxidation of 5-OH group and introduction of a double bond in C ring. Ring D is now opened through a hydrolytic reaction mediated by a crotonase yielding 2-(2-carboxyethyl)-4-hydroxy-3-methyl-6-oxocyclohex-1-ene-1-carboxyl-CoA which suffers the hydrolytic cleavage of ring C. After the opening of both CD rings the intermediate is the substrate of a thiolase yielding a molecule of acetyl-CoA and 4-methyl-5-oxo-octanedioyl-CoA. This last compound, after a β-oxidative process, releases another molecule of acetyl-CoA and 2-methyl-β-ketoadipyl-CoA, which is metabolically resolved as propionyl-CoA and succinyl-CoA [26–29] (Fig. 2).

2 Materials

2.1 Isolation of P. putida DOC21 Mutants Unable to Catabolize BA by Tn5 Transposon Mutagenesis

1. Luria-Bertani broth (LB): 10 g/L of Bacto Tryptone, 5 g/L of yeast extract, and 10 g/L of NaCl. For solid LB medium, supply with 2% (w/v) of agar.
2. MM solid media: 13.6 g/L of KPO_4H_2, 2.0 g/L of $(NH_4)_2SO_4$, 0.25 g/L of $MgSO_4.7H_2O$, and 0.005 g/L of $FeSO_4.7H_2O$, agar 2% (w/v), pH 7.0. It can be supplied with different carbon sources.
3. Kanamycin (Km), chloramphenicol (Cm), and ampicillin (Ap).
4. *E. coli* DH10B pGS9 [30].
5. *E. coli* HB101 pRK600 [31].
6. *P. putida* DOC21 [8].
7. 4-Hydroxyphenylacetic acid.
8. Shaker incubator.
9. Laboratory heaters (30 and 37 °C).
10. Spectrophotometer.
11. Centrifuge (e.g., Eppendorf 5810R).
12. Microcentrifuge (e.g., Eppendorf MiniSpin).
13. Polypropylene tubes 15 and 50 mL.
14. Microtubes 1.5 and 2 mL.
15. Filters 0.45 μm pore size.
16. Sterile toothpicks.

2.2 Identification of the Target Tn5 Insertion Point and Surrounding Sequences in the Genome of P. putida Strain DOC21

1. Plasmid pJQ200KS [32].
2. Plasmid pGEM-T *Easy* (Promega).
3. Specific oligonucleotide primers (*see* Table 1).
4. QIAquick PCR Purification Kit (QIAGEN) or QIAquick Gel Extraction Kit (QIAGEN).
5. Gentra Puregene Yeast/Bacteria Kit (QIAGEN).

Table 1
Specific oligonucleotide primers used for identification of the target Tn*5* insertion point and surrounding sequences in the genome of mutants obtained from *P. putida* strain DOC21

Primer	Sequence	Target
Tn5-ext	5′-GGGTGATCCTCGCCGTACTGCC-3′	Tn*5* transposon
Tn5-int	5′-GCGGACTGGCTTTCTACGTGTTC-3′	
Tn5-2	5′-CCGCCGAAGAGAACACAG 3′	
M13 forward (−47)	5′-CGCCAGGGTTTTCCCAGTCACGAC 3′	Universal primers of the plasmid pJQ200KS
M13 reverse (−48)	5′-AGCGGATAACAATTTCACACAGGA 3′	

6. Thermal cycler.
7. Polymerase chain reaction reagents.
8. Restriction endonucleases and T4 DNA ligase.
9. *E. coli* DH10B chemically competent cells.
10. LB broth and solid media.
11. Kanamycin (Km), chloramphenicol (Cm), ampicillin (Ap), and gentamycin (Gm).
12. Shaker incubator.
13. Laboratory heaters (30 and 37 °C).
14. Spectrophotometer.
15. Centrifuge (e.g., Eppendorf 5810R).
16. Microcentrifuge (e.g., Eppendorf MiniSpin).
17. Polypropylene tubes 15 and 50 mL.
18. Microtubes 1.5 and 2 mL.
19. Filters 0.45 μm pore size.
20. Sterile toothpicks.
21. Sodium acetate 3 M.
22. Ethanol.
23. QIAprep Spin Miniprep Kit (QIAGEN).

2.3 Genome Editing Through Deletion of Specific Genes Involved in Bile Acid Catabolism in *P. putida* DOC21

1. Plasmid pJQ200KS [32].
2. Plasmid pGEM-T *Easy* (Promega).
3. QIAquick PCR Purification Kit (QIAGEN) or QIAquick Gel Extraction Kit (QIAGEN).
4. Gentra Puregene Yeast/Bacteria Kit (QIAGEN).
5. Thermal cycler.
6. Polymerase chain reaction reagents.
7. Restriction endonucleases and T4 DNA ligase.
8. *E. coli* DH10B chemically competent cells.
9. LB broth and solid media.
10. Sucrose.
11. Kanamycin (Km), chloramphenicol (Cm), ampicillin (Ap), and gentamycin (Gm).
12. Shaker incubator.
13. Laboratory heaters (30 and 37 °C).
14. Spectrophotometer.
15. Centrifuge (e.g., Eppendorf 5810R).
16. Microcentrifuge (e.g., Eppendorf MiniSpin).

17. Polypropylene tubes 15 and 50 mL.
18. Microtubes 1.5 and 2 mL.
19. Filters 0.45 μm pore size.
20. Sterile toothpicks.

2.4 Identification of 4-Androstene-3,17-dione and 1,4-Androstadiene-3,17-dione as Metabolic Intermediates Using Specific Mutants of P. putida DOC21

1. Succinate.
2. LCA (Sigma-Aldrich).
3. 500 mL flasks.
4. Rotary shaker.
5. Microfuge.
6. Filters 0.22 μm pore size.
7. TLC Silica gel 60, aluminum sheets 20 × 20 cm (Merck KGaA).
8. Drummond glass micropipettes (10 μL).
9. Soft pencil, ruler.
10. Organic solvents: benzene, acetone.
11. TLC developing chamber.
12. Filter paper.
13. Hair dryer.
14. Exhaust hood.
15. Benzene.
16. Acetone.
17. AD (Sigma-Aldrich).
18. ADD (Sigma-Aldrich).
19. H_2SO_4 solution (30% by vol.).
20. Compressed gas sprayer for TLC.
21. Oven.
22. Gloves.

2.5 Identification of StdA1 and StdA2 from P. putida DOC21 as ATP-Dependent Acyl-CoA Synthetases Involved in the Catabolism of the C-17 BA-Acyl Chain

1. Erlenmeyer flasks (2000 mL).
2. Succinate.
3. Cholic acid (CA) (Sigma-Aldrich).
4. Shaker incubator.
5. Filters 0.22 μm pore size.
6. HCl.
7. Ethyl acetate.
8. Separation funnels.
9. Avanti J-30I High Performance Beckman centrifuge equipped with an Avanti JA-10 rotor (Beckman Coulter Ltd.).

10. HPLC system consisting of two high-pressure pumps (Shimadzu LC-10ATvp), an autoinjector (Gilson 234), a C18-reversed phase column, and a UV-visible light diode array detector (Shimadzu SPD-M 10) with a scanning range from 190 to 500 nm.
11. Semipreparative reversed-phase column 250 by 8 mm, Eurospher II, 100-5 C18 H (Knauer).
12. K–Na phosphate buffer 10 mM, pH 7.1.
13. Acetonitrile HPLC grade.
14. 50 mM Tris–HCl pH 8.0.
15. $NAD(P)^+$.
16. $K_3Fe[CN]_6$.
17. α,β-Hydroxysteroid dehydrogenase from *Pseudomonas testosteroni* (Sigma-Aldrich).
18. Spectrophotometer.
19. MOPS buffer pH 7.8.
20. CoA.
21. ATP.
22. $MgCl_2$.
23. French press (Amico).
24. Sephadex G-25 PD-10 columns.
25. Bradford protein assay reagents [31].

2.6 Identification of ATP-Dependent Acyl-CoA Synthetase StdA3 from *P. putida* DOC21 Involved in the Degradation of the C and D Rings of the Steroid Core

1. Erlenmeyer flasks (2000 mL).
2. Shaker incubator.
3. Avanti J-30I High Performance Beckman centrifuge equipped with an Avanti JA-10 rotor (Beckman Coulter Ltd.).
4. Separation funnels.
5. HPLC system consisting of a Waters binary pump 1525 HPLC module (Milford), a Luna 3 μm PFP(2) 50 × 4.6 mm column (Phenomenex), and a Waters 2996 photodiode array detector.
6. Eluents: ethyl acetate, hexane, and acetic acid; all of them HPLC grade.
7. Bradford protein assay reagents [33].
8. HEPES buffer pH 7.3.
9. $MgCl_2$.
10. ATP.
11. CoA.
12. Pyruvate.
13. HCl.
14. Ethyl acetate.

3 Methods

3.1 Isolation of Mutants of *P. putida* DOC21 Unable to Catabolize BAs by *Tn5* Transposon Mutagenesis

Tn*5* transposon mutagenesis has been proven to be a very useful technique in bacterial genetics [8, 34–36]. Tn*5* insertion into a gene leads to its inactivation, the insertion being stable and non-leaky. However, although the insertion of Tn*5* transposon occurs randomly, the existence of hot spots in the acceptor genome has been observed. Moreover, the insertion of the Tn*5* into a bacterial genome implies the creation of a new genetic and physical marker that could be readily mapped, allowing the sequencing and identification of the insertion point and surrounding DNA from kanamycin-resistant transconjugants. However, sometimes the insertion of the transposon on a set of cotranscribed genes (an operon) exerts polar effects silencing the expression genes downstream from the insertion point. This effect has been attributed to the presence of transcriptional and translational termination signals into the transposon [37].

1. Seed three different cultures in sterile 50 mL polypropylene tubes: (i) *E. coli* DH10B containing pGS9 [30] in a 10 mL aliquot of LB broth supplemented with kanamycin (Km, 25 μg/mL) and chloramphenicol (Cm, 30 μg/mL); (ii) *E. coli* HB101 pRK600 [31] in other 10 mL of LB broth supplemented with Cm (30 μg/mL); and (iii) *P. putida* DOC21 in 10 mL of LB broth with ampicillin (Ap, 100 μg/mL). Incubate *E. coli* strains at 37 °C and *P. putida* strain at 30 °C during 8–10 h.
2. Determine the growth of each culture as a function of their absorbance at 540 nm.
3. Collect aliquots of each culture in a unique 15 mL sterile polypropylene tube and mix them in a proportion of 5:1:1 (five times more cells of the acceptor *P. putida* DOC21 than the donor *E. coli* pGS9 and the helper *E. coli* pRK600, in function of the absorbance reached for each individual culture). Use, at least, 1 mL of *E. coli* pGS9 culture and the share of the other ones. Preserve the surplus of the cultures for future actions (*see* **step 11**).
4. Centrifuge the bacterial mixture of cells at 3220 × *g* for 10 min.
5. Discard the supernatant and suspend the bacteria in 1 mL of fresh, sterile LB broth. Transfer them to a 1.5 mL Eppendorf tube.
6. Centrifuge the microbial mixture at 12,700 × *g* for 2 min.
7. Suspend the bacteria in 1 mL of fresh, sterile LB broth.
8. Repeat **step 6**.
9. Suspend the mixed cultures in 50–80 μL of LB broth.

10. Dispose, using sterile forceps, a sterile Millipore bacterial filter (0.45 μm pore size) on the surface of a plate containing solid LB medium without any antibiotic.
11. Apply the cell suspension from **step 9** onto the sterile filter. At the same time, using 10–20 μL from the remaining cultures from **step 3**, inoculate LB plates containing Km, Cm, Ap, or a combination thereof, at the concentrations shown in **step 1**. They are used as control plates to verify the lack of microbial contamination. Incubate all the plates at 30 °C for 12 h (*see* **Notes 1** and **2**).
12. Once the validity of the strains (through examination of the control plates) is verified, place the filter with the mixture of strains onto a 15 mL polypropylene tube containing 3 mL of sterile LB broth. Discard the control plates.
13. Shake the tube until complete suspension of the bacteria.
14. Make serial dilutions from the suspension and seed 100 μL of each dilution onto LB plates supplied with Ap and Km (at the concentrations indicated in **step 1**).
15. Incubate the plates at 30 °C for 48 h, until rising of transconjugant colonies (*see* **Notes 3** and **4**). Use those plates where the higher number of isolated colonies arises.
16. Select transconjugants by seeding with a sterile toothpick from the LB plates supplied with Km and Ap to paired MM plates containing (i) 10 mM 4-hydroxyphenylacetic acid (or other steroid nonrelated carbon source, i.e., succinate), Ap, and Km, and (ii) CA (5 mM) plus Km (not Ap) (*see* **Note 5**). Incubate the plates containing MM for 12–20 h. Those transconjugants that do not grow in MM containing CA as a sole carbon source but that are able to grow in the counterselection MM plates (those containing a non-steroid carbon source, i.e., 4-hydroxyphenylacetic acid) should be considered as mutants specifically affected in BA metabolism.

    The full process is schematized in Fig. 3.

3.2 Identification of the Tn5 Insertion Point in the Genome of P. putida Strain DOC21 Mutants and Location of the Genomic Sequences Surrounding the Insertion Point

A method for the identification of the surrounding sequences around the insertion point of Tn5 has been developed in our laboratory. This method is based on the homologous recombination of a specifically designed suicide plasmid (derived from pJQ200KS) [32] containing a DNA fragment corresponding to the inverted repeated sequences flanking the Tn*5* transposon (construct pJQ-Tn*5*) [36]. Later, the genomic DNA from the recombinants is isolated, and after restriction of this DNA with specific endonucleases, and ligation of the constructs, the plasmids containing the adjacent sequences are recovered. The transformation of *E. coli* competent cells with the ligated DNA allows to obtain

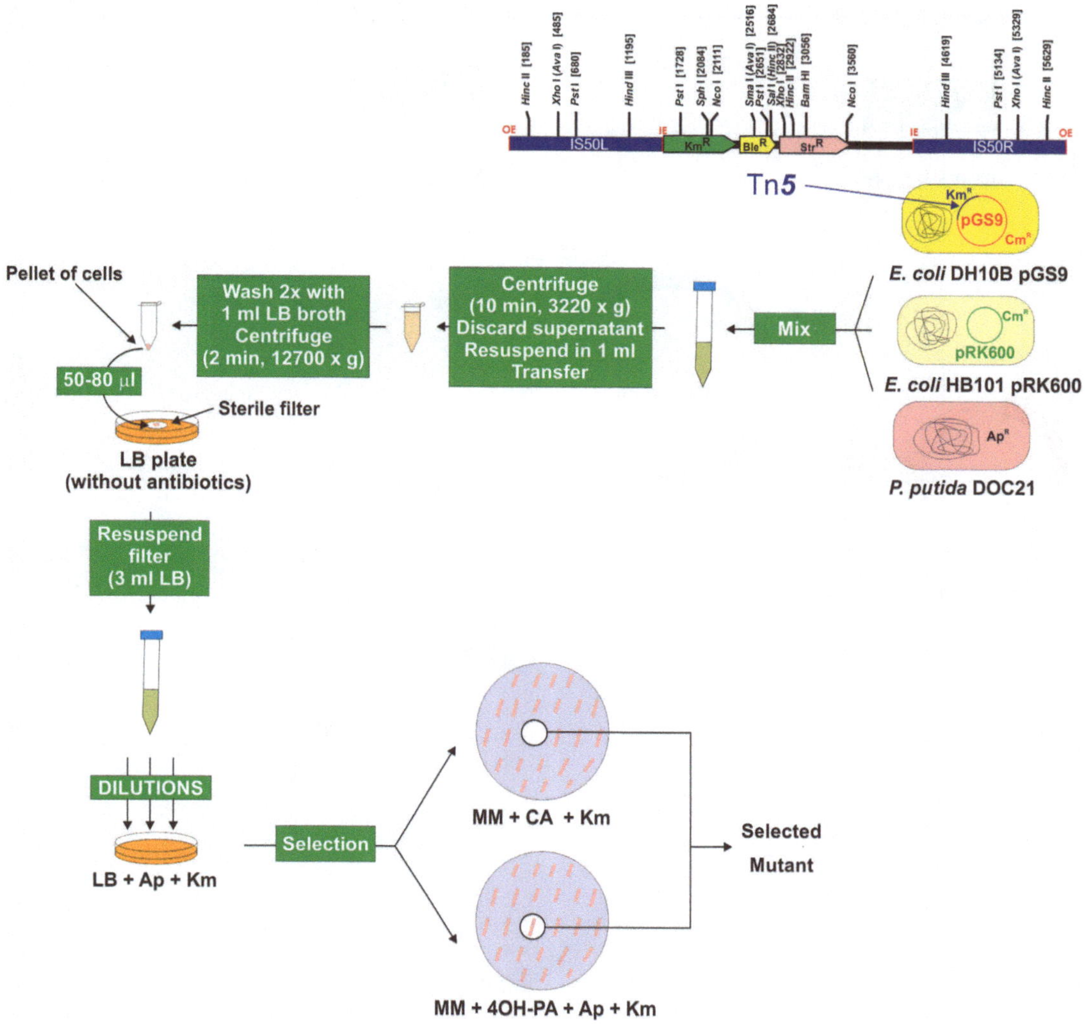

Fig. 3 Schematic representation of the procedure followed for collecting Tn*5* insertional mutants from *P. putida* DOC21

colonies containing the plasmid (and thus, the replicative origin contained in it) together with the genomic DNA adjacent to one of the IS*50* from Tn*5*.

3.2.1 Construction of pJQ-Tn5 (Fig. 4) [36]

1. Amplify by PCR a 1113 bp fragment present in both the IS*50*L and IS*50*R sequences from Tn*5* transposon, using the oligonucleotide primers Tn*5*-ext and Tn*5*-int (*see* **Note 6**).
2. Clone the amplified fragment into pGEM-T *Easy* according to the manufacturer instructions.
3. Transform *E. coli* DH10B chemically competent cells. Plate transformed cells on LB plates supplied with Ap (100 µg/mL). Select clones.

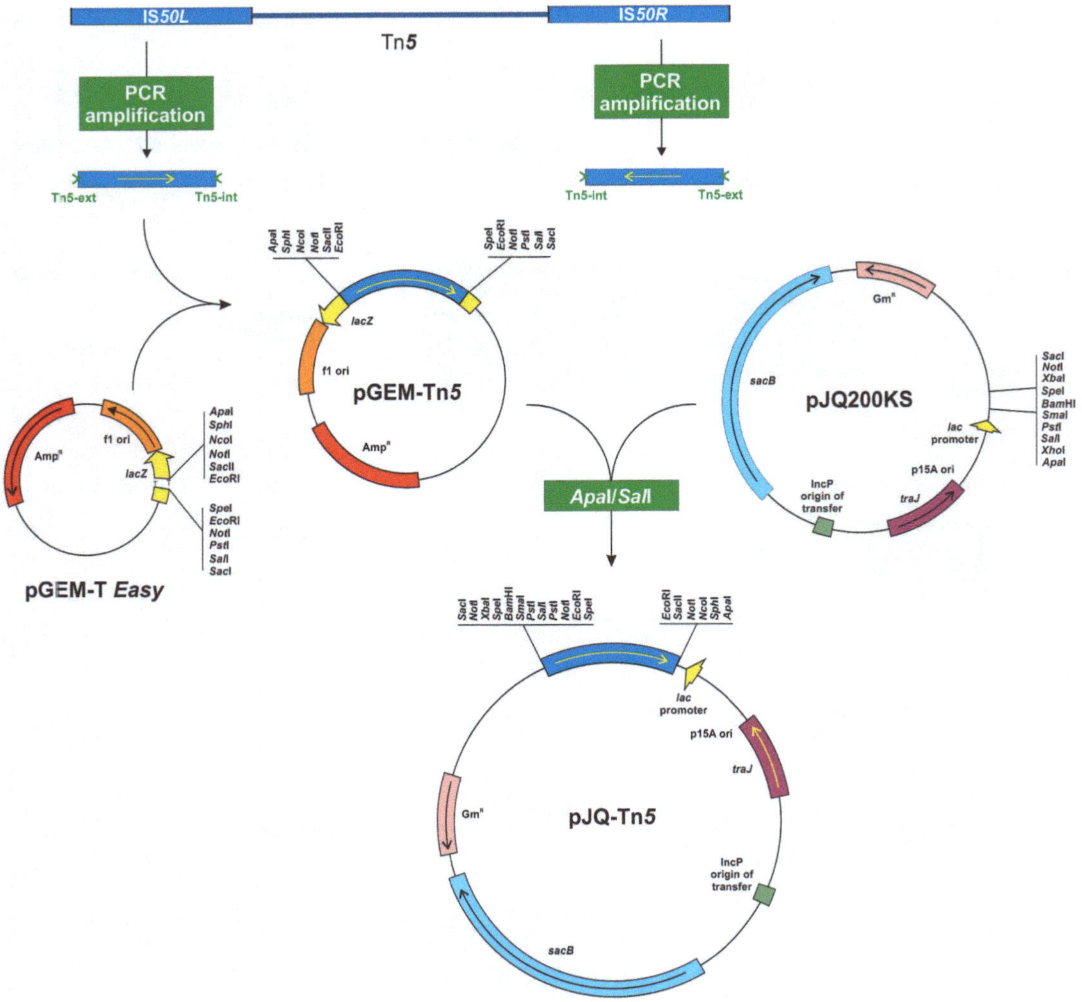

Fig. 4 Schematic representation of the process followed for the collection of the plasmid pJQ-Tn5 needed for identification of both the insertion points of Tn*5* transposon in the genome of *P. putida* strain DOC21 and the genomic sequences surrounding this point

4. Culture the selected clones in 15 mL sterile polypropylene tubes containing LB broth with Ap (100 μg/mL).
5. Collect the cells by centrifugation at maximum speed in a microcentrifuge for 5 min.
6. Extract the plasmids by using QIAprep Spin Miniprep Kit.
7. Rescue the cloned fragment by *Apa*I and *Sal*I restriction of the construction.
8. Ligate the fragment into a pJQ200KS [32] previously restricted with *Apa*I and *Sal*I.
9. Transform competent *E. coli* DH10B cells with the ligation reaction. Plate the transformed cells onto LB plates supplied with Gm (30 μg/mL).

10. Select the appropriate clone containing the construct pJQ-Tn*5*.

3.2.2 Transferring of pJQ-Tn5 into the P. putida *DOC21 Tn5 Mutant by Triparental Mating (Fig. 5)*

1. Run three different cultures in sterile 50 mL polypropylene tubes: (i) *E. coli* DH10B containing pJQ-Tn*5* in 10 mL of LB broth supplemented with Gm (30 μg/mL), (ii) *E. coli* HB101 pRK600 [31] in other 10 mL of LB broth supplemented with Cm (30 μg/mL), and (iii) *P. putida* DOC21 mutant in 10 mL of LB broth supplied with Ap (100 μg/mL) and Km (25 μg/mL). Incubate *E. coli* strains and *P. putida* for 8–10 h at 37 °C and 30 °C, respectively.
2. Determine the growth of each culture as a function of their absorbance at 540 nm.
3. Considering the absorbance reached by the cultures, collect, in a 15 mL sterile polypropylene tube, aliquots of *E. coli* cultures containing the same number of cells. Collect in the same tube the amount of *P. putida* DOC21 Tn*5* mutant culture corresponding to five times more cells of the acceptor (*P. putida* DOC21) than the donor (*E. coli* pJQ-Tn*5*) and the helper (*E. coli* pRK600). Use, at least, 1 mL of *E. coli* pJQ-Tn*5* culture and the share of the other ones. Preserve the surplus of the cultures to use them later for establishing controls of the process (*see* **step 11**).
4. Centrifuge the bacterial mixture at 3220 × *g* for 5 min.
5. Discard the supernatant and suspend the bacteria in 1 mL of fresh, sterile LB broth without any antibiotic. Transfer them to a 1.5 mL Eppendorf tube.
6. Centrifuge the mixture at 12,700 × *g* for 2 min.
7. Suspend the bacteria in 1 mL of fresh, sterile LB broth.
8. Repeat **step 6**.
9. Suspend the mixed cultures in 50–80 μL of LB broth.
10. Dispose (by using sterile forceps) a sterile Millipore bacterial filter (0.45 μm pore size) on the surface of a plate containing solid LB medium without any antibiotic.
11. Apply the cell suspension from **step 9** onto the sterile filter. At the same time, using 10–20 μL from the rest of the cultures from **step 3**, inoculate LB plates containing Km, Gm, Cm, Ap, or combinations thereof, at the concentrations indicated in **step 1** (control plates to verify no contamination of the strains). Incubate these plates at 30 °C for 12 h (*see* **Notes 1** and **2**).
12. Once the suitability of the strains (through examination of the control plates) is verified, place the filter with the mixture of strains (using sterile forceps) onto a 15 mL polypropylene tube

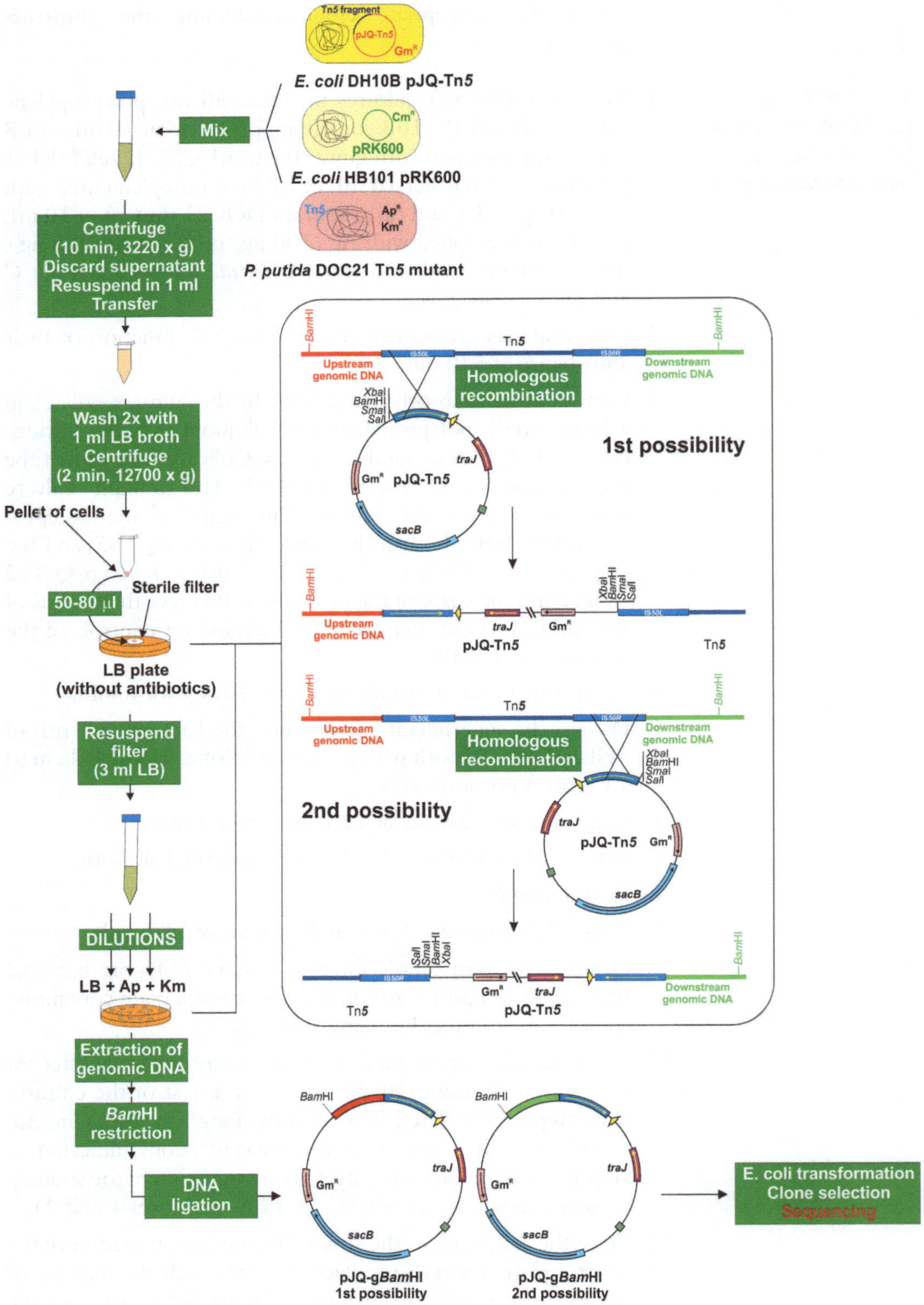

Fig. 5 Schematic representation of the procedure followed to get the identification of (i) the point in which the transposon Tn*5* is nested in the genome of *P. putida* DOC21 mutants as well as (ii) the genomic surrounding

containing 3 mL of sterile LB broth and shake the tube until complete suspension of the bacteria. Discard the control plates.

13. Make serial dilutions from the suspension and seed 200 μL of each dilution onto LB plates supplied with Ap, Gm, and Km (at the concentrations indicated in **step 1**).
14. Incubate the plates at 30 °C for 48 h or the time required (until rising of transconjugant colonies).
15. Select isolated colonies streaking them on LB plates containing Ap, Gm, and Km.

3.2.3 Cloning of the DNA Fragments in Which Tn5 Has Been Inserted (Fig. 5)

1. Grow the selected pJQ-Tn*5* recombinant clones (*see* **Note 7**) in sterile 50 mL polypropylene tubes containing 10 mL of LB broth with Ap, Gm, and Km (concentrations indicated in Subheading 3.2.2, **step 1**). Incubate the cultures at 30 °C for 8–10 h with vigorous shaking (250 rpm).
2. Transfer 1 mL from each culture into 1 mL sterile Eppendorf tubes on ice.
3. Centrifuge at 12,700 × *g* for 1 min.
4. Purify the genomic DNA using the Gentra Puregene Yeast/Bacteria Kit.
5. Digest the 50 μL of the genomic DNA obtained from **step 4** with the endonucleases *Bam*HI, *Xba*I, *Sma*I, or *Sal*I (*see* **Note 8**).
6. Precipitate the restriction reactions by adding 1/10 volume sodium acetate 3 M, pH 5.2 and 2.5 volumes of cold 100% ethanol. Place at −20 °C for >20 min.
7. Centrifuge at maximum speed on a microfuge for 15–20 min.
8. Decant the supernatant.
9. Wash the DNA pellet with 1 mL 70% ethanol. Spin at maximum speed on a microcentrifuge for 5 min. Decant the supernatant.
10. Air-dry the pellet for 5 min.
11. Resuspend the pellet in 35 μL of sterile Milli-Q water by incubating at 65 °C, 5–10 min, and at room temperature, for 10–15 additional min.

Fig. 5 (continued) sequences. The knowledge of the Tn*5* insertion points is based on the homologous recombination of pJQ-Tn*5* over both IS*50* fragments flanking the transposon. In the scheme are indicated the two different recombination possibilities. The use of *Bam*HI restriction of the genomic DNA from the recombinant strains in which pJQ-Tn*5* has been nested is only for didactical purposes. This restriction could be performed with *Bam*HI, *Xba*I, *Sal*I, or *Sac*I, according to the presence of these targets in the genomic DNA

12. Circularize the DNA fragments: use 17 μL of the DNA from **step 11** in a 20 μL ligation reaction (2 μL of 10× T4 DNA ligase ligation buffer, 1 μL of T4 DNA ligase, and 17 μL of DNA).
13. Use 10 μL from the ligation reaction to transform chemically competent *E. coli* DH10B cells.
14. Seed the transformed bacteria on LB plates containing Gm (30 mg/mL).
15. Select clones.
16. Culture the selected clones in 3 mL of LB broth at 37 °C during 8–12 h.
17. Collect the bacteria by centrifuging in Eppendorf tubes.
18. Extract the plasmids by using QIAprep Spin Miniprep Kit.
19. Check the plasmids through restriction with the same endonuclease used in **step 5** (*see* **Note 9**).
20. Sequence the selected constructions from universal primers of the plasmid (M13 forward −47 and M13 reverse −48) and Tn*5*-2 (oligonucleotide primer present at 152 bp from the end of IS*50* sequences) (*see* **Note 10**).

3.3 Genome Editing Through Deletion of Specific Genes Involved in Bile Acid Catabolism in P. putida DOC21

In order to study the metabolic role of specific genes and enzymes, or to obtain catabolic intermediates belonging to the BA degradative pathway, a procedure of genome editing based on the specific deletion of selected genes is proposed. This technique consists, summarily, in force a double homologous recombination of a suicide plasmid (pJQ200KS) [32] into the genome, using DNA fragments upstream and downstream of the fragment that is desired to be deleted as homologous recombinational places (Fig. 6). The first recombination is triggered by the selective pressure to conserve the Gm resistance through culture of the clones in LB plates containing Gm, and it inserts the plasmid into the chromosome; the second recombinational process is induced by culturing the selected clones in LB plates without Gm but with sucrose. Plasmid pJQ200KS harbors the *sacB* gene, whose enzymatic product synthesize high-molecular-weight, β-(2,6)-linked levan from sucrose by transfer of fructosyl units in the periplasmic space, having a strong deleterious effect over the cell. To avoid cell death, some of the clones suffer a second homologous recombinational event that eliminates the inner fragment of the target gene as well as the suicide plasmid.

1. Amplify by PCR the fragments (>400 bp) adjacent (upstream and downstream) of the DNA fragment to be deleted. The oligonucleotides in 3′ (for the fragment located upstream) and 5′ (for the fragment located downstream) should include a restriction target sequence that is not present inside of the amplified fragments nor *Not*I.

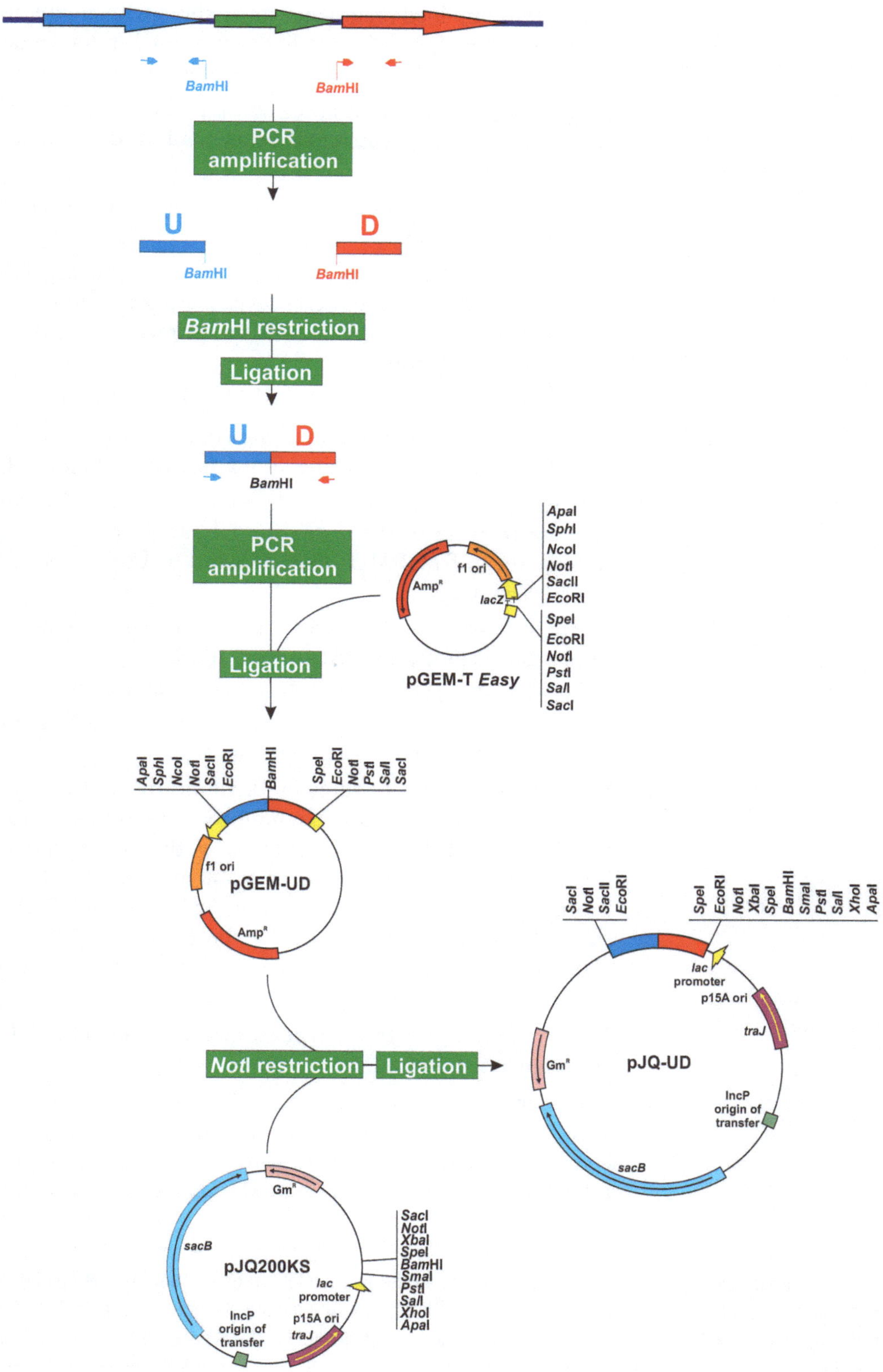

Fig. 6 Schematic representation of the construction of a theoretical plasmid derived from pJQ200KS illustrating the process for obtaining a construction to delete genes in *P. putida* DOC21. U (from "upstream")

2. Purify the amplified fragments from the PCR amplification reaction (using QIAquick PCR Purification Kit) or from an agarose gel (using QIAquick Gel Extraction Kit).
3. Digest both fragments with the endonuclease whose target has been included in the upstream 3′ primer and the downstream 5′ primer.
4. Using equal amounts of each fragment, ligate them in a T4 DNA ligase reaction.
5. Using the T4 DNA ligation reaction as a template DNA, amplify by PCR both fragments ligated using the primer in 5′ from the fragment upstream and the primer in 3′ from the fragment downstream.
6. Ligate the PCR amplicon into pGEM-T *Easy* plasmid.
7. Transform *E. coli* DH10B with the constructs and select the appropriate clones in LB plates containing Ap (100 μg/mL).
8. Rescue the fragment from the selected clones using *Not*I endonuclease (cutting to the restriction targets from the multiple cloning sites of pGEM-T *Easy* disposed at both sites of the cloned fragment).
9. Ligate the fragment into *Not*I restricted and dephosphorylated (using calf alkaline phosphatase) pJQ200KS.
10. Transform *E. coli* DH10B with the constructs and select the appropriate clones in LB plates containing Gm (30 μg/mL). The full process is schematized in Fig. 6.
11. Seed three different cultures in sterile 50 mL polypropylene tubes: (i) *E. coli* DH10B containing the pJQ200KS derivative in 10 mL of LB broth supplemented with gentamycin (Gm, 30 μg/mL), (ii) *E. coli* HB101 pRK600 in other 10 mL of LB broth supplemented with Cm (30 μg/mL), and (iii) *P. putida* strain DOC21 in 10 mL of LB broth supplied with Ap (100 μg/mL). Incubate the *E. coli* strains at 37 °C and *P. putida* at 30 °C for 8–10 h.
12. Determine the growth of each culture as a function of their absorbance at 540 nm.
13. Considering the absorbance reached by the cultures, collect, in a 15 mL sterile polypropylene tube, aliquots of each *E. coli* cultures containing equivalent number of cells. Add to this tube a volume of the *P. putida* DOC21 culture containing

Fig. 6 (continued) and D (from "downstream") indicated the two flanking regions used for the deletion of a gene located between these two fragments. The *Bam*HI target in the oligonucleotides in 3′ from U and in 5′ from D is used for didactical reasons. The endonuclease target to include in these degenerate oligonucleotides will be dependent of the restriction sites inside the fragments to be amplified, avoiding the use of the targets that would be present as well as *Not*I, due to its use in the process of cloning

five times more cells than the donor *E. coli* containing the plasmid pJQ200KS and that of the helper strain (*E. coli* pRK600). Use, at least, 1 mL of donor *E. coli* pJQ200KS culture and the share of the other ones. Preserve the surplus of the cultures; they will be used later to establish controls of the process (*see* **step 21**).

14. Centrifuge the bacterial mixture at 3220 × *g* for 5 min.
15. Discard the supernatant and suspend the cells in 1 mL of fresh, sterile LB broth without any antibiotic. Transfer the cells to a 1.5 mL Eppendorf tube.
16. Centrifuge them at 12,700 × *g* for 2 min.
17. Suspend the bacteria in 1 mL of fresh, sterile LB broth.
18. Repeat **step 16**.
19. Suspend the mixed cultures in 50–80 μL of LB broth.
20. Dispose, with sterile forceps, a sterile Millipore bacterial filter (0.45 μm pore size) on the surface of a plate containing solid LB medium without any antibiotic.
21. Apply the cell suspension from **step 19** onto the sterile filter. At the same time, using 10–20 μL from the surplus of the cultures from **step 3**, inoculate LB plates containing Gm, Cm, and Ap or a combination thereof at the concentrations indicated in **step 1**, and incubate them at 30 °C for 12 h (*see* **Notes 1** and **2**).
22. Once the suitability of the strains (through examination of the control plates) is verified, place the filter with the mixture of strains (using sterile forceps) onto a 15 mL polypropylene tube containing 3 mL of sterile LB broth and shake the tube until complete suspension of the bacteria. Discard the control plates.
23. Make serial dilutions from the suspension and seed 200 μL of each dilution onto LB plates supplied with Gm (30 μg/mL).
24. Incubate the plates at 30 °C, until rising of transconjugant colonies (about 48 h). This medium only allows the growth of these *P. putida* DOC21 derivatives in which the pJQ200KS-derived plasmid has been inserted onto the chromosome.
25. Pick isolated colonies with a sterile toothpick and suspend each colony on an Eppendorf tube containing 200 μL of sterile Milli-Q quality water.
26. Make serial dilutions of the cells and plate them in LB plates supplied with 10% sucrose without Gm.
27. Incubate the plates at 30 °C until colonies arise (about 48 h).
28. Select isolated clones by streaking them in LB plates containing 10% sucrose and LB plates supplied with Gm. The clones that

have suffered a deletion should be fully sensitive to Gm (*see* **Note 11**).

29. Check that the deletion has occurred in the selected clones. Thus, amplify by PCR using specific oligonucleotide primers designed over genomic DNA (upstream and downstream) of the fragments that have been cloned for recombinations. The colonies in which the deletion has taken place should present an amplification product with a reduced molecular size (it has lost the deleted fragment) than the amplification observed when the wild type strain is used (*see* **Note 12**).

The procedure is schematized in Fig. 7.

3.4 Identification of 4-Androstene-3,17-dione and 1,4-Androstadiene-3,17-dione as Metabolic Intermediates Using Specific Mutants of P. putida DOC21

Using the technique described in Subheading 3.3, the genes encoding the 3-ketosteroid-Δ^1-dehydrogenase (*stdH*) and the reductase component of the 3-ketosteroid-9α hydroxylase (*stdJ*) from *P. putida* DOC21 were deleted, collecting the mutant strains *P. putida* DOC21 Δ*stdH* and *P. putida* DOC21 Δ*stdJ*. As expected, mutant Δ*stdH* accumulated AD in the culture broths, whereas *P. putida* DOC21 Δ*stdJ* accumulated ADD when cultured in a medium containing testosterone or LCA (5 mM) as carbon sources. The accumulation of both compounds in the culture broths could be easily observed by thin-layer chromatography (TLC).

1. Culture the mutants *P. putida* DOC21 Δ*stdH* and *P. putida* DOC21 Δ*stdJ* in 500 mL Erlenmeyer flasks containing 100 mL of MM supplied with succinate (10 mM) and LCA (5 mM) as carbon sources (*see* **Note 13**).
2. Incubate the flasks on a rotary shaker (250 rpm) at 30 °C until 80 h after reach stationary phase of growth (*see* **Note 14**). Take aliquots (0.5 mL) of the cultures at the desired times.
3. Centrifuge the samples of the cultures on a microfuge at maximum speed for 3 min. Recover the supernatant and discard the cell pellet.
4. Filtrate the supernatants through 0.22 μm pore size filters for the complete removal of cells and salt particles.
5. Previously, prepare a TLC developing chamber by disposing a filter paper lying against one of the major walls of the container. Add 80–100 mL of the mobile phase (benzene/acetone 5:3; v:v) into the chamber and let it soak the blotting paper. Close the chamber and leave it for 1–2 h to saturate it.
6. Draw with a pencil a parallel line at 1.5 cm above the lower border of the TLC plate. Divide this line with separate marks of 1 cm of distance between them, and separate 1 cm of both sides of the plate. Write with a pencil the initials of the sample to apply below of each mark. Draw another line of 0.5 cm in the

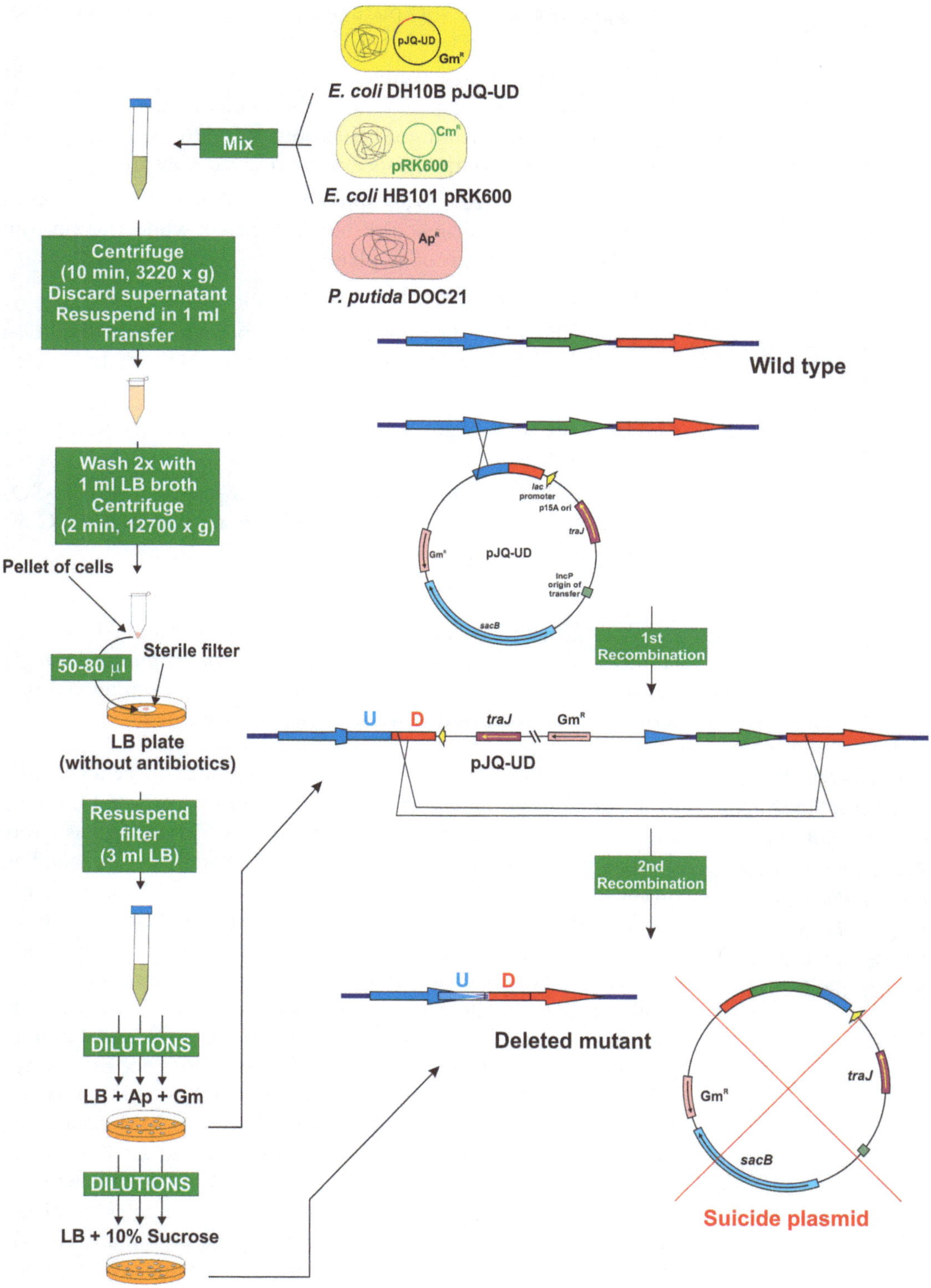

Fig. 7 Schematic representation of the procedure followed to delete a specific gene in *P. putida* DOC21. The two homologous recombinations that occur during the process are shown. The suicide plasmid used in the representation (pJQ-UD) is the same whose construction is indicated in Fig. 6

upper border of the plate to mark the end of the chromatographic step (*see* **Note 15**).

7. Drop 10 μL of each sample and patterns (AD and ADD 5 mM each solved in sterile MM) in the marks of the bottom line using a glass Drummond micropipette. After spotting, use a hair dryer to dry the spots evaporating the solvent.
8. Introduce the TLC plate into the chamber and let the top of the plate lean against the blotting paper while the bottom contacts with the mobile phase.
9. Seal the chromatographic chamber with its lid.
10. Leave the TLC plate inside the chamber allowing the elution until the solvent reaches the line on the top of the plate (*see* **step 6**).
11. Take the TLC plate out of the chamber and let it dry in a ventilated place (exhaust hood).
12. Reveal the spots in the plate by spraying it with 30% H_2SO_4 (w/v) (*see* **Note 16**) and heating it on an oven at 100 °C for 10 min.
13. Compare the sample stains with the pattern ones.

A TLC plate showing the accumulation of AD and ADD in the culture broths of *P. putida* DOC21 Δ*stdH* and *P. putida* DOC21 Δ*stdJ* mutants, respectively, is shown in Fig. 8. Commercial patterns of these compounds are also chromatographed.

3.5 Identification of StdA1 and StdA2 from *P. putida* DOC21 as ATP-Dependent Acyl-CoA Synthetases Involved in the Catabolism of the C-17 BA-Acyl Chain

Between the pool of *std* genes (steroid degradation genes) found in the genome of *P. putida* DOC21, three of them (*stdA1*, *stdA2*, and *stdA3*) encode acyl-CoA synthetases. These genes were deleted in *P. putida* DOC21 using the procedure described in Subheading 3.3, yielding the mutant strains Δ*stdA1*, Δ*stdA2*, and Δ*stdA3*. When cultured in MM containing BA, testosterone, or AD, their behavior was significantly different. Thus, Δ*stdA1* and Δ*stdA2* strains were unable to grow with any BA, but they showed a similar growth to the wild type when cultured in MM containing testosterone or AD as carbon sources. However, Δ*stdA3* mutant grew poorly in media supplied with BA, and it does not grow when cultured in MM containing testosterone or AD as carbon sources. Moreover, when these mutants were cultured in MM containing succinate (10 mM) and CA, Δ*stdA1* accumulated Δ^1-3-ketocholate, and/or Δ^4-3-ketocholate, and $\Delta^{1,4}$-3-ketocholate, whereas Δ*stdA2* accumulated 7α,12α-dihydroxy-3-oxopregna-1,4-diene-20-carboxylate (DHOPDC). Likewise, when the mutant Δ*stdA3* was cultured in MM containing testosterone or bile acids, 3aα-*H*-4α(3′propanoate)7aβ-methylhexahydro-1,5-indanedione (HIP) (or the corresponding hydroxylated derivatives) was accumulated in the culture broths [17].

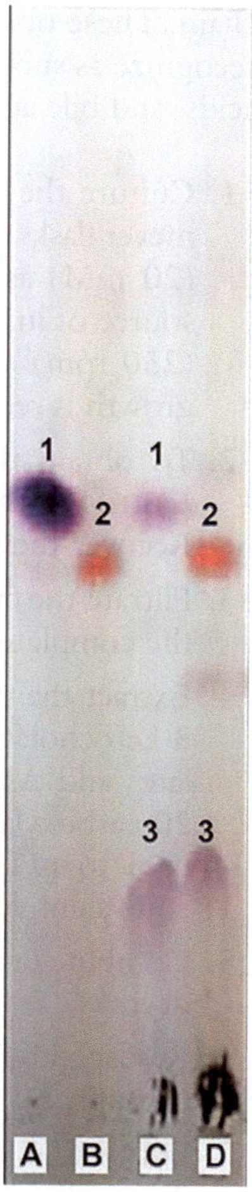

Fig. 8 Thin-layer chromatography (TLC) of commercial AD (A) and ADD (B) and the products accumulated in the culture broth of the mutants *P. putida* Δ*stdH* (C) and *P. putida* Δ*stdJ* (D) when these strains are cultured in MM containing succinate (10 mM) as carbon and energy source and LCA (5 mM) as precursor of intermediates. (1) AD spots; (2) ADD spots; (3) LCA remaining in the culture broths

The assay of the enzymatic activities revealed that StdA1 was able to activate, in an ATP-dependent reaction, CA, 3-ketocholate, Δ^1-3-ketocholate, and/or Δ^4-3-ketocholate, and $\Delta^{1,4}$-3-ketocholate to their CoA thioester derivatives; StdA2 activated DHOPDC to DHOPDC-CoA; and StdA3 catalyzed the CoA thioesterification of HIP and its hydroxylated derivatives. Thus, it could be concluded that StdA1 and StdA2 are involved in the degradation of the C-17 acyl chain, whereas StdA3 initiates the degradation of the C and D rings of the steroid molecule (*see* Fig. 2). Furthermore, the results obtained with mutant Δ*stdA1* suggest that the introduction of unsaturations in C-1 and C-4 in the A ring by StdH and by StdI could be accomplished without shortening of the C-17 acyl chain. These observations suggest that StdH and StdI are able to recognize as substrates testosterone, AD, 3-keto-derivatives of bile acids, and bile acids [17].

3.5.1 Preparation of the Substrates for StdA1 and StdA2 Assays

1. Culture the mutants Δ*stdA1* and Δ*stdA2* in 2000 mL Erlenmeyer flasks containing 500 mL of MM supplied with succinate (20 mM) as carbon and energy source and CA (5 mM) as source of intermediates. Incubate the flasks on a rotary shaker (250 rpm) at 30 °C until 80 h after the stationary phase of growth is reached.
2. To obtain the culture broths containing the intermediates, centrifuge the cultures at 12,400 × *g* for 10 min. at 4 °C. Recover the supernatant and discard de cell pellet.
3. Filtrate the culture broths through 0.22 μm pore size filters for the complete removal of cells and salt particles.
4. Extract the metabolites accumulated by mutants Δ*stdA1* (Δ^1-3-ketocholate, and/or Δ^4-3-ketocholate, and $\Delta^{1,4}$-3-ketocholate) and Δ*stdA2* (7α,12α-dihydroxy-3-oxopregna-1,4-diene-20-carboxylate, DHOPDC). The supernatant should be acidified to pH ~ 1 with HCl and extracted three times with equivalent volumes of ethyl acetate.
5. Combine organic fractions from each mutant and evaporate to dryness.
6. Resuspend the residues in 2 mL of ethyl acetate and filter through a 0.22 μm filter.
7. Purify Δ^1-3-ketocholate, and/or Δ^4-3-ketocholate, and $\Delta^{1,4}$-3-ketocholate by HPLC using a semipreparative reversed-phase column (250 by 8 mm, Eurospher II, 100-5 C18 H) with a total flow rate of 0.8 mL/min using as eluents K–Na phosphate buffer 10 mM pH 7.1 (eluent A) and acetonitrile (eluent B). Use a gradient method, starting with 20% eluent B for 2 min, increasing to 34% eluent B within 9 min, and returning to 20% eluent B within 1 min, followed by an equilibration of 6 min.
8. For production of 3-ketocholate as a substrate for StdA1, perform a reaction containing 50 mM Tris–HCl pH 8.0, 1 mM $NAD(P)^+$ or $K_3Fe[CN]_6$, and 0.3 U of α,-β-hydroxysteroid dehydrogenase from *Pseudomonas testosteroni*. Start the reaction by adding 1 mM cholate. Cholate-dependent reduction of $NAD(P)^+$ and $K_3Fe[CN]_6$ is measured spectrophotometrically at 365 nm ($\varepsilon = 3.4\ mM^{-1}\ cm^{-1}$) and at 436 nm ($\varepsilon = 0.7\ mM^{-1}\ cm^{-1}$), respectively [16]. Extract the reaction products as indicated in **step 4**, and quantify residual cholate by HPLC, using an isocratic method with 70% eluent A and 30% eluent B.
9. Purify DHOPDC extracted from Δ*stdA2* by HPLC using a semipreparative reversed-phase column (250 by 8 mm,

Eurospher II, 100-5 C18 H) with a total flow rate of 0.8 mL/min. The gradient method starts with 20% eluent B for 2 min and increases to 70% eluent B in 1 min, followed by an equilibration of 6 min.

10. Analyze the concentration of Δ^1-3-ketocholate, and/or Δ^4-3-ketocholate, $\Delta^{1,4}$-3-ketocholate, and DHOPDC in the resulting solutions by measuring their absorbance at 245 nm, which is the characteristic maximum for steroid compounds containing a Δ^1-, Δ^4-, and $\Delta^{1,4}$-3-keto structure. Calculations are based on an averaged molar extinction coefficient of $\varepsilon_{245\ \mathrm{nm}}$ of 14.7 cm^{-1} mM^{-1} [38].

3.5.2 StdA1 and StdA2 ATP-Dependent Acyl-CoA Synthetase Assays

1. Culture *P. putida* DOC21 and the Δ*stdA1* and Δ*stdA2* mutants in 2000 mL Erlenmeyer flasks containing 500 mL of MM containing succinate (10 mM) and CA (5 mM). Incubate on a rotary shaker (250 rpm, 30 °C) until the late exponential phase of growth.
2. Harvest the bacteria by centrifugation (9000 × *g*, 10 min at 4 °C).
3. Wash them with 50 mM K–Na–PO_4 buffer pH 7.0 and centrifuge (9000 × *g*, 10 min at 4 °C).
4. Suspend the bacterial pellet in 3 mL of the same buffer.
5. Disrupt the bacteria by three passages through a French press at 138 MPa.
6. Remove the cell debris by centrifugation (17,900 × *g*, 30 min at 4 °C).
7. Desalt the homogenates through Sephadex G-25 PD-10 columns to eliminate low-molecular-weight compounds.
8. Froze at −20 °C until use.
9. Analyze protein concentration by Bradford protein assay using bovine serum albumin as a standard.
10. Assay StdA1 and/or StdA2 activities by measuring the formation of different acyl-CoA thioesters in a mixture of reaction containing 50 mM MOPS buffer pH 7.8, 1 mM CoA, 2 mM ATP, 3 mM $MgCl_2$, and 0.6–1.0 mg/mL of cell-free extracts from the different strains, starting the reaction by adding 0.5 mM of the tested substrate (CA, 3-ketocholate, Δ^1-3-ketocholate, Δ^4-3-ketocholate, $\Delta^{1,4}$-3-ketocholate, or DHOPDC).
11. Withdraw samples immediately after the starting of the reaction and at defined time intervals (15 min to 2 h).
12. Analyze the CoA thioesters produced by StdA1(cholyl-CoA, 3-ketocholyl-CoA, $\Delta^{1,4}$-3-ketocholyl-CoA) and StdA2 (DHOPDCCoA) by HPLC using a gradient method, starting with 10% acetonitrile and 90% K–Na phosphate buffer 10 mM,

pH 7.1 for 2 min, increasing to 56% acetonitrile within 23 min, and returning to 10% acetonitrile within 1 min, followed by an equilibration of 6 min [21, 22] (*see* **Note 17**).

3.6 Identification of ATP-Dependent Acyl-CoA Synthetase StdA3 from P. putida DOC21 Involved in the Degradation of the C and D Rings of the Steroid Core

P. putida DOC21 Δ*stdA3* accumulates 3aα-*H*-4α(3′propanoate) 7aβ-methylhexahydro-1,5-indanedione (HIP) when cultured in MM containing testosterone, AD, or LCA as source of precursors. When this mutant is cultured in the presence of other bile acids, the respective hydroxylated derivatives of HIP are accumulated in the culture broth [3aα-*H*-4α(3′(*R*)-hydroxy-3′-propanoate)-7a-β-methylhexahydro-1,5-indanedione from CDCA or 3aα-*H*-4α(3′(*S*)-hydroxy-3′-propanoate)-7aβ-methylhexahydro-1,5-indanedione from UDCA -3′-OH-HIP-; 3aα-*H*-4α(3′-propanoate)-7-α-hydroxy-7aβ-methylhexahydro-1,5-indanedione-7-OH-HIP- from DCA; and 3aα-*H*-4α(3′(*R*)-hydroxy-3′-propanoate)-7-α-hydroxy-7aβ-methylhexahydro-1,5-indanedione -3′,7-diOH-HIP- formed from CA]. StdA3 has been characterized as the ATP-dependent acyl-CoA synthetase catalyzing the activation of the HIPs to their CoA derivatives for its further catabolism [17]. A similar activity has been identified in actinobacteria when the metabolism of cholesterol was studied [25].

3.6.1 Obtaining of the Substrates for StdA3 Assay

1. Culture the mutant Δ*stdA3* in 1 L of MM supplied with 20 mM of pyruvate (2 × 2000 mL Erlenmeyer flasks containing 500 mL MM each). Incubate the flasks on a rotary shaker (250 rpm) at 30 °C.
2. Harvest the cells by centrifugation (12,400 × *g* during 10 min at 4 °C) at the mid-exponential phase of growth (OD_{540} = 1.2–1.4).
3. Wash the cells with 400 mL of MM without carbon source; recover the cells by centrifugation at 12,400 × *g* during 10 min at 4 °C.
4. Suspend the cells in 400 mL MM containing 1–2 mM of the required steroid (BA or testosterone).
5. Incubate the bacterial cell suspensions with shaking (250 rpm) for 80–100 h at 30 °C.
6. Harvest the cells by centrifugation at 12,400 × *g* during 10 min at 4 °C and recover the supernatants.
7. Extract the metabolites accumulated by mutant Δ*stdA3* (HIP, or the corresponding hydroxylated derivatives) by acidification of the supernatants to pH ~ 3 with concentrated HCl and followed by three successive organic extractions with one equivalent of ethyl acetate.
8. Combine the organic fractions and evaporate the solvent to dryness.

9. Suspend the solid residue in 2 mL of ethyl acetate and pass through a 0.22 μm filter.
10. Purify by HPLC the extracted metabolites by loading the sample onto a 250 × 10 mm Luna 5 mm silica (2) column. Operate the HPLC system at 11.1 mL/min.
11. For HPLC separation of the HIPs, use a 60 min gradient of 0–100% ethyl acetate in hexane and 0.5% acetic acid.
12. Monitor the eluates at 290 nm.
13. Recover the fractions containing HIP (or the hydroxylated derivatives), evaporate them to dryness under nitrogen, and store until use at −80 °C.

3.6.2 StdA3 ATP-Dependent Acyl-CoA Synthetase Assay

1. Prepare cell-free extracts of wild-type strain and Δ*stdA3* mutant according to the **steps 1–9** from Subheading 3.6.2.
2. Assay StdA3 activity by measuring the formation of different acyl-CoA thioesters in 100 μL of mixture of reaction containing 100 mM HEPES pH 7.3, 5 mM $MgCl_2$, 2 mM ATP, 1.2 mM CoA, 2 μM of StdA3, and 1 mM of the tested substrate (3′,7-diOH-HIP, 3′-OH-HIP, 7-OH-HIP, or HIP).
3. Incubate the reactions for 2 h at 22 °C.
4. Analyze the thioester formation by HPLC by injecting 1 mL of samples on a Luna 3 μm PFP(2) 50 × 4.6 mm column equilibrated with 0.1 M ammonium acetate, pH 4.5. Elute the CoA thioesters using a 20 mL gradient from 0% to 90% methanol in 0.1 M ammonium acetate, pH 4.5, monitoring the eluate at 258 nm.

4 Notes

1. Incubate the filter containing the mixture of strains. The temperature selected should be the optimal for the more restrictive strain (acceptor strain).
2. Avoid the formation of bubbles when disposing the mixture of cells over the filter; these bubbles could explode spreading the strains over the plate.
3. Considering that none of the parental strains carries both antibiotic resistances used for selection (Ap and Km), in these plates only those *P. putida* DOC21 (Ap resistance) transconjugants could grow, in which the transposon (giving the Km resistance) has been inserted into the genome.
4. If the acceptor strain would have a Km resistance, the selection should be made using these other antibiotic markers (bleomycin and streptomycin) also present in the transposon (Fig. 3).

5. In *P. putida* DOC21 a deleterious effect inhibiting the bacterial growth has been observed when this strain is cultured in media containing Ap and BA. Although this effect has not been investigated, it could be speculated that a mixture of bile acids (acting as detergents) and ampicillin could have a synergistic effect, sensitizing the bacteria against the antibiotic.
6. Remember that IS*50*L and IS*50*R are repeated and inverted sequences. Thus, the recombination through the sequence cloned in pJQ-Tn5 could occur with the same probability over both IS*50* sequences but in an opposite direction, allowing the cloning of both the sequences upstream and downstream from the Tn*5* insertion point (Fig. 5).
7. Use various clones. Considering that the recombination event could happen with the same probability over the two IS*50* sequences, the resulting recombinants will correspond to both events.
8. The use of these restriction endonucleases (*Bam*HI, *Xba*I, *Sma*I, or *Sal*I) is due to the presence of target sequences in the multiple cloning site of the plasmid but the lack of these targets inside the IS*50* sequences.
9. After the restriction of the clones with the same endonuclease used to rescue the fragment adjacent to Tn*5*, only a band in an agarose gel should be observed. It corresponds to the pJQ200KS plus the insert (1269 bp from the IS*50* fragment plus the adjacent genomic DNA). If more restriction bands are observed, this could correspond to the illegitimate ligation of fragments from the genome. Discard these clones.
10. After sequencing of the DNA placed at both sides of the Tn*5* insertion point, a directed repeat sequence of 9 bp will be observed. This repetition is due to the insertion of the transposon and should be corrected.
11. Never select a strain that showed Gm resistance even though it grew in the sucrose selective plates. These strains could grow due to mechanisms different to the deletion of the desired gene, as in example inactivation of the *sacB* gene or the promoter of this gene.
12. Other useful techniques for determination of the deletion of the fragment could be Southern blot using the genomic DNA from the mutant or Northern blot or real-time Q-PCR, using RNA extracted from induced bacteria.
13. Both mutants were unable to grow using BA, testosterone, or ADD as sole carbon sources. MM was supplied with 10 mM succinate (to support the bacterial growth) and 5 mM LCA or testosterone as a source of the intermediates (AD or ADD). Catabolism of succinate causes a strong repression over the use

of steroids; thus, other non-steroid carbon sources to maintain the cellular growth could be profitable.

14. When succinate (10 mM) and a steroid compound (5 mM) are used as carbon sources, the bacteria do not start the biotransformation of the steroid compounds into the precursors until succinate depletion occurs.
15. Handle the TLC plates with gloves on. Never touch the surface of the plate, even with gloves.
16. Consider that commercial gas sprayers could be made of plastic non-resistant to sulfuric acid (mainly the tubes and the head of the sprayer). Prepare the 30% H_2SO_4 (w/v) solution for revealing, just in the moment of use, and clean carefully the sprayer just after use. Alternatively, the H_2SO_4 solution could be disposed with a roller made with H_2SO_4-resistant materials but carefully avoiding the leakage of sulfuric acid.
17. Alternatively, analyze the products of the reaction assays by LC–tandem mass spectrometry (LC-MS/MS) as described [16].

Acknowledgments

This research was funded by the Ministerio de Economía y Competitividad (Madrid, España, grants BFU2009-11545-C03-01, BIO2012-39695-C02-02, and BIO2015-66960-C3-3R), by a CENIT Project RTC-2014-2249-1 (CDTI, Ministerio de Economía y Competitividad, Madrid, España), and by a grant from the Junta de Castilla y León (Consejería de Educación, Valladolid, España) LE246A11-2. The authors also want to thank the support to their actual research by the Horizon Europe Framework Programme (call: HORIZON-CL4-2021-RESILIENCE-01-11) through the ESTELLA project ("DESign of bio-based Thermoset polymer with rEcycLing capabiLity by dynAmic bonds for bio-composite manufacturing") (Project no. 101058371), the Ministerio de Ciencia e Innovación (grant TED2021-132593B-I00 belonging to the 2021 convocatory "Proyectos Estratégicos Orientados a la Transición Ecológica y a la Transición Digital" and RTI2018-095584-B-C43 from Proyectos de I+D+i RETOS INVESTIGACION), and the Junta de Castilla y León, grant LE250P20.

References

1. Russell DW (2003) The enzymes, regulation, and genetics of bile acid synthesis. Annu Rev Biochem 72:137–174
2. di Gregorio MC, Cautela J, Galantini L (2021) Physiology and physical chemistry of bile acids. Int J Mol Sci 22:1780

3. Hofmann AF, Hagey LR, Krasowski MD (2010) Bile salts of vertebrates: structural variation and possible evolutionary significance. J Lipid Res 51:226–246
4. Hofmann AF, Hagey LR (2008) Bile acids: chemistry, pathochemistry, biology, pathobiology, and therapeutics. Cell Mol Life Sci 65: 2461–2483
5. Durník R, Šindlerová L, Babica P et al (2022) Bile acids transporters of enterohepatic circulation for targeted drug delivery. Molecules 27: 2961
6. Collins SL, Stine JG, Bisanz JE et al (2023) Bile acids and the gut microbiota: metabolic interactions and impacts on disease. Nat Rev Microbiol 21(4):236–247. https://doi.org/10.1038/s41579-022-00805-x
7. Hagey LR, Vidal N, Hofmann AF et al (2010) Evolutionary diversity of bile salts in reptiles and mammals, including analysis of ancient human and extinct giant ground sloth coprolites. BMC Evol Biol 10:133
8. Merino E, Barrientos A, Rodríguez J et al (2013) Isolation of cholesterol- and deoxycholate-degrading bacteria from soil samples: evidence of a common pathway. Appl Microbiol Biotechnol 97:891–904
9. Horinouchi M, Hayashi T, Kudo T (2012) Steroid degradation in *Comamonas testosteroni*. J Steroid Biochem Mol Biol 129:4–14
10. van der Geize R, Yam K, Heuser T et al (2007) A gene cluster encoding cholesterol catabolism in a soil actinomycete provides insight into *Mycobacterium tuberculosis* survival in macrophages. Proc Natl Acad Sci 104:1947–1952
11. van der Geize R, Grommen AWF, Hessels GI et al (2011) The steroid catabolic pathway of the intracellular pathogen *Rhodococcus equi* is important for pathogenesis and a target for vaccine development. PLoS Pathog 7: e1002181
12. Drzyzga O, de las Heras LF, Morales V et al (2011) Cholesterol degradation by *Gordonia cholesterolivorans*. Appl Environ Microbiol 77: 4802–4810
13. Bragin EY, Shtratnikova VY, Dovbnya DV et al (2013) Comparative analysis of genes encoding key steroid core oxidation enzymes in fast-growing *Mycobacterium* spp. strains. J Steroid Biochem Mol Biol 138:41–53
14. Uhía I, Galán B, Kendall SL et al (2012) Cholesterol metabolism in *Mycobacterium smegmatis*. Environ Microbiol Rep 4:168–182
15. Mohn WW, Wilbrink MH, Casabon I et al (2012) Gene cluster encoding cholate catabolism in *Rhodococcus* spp. J Bacteriol 194:6712–6719
16. Birkenmaier A, Holert J, Erdbrink H et al (2007) Biochemical and genetic investigation of initial reactions in aerobic degradation of the bile acid cholate in *Pseudomonas* sp. strain Chol1. J Bacteriol 189:7165–7173
17. Barrientos Á, Merino E, Casabon I et al (2015) Functional analyses of three acyl-CoA synthetases involved in bile acid degradation in *Pseudomonas putida* DOC21. Environ Microbiol 17:47–63
18. Feller FM, Holert J, Yücel O et al (2021) Degradation of bile acids by soil and water bacteria. Microorganisms 9:1759
19. Olivera ER, Luengo JM (2019) Steroids as environmental compounds recalcitrant to degradation: genetic mechanisms of bacterial biodegradation pathways. Genes (Basel) 10:512
20. Olivera ER, de la Torre M, Barrientos Á et al (2018) Steroid catabolism in bacteria: genetic and functional analyses of *stdH* and *stdJ* in *Pseudomonas putida* DOC21. Can J Biotechnol 2:88–99
21. Holert J, Jagmann N, Philipp B (2013) The essential function of genes for a hydratase and an aldehyde dehydrogenase for growth of *Pseudomonas* sp. strain Chol1 with the steroid compound cholate indicates an aldolytic reaction step for deacetylation of the side chain. J Bacteriol 195:3371–3380
22. Holert J, Kulic Z, Yucel O et al (2013) Degradation of the acyl side chain of the steroid compound cholate in *Pseudomonas* sp. strain Chol1 proceeds via an aldehyde intermediate. J Bacteriol 195:585–595
23. Horinouchi M, Taguchi K, Arai H et al (2001) Meta-cleavage enzyme gene *tesB* is necessary for testosterone degradation in *Comamonas testosteroni* TA441. Microbiology (Reading) 147:3367–3375
24. Yam KC, D'Angelo I, Kalscheuer R et al (2009) Studies of a ring-cleaving dioxygenase illuminate the role of cholesterol metabolism in the pathogenesis of *Mycobacterium tuberculosis*. PLoS Pathog 5:e1000344
25. Casabon I, Crowe AM, Liu J et al (2013) FadD3 is an acyl-CoA synthetase that initiates catabolism of cholesterol rings C and D in actinobacteria. Mol Microbiol 87:269–283
26. Horinouchi M, Hayashi T (2021) Identification of the coenzyme A (CoA) ester intermediates and genes involved in the cleavage and degradation of the steroidal C-ring by *Comamonas testosteroni* TA441. Appl Environ Microbiol 87:e0110221
27. Horinouchi M, Koshino H, Malon M et al (2019) Steroid degradation in *Comamonas testosteroni* TA441: identification of the entire

β-oxidation cycle of the cleaved B ring. Appl Environ Microbiol 85:e01204-19
28. Crowe AM, Casabon I, Brown KL et al (2017) Catabolism of the last two steroid rings in *Mycobacterium tuberculosis* and other bacteria. mBio 8:e00321-17
29. Horinouchi M, Malon M, Hirota H et al (2019) Identification of 4-methyl-5-oxo-octane-1,8-dioic acid and the derivatives as metabolites of steroidal C,D-ring degradation in *Comamonas testosteroni* TA441. J Steroid Biochem Mol Biol 185:277–286
30. Selvaraj G, Iyer VN (1983) Suicide plasmid vehicles for insertion mutagenesis in *Rhizobium meliloti* and related bacteria. J Bacteriol 156:1292–1300
31. Herrero M, de Lorenzo V, Timmis KN (1990) Transposon vectors containing non-antibiotic resistance selection markers for cloning and stable chromosomal insertion of foreign genes in gram-negative bacteria. J Bacteriol 172: 6557–6567
32. Quandt J, Hynes MF (1993) Versatile suicide vectors which allow direct selection for gene replacement in gram-negative bacteria. Gene 127:15–21
33. Bradford M (1976) A rapid and sensitive method for the quantitation of microgram quantities of protein utilizing the principle of protein-dye binding. Anal Biochem 72:248–254
34. Ruvkun GB, Sundaresan V, Ausubel FM (1982) Directed transposon Tn*5* mutagenesis and complementation analysis of *Rhizobium meliloti* symbiotic nitrogen fixation genes. Cell 29:551–559
35. Olivera ER, Miñambres B, García B et al (1998) Molecular characterization of the phenylacetic acid catabolic pathway in *Pseudomonas putida* U: the phenylacetyl-CoA catabolon. Proc Natl Acad Sci U S A 95:6419–6424
36. Arcos M, Olivera ER, Arias S et al (2010) The 3,4-dihydroxyphenylacetic acid catabolon, a catabolic unit for degradation of biogenic amines tyramine and dopamine in *Pseudomonas putida* U. Environ Microbiol 12:1684–1704
37. Berg DE, Weiss A, Crossland L (1980) Polarity of Tn*5* insertion mutations in *Escherichia coli*. J Bacteriol 142:439–446
38. Tenneson ME, Baty JD, Bilton RF et al (1979) The degradation of cholic acid by *Pseudomonas* spp. N.C.I.B. 10590. Biochem J 184:613–618

Chapter 5

Targeted Mutagenesis of *Mycobacterium* Strains by Homologous Recombination

Shikui Song and Zhengding Su

Abstract

Targeted mutagenesis by homologous recombination (TMHR) is an efficient allelic exchange mutagenesis for bacterial genome engineering in synthetic biology. Unlike other allelic exchange methods, TMHR does not require a heterologous recombinase to insert or excise a selectable marker from the genome. In contrast, positive and negative selection is achieved solely by suicide vector-encoded functional and host cell proteins. Here we describe a concise protocol to knock out and knock in a 3-ketosteroid-1,2-dehydrogenase gene (*kstd*) in *Mycobacterium neoaurum* HGMS2 using TMHR approach. The homology arms flanking the *kstd* gene are amplified by PCR in vitro and then subcloned into a common homologous recombination vector. The vector is then electroporated into the HGMS2 competent cells. The replacement of the *kstd* gene by homologous recombination produces antibiotic-resistant single-crossover recombination via the first allelic exchange. Double-crossover markerless mutants are directly separated using sucrose-mediated counterselection. These two steps can generate seamless mutations down to a single DNA base pair. The whole process takes less than 2 weeks.

Key words Mutagenesis, Homologous recombination, Gene knockout, Knockin, *Mycobacterium*, PCR, 3-Oxosteroid 1-dehydrogenase

1 Introduction

Mycobacteria have been used as important industrial strains for efficiently converting phytosterols to sterols [1–5]. So far, a large numbers of mycobacterial genomic sequence data have become available [6–8], providing feasibility for engineering the phytosterol metabolic pathways in *Mycobacteria* [9–11]. Consequently, many key enzymes that play important roles in the accumulation of steroid intermediates in the phytosterol metabolic pathway have been identified [12–14]. Nonpathogenic *Mycobacterium* strains are not only important industrial strains but also suitable models for studying pathogenic *Mycobacteria* [15–17].

Carlos Barreiro and José-Luis Barredo (eds.), *Microbial Steroids: Methods and Protocols*, Methods in Molecular Biology, vol. 2704, https://doi.org/10.1007/978-1-0716-3385-4_5,

Gene knockout and knockin strategies have been increasingly used for engineering *Mycobacteria* that efficiently accumulate steroid intermediates [12, 14, 18]. The state-of-the-art CRISPR system has been successfully used to generate marker-free deletions in *M. smegmatis* [19]. However, its off-target issue causes low recombination efficiency [19, 20], compared with a conventional mutagenesis that uses homologous recombination to target an endogenous gene and substitute it with a designed mutant gene.

Targeted mutagenesis by homologous recombination (TMHR) has been used for engineering a wide range of bacterial species [14, 21–23]. These mutant alleles are not disturbed by antibiotic resistance markers, precisely enabling genome engineering in single-base and significantly eliminating off-target recombination. TMHR uses *Bacillus subtilis* fructansucrase gene (*sacB*) as a counterselection marker that is lethal to *Mycobacteria* at a concentration of 10% sucrose [24]. Thus, the expression vector harboring the *sacB* gene exhibits significant efficiency in sucrose-containing media, ensuring a successful homologous recombination via a two-step selection [25]. Here, we demonstrate to use TMHR approach to manipulate gene engineering in *Mycobacterium* sp. HGMS2, an industrial strain that accumulates 4-androstenedione (also known as androst-4-ene-3,17-dione; 4-AD) [7, 14], to improve the bioconversion yield of phytosterols to 4-AD, providing a comprehensive and up-to-date protocol for genome engineering of *Mycobacterium* strains.

2 Materials

Prepare all solutions using Milli-Q water (18 MΩ-cm at 25 °C) and analytical grade reagents. Prepare and store all reagents at 4 °C (unless indicated otherwise). Cautiously follow all waste disposal regulations when disposing waste materials.

2.1 Microbial Strains

1. *Mycobacterium neoaurum* HGMS2, a nonpathogenic industrial *Mycobacterium* strain deposited at China Center for Type Culture Collection (CCTCC No: M2012522) [7].
2. *E. coli* DH5α competent cells (Thermo Fisher Scientific).

2.2 Genetic Manipulation Material

1. Plasmid pT18mobsacB and Plasmid p2NIL (AddGene).
2. Primer *kstd*-U-F (5′-GCAGTAggatccCCACAATCTCGGCATACC-3′).
3. Primer *kstd*-U-R (5′-CCACAActcgagGGTGAGGGCGGCGACCATG-3′).
4. Primer *kstd*-D-F (5′-CCACAActcgagaTGACGTTCGGCTACCTCGC-3′).

5. Primer *kstd*-D-R (5′-ACGAGCaagcttCCGACAGTCTCCCA CTCGG-3′).
6. Primer *sacB*-F (5′-CCGGAATTCCACATAACCTGCC GTTC-3′).
7. Primer *sacB*-R (5′-CCGGAATTCTTATTTGTTAACTGTTA ATTGT-3′).
8. Enzymes including *BamH* I, *Hind* III, *EcoR* I and *Xba* I, and T4 DNA ligase.
9. KOD DNA Polymerase, dNTPs, and $MgSO_4$.
10. Bacterial Genomic DNA Extraction Kit (Tiangen) to extract the genomic DNAs of *Mycobacterium* sp. HGMS2 and its mutants.
11. Plasmid Purification Kit (Tiangen).
12. PCR Cleanup Kit (Omega BioTek) to purify PCR fragments.

2.3 Agarose Gel

1. Agarose and EDTA.
2. TAE buffer (50×): Prepare 2 M Tris–acetate, 1 M sodium acetate, and 50 mM EDTA, and bring up the volume to 500 mL following adjustment of pH to 8.0 as a stock solution. Store at room temperature and produce 1× working solutions by diluting with Milli-Q water.
3. 1% (w/v) agarose gel: Dissolve 1 g of agarose in 100 mL of 1× TAE buffer. Heat until completely dissolved and mix well before use. Store at room temperature.

2.4 Media and Other Materials

1. 10% (vol/vol) glycerol solution: Mix 10 mL of glycerol with 90 mL of Milli-Q water. Autoclave at 121 °C for 15 min and store at 4 °C.
2. 10% (vol/vol) Tween 80 solution: Mix 10 mL of Tween 80 with 90 mL of Milli-Q water. Autoclave at 121 °C for 15 min and store at 4 °C.
3. Kanamycin solution (30 mg/mL): Dissolve 0.6 g of kanamycin in Milli-Q water to a final volume of 20 mL. Sterilize it by filtration through a 0.22 μm membrane. Aliquot and store at − 25 °C. Dissolve completely and add at a ratio of 1:1000 (v/v).
4. LB medium: Dissolve 10 g of tryptone, 5 g of yeast extract, and 10 g of NaCl in 1 L of Milli-Q water, and adjust the pH to 7.4 with HCl. Autoclave at 121 °C for 15 min and store at room temperature.
5. LBT medium: Mix 0.5 mL of 10% Tween 80 with 100 mL of sterile LB medium to make a final concentration of 0.05% (v/v). Filter-sterilize and store at room temperature.

6. LBK medium: Mix 100 μL (30 mg/mL) of sterilized kanamycin solution with 100 mL of sterile LB medium to make a final concentration of 30 μg/mL. Store at 4 °C for short term.
7. LB solid medium: Dissolve 1 g of tryptone, 0.5 g of yeast extract, 1 g of NaCl, and 15–20 g of agar in 100 mL of Milli-Q water and adjust the pH to 7.4 with HCl. Autoclave at 121 °C for 15 min and store at room temperature.
8. LBS solid medium: Dissolve 10 g of sucrose and 1.8–2 g of agar in 100 mL of LB medium. Autoclave at 115 °C for 15 min and store at room temperature.

2.5 Equipment

1. Shaker and incubator.
2. Biological Safety Cabinet.
3. Veriti™ 96-Well Thermal Cycler, laboratory centrifuge, and NanoDrop 2000c spectrophotometer (Thermo Fisher Scientific).
4. Scientz-2C gene electroporator (Ningbo Scientz Biotechnology Co. Ltd.).

3 Methods

Unless otherwise specified, all procedures are performed in a biosafety hood.

3.1 Primer Design

1. Use the NCBI database to retrieve the genome sequence of *Mycobacterium* sp. HGMS2. Select a targeted-mutation gene, for example, 3-ketosteroid-1,2-dehydrogenase gene (*kstd*), and rationally design two PCR primers carrying restriction enzyme sites using DNA sequence analysis software.
2. Digest the amplified upstream and downstream fragments for each target gene with two pairs of restriction enzymes, *Bam*H I/*Xba* I and *Xba* I/*Hind* III, respectively (*see* **Note 1**). The nucleotide sequences of the primers are shown in Subheading 2.2 (*see* **Note 2**).

3.2 Vector Construction

To construct a TMHR vector based on p2NIL plasmid, use pT18mobsacB as a template to amplify the sucrose lethality gene *SacB* that is originally from *Bacillus subtilis* 168.

1. Amplify *SacB* and its promotor and terminator by PCR introducing *EcoR* I sites at both ends using the primers *sacB*-F and *sacB*-R.
2. Digest individually the PCR fragment and the p2NIL plasmid with *EcoR* I.

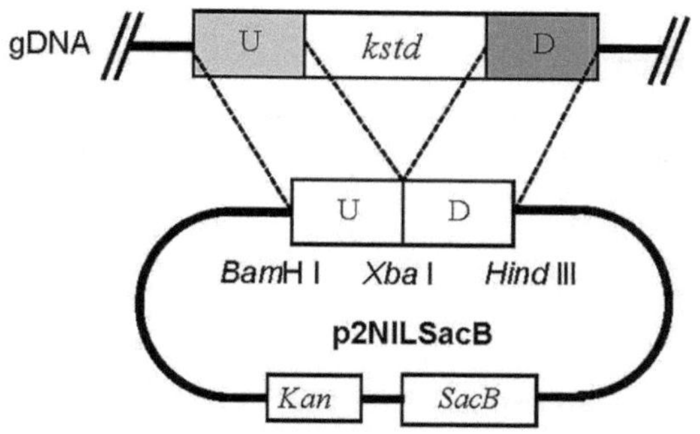

Fig. 1 Structure of p2NIL–SacB homologous recombination vector. U the upstream arm, D the downstream arm, Kan kanamycin gene, SacB sucrose lethality gene

3. Recover the digested *SacB* PCR fragment and p2NIL plasmid by agarose gel purification.
4. Ligate digested *SacB* DNA fragment to the digested p2NIL plasmid by T4 DNA ligase. The resultant plasmid is named p2NILSacB (Fig. 1).

3.3 Amplification of the Upstream and Downstream Homology Arms of Targeted Gene

1. Dissolve primers in sterile water to generate primer stock solutions in the concentration of 100 μM and store at −25 °C (*see* **Note 3**). Dilute the stock solutions to 10 μM concentration as a working solution (i.e., add 5 μL of 100 μM stock solution to mix with 45 μL sterile water).
2. Individually amplify the upstream and downstream homology arms of a target gene by PCR. In this study, *Mycobacterium* sp. HGMS2 genomic DNA was used as a template to establish optimal PCR amplification system. The 25 μL system contains 50–100 ng of template DNA, 1.5 mmol/L of Mg^{2+}, 0.2 mmol/L of dNTPs, 0.3 μmol/L of each primer, 0.01 U of KOD DNA polymerase, and 10× PCR buffer (Table 1).
3. Set up the PCR reaction (based on optimal conditions for DNA polymerase) as follows: denaturation at 95 °C for 5 min, hybridization at 65 °C for 30 s, and extension at 68 °C for 1.5 min for 35 cycles (*see* **Note 4**).
4. Store the PCR product temporarily at 4 °C (*see* **Note 3**, Table 2).
5. DNA agarose gel electrophoresis: Load 10 μL of PCR product into gel wells mixed with dye. Use a DNA marker as a reference. Run gel electrophoresis at 110 volts for 15 min and then image the gel with a UV transilluminator equipped with a digital camera.

Table 1
PCR amplification reaction

Step	Reagent	Volume (μL)	Remark
1	ddH_2O	17	For volume up to 25 μL
2	10× PCR buffer	2.5	Must be completely defrosted
3	dNTPs	2.5	10 mM stock solution
4	Forward primer	1	10 μM stock solution
5	Revise primer	1	10 μM stock solution
6	*Pfu* or KOD DNA polymerase	0.5	
7	Template	0.5	Appropriate concentration
8	Total volume	25	On ice

Table 2
Parameters for PCR amplification

Step	Temperature	Time	Cycle
1	95 °C	5 min	1
2	95 °C 68 °C 72 °C	30 s 30 s 1 kbp/min	35×
3	72 °C	10 min	1
4	10 °C	–	1

6. Purification of PCR products: Determine the concentration and purity of eluted DNAs by measuring absorbance at 260 nm and 280 nm ($A_{260\ nm}$ and $A_{280\ nm}$) on a UV spectrophotometer.
7. Store purified PCR products at 4 °C (*see* **Note 3**).

3.4 Construction of Knockout Plasmids

1. Digest purified PCR products and p2NIL by incubation with selected restriction enzymes at 37 °C for 2 h (*see* **Note 5**) (Table 3). Double-enzyme digestion reaction in 50 μL: 1600 ng of vector or PCR product, 2 μL of each restriction endonuclease (*Bam*H I/*Hind* III, *Bam*H I/*Xba* I, or *Xba* I/*Hind* III), 5 μL of 10× buffer, and add ddH_2O to a total volume of 50 μL (*see* **Note 6**).
2. Purify the digested products (*see* Subheading 3.2).
3. Set up the ligation reaction of the homology arms and vehicle, and incubate at 16 °C for 5 h (*see* **Note 5**) (Table 4). Ligation

Table 3
Double digestion

Component	Volume (μL)
ddH_2O	36
10× buffer	5
Restriction endonuclease 1	2
Restriction endonuclease 2	2
Vector or PCR product	2
Total volume	50

Table 4
DNA ligation reaction

Component	Volume (μL)
Upstream arm DNA	3
Downstream arm DNA	3
Vector	2
T4 ligase	1
10× T4 ligase buffer	1
Total volume	10

reaction in 10 μL: 3 μL of upstream or downstream arm DNA each, 2 μL of vector, 1 μL of T4 ligase, and 10× T4 ligase buffer each (*see* **Note** 7).

4. To transform ligated plasmids into *E. coli* DH5α competent cells, add the ligation mixture to a 100 μL of pre-thawed competent cells, and then incubate the cells on ice for 30 min.
5. Add 0.5 mL of LB medium and transfer the cells to a sterile culture tube. Incubate the cells by shaking at 200 rpm for 1 h at 37 °C.
6. Plate 300 μL cells on LB agar plate with the appropriate antibiotic selection and incubate for 20 h at 37 °C.
7. Screening and identification of positive recombinants. Set up a colony PCR system according to Subheading 3.2 (using single colony instead of plasmid as DNA template) and verify by agarose gel electrophoresis.
8. Select at least four positive colonies and culture in LB media containing appropriate antibiotics for 12 h.

9. Extract each plasmid, using the plasmid extraction kit, and verify the correct recombinant plasmid by DNA sequencing.
10. Determine the concentration and purity of the eluted DNA by measuring the $A_{260\ nm}$ and $A_{280\ nm}$ values on a UV spectrophotometer. The $A_{260\ nm}/A_{280\ nm}$ ratio of each plasmid is generally 1.8–2.0.
11. Use the samples directly for the next step or store at −25 °C for future use (*see* **Note 3**).

3.5 Preparation of Mycobacterial Competent Cells

1. Take 100 μL of *Mycobacterium* strain preserved in glycerol at − 80 °C and transfer it to 50 mL of LBT medium.
2. Incubate at 200 rpm and 30 °C until the $OD_{600\ nm}$ of the cells reach 0.4–0.6 (about 10 h) (*see* **Note 8**).
3. Collect the *Mycobacterium* cells by centrifugation and wash three times with 10% glycerol solution containing 0.05% Tween 80. Carry out the washing process in a biosafety hood (*see* **Note 9**).
4. Resuspend cell pellets in 100 μL of 10% glycerol solution and use it directly as competent cells.

3.6 Electroporation of Knockout Plasmids into Mycobacterial Competent Cells and Twice Recombination

1. Aliquot 5–10 μg of the knockout plasmid constructed in Subheading 3.4 in autoclaved EP tubes and concentrate by freeze drying (*see* **Note 10**).
2. Dissolve the dried plasmid in 5 μL of autoclaved Milli-Q H_2O for DNA electroporation (*see* **Note 11**).
3. Transfer 100 μL of mycobacterial competent cells to a 0.2 cm precooled electroporation cuvette (*see* **Note 12**).
4. Add concentrated plasmids to the cuvette and gently mix the competent cells by pipetting.
5. Cover the cuvette with a clean lid to prevent contamination (*see* **Note 12**).
6. Place the electroporation cuvette in the cassette of the gene transfer instrument and apply a constant wave composed of single pulse of 2.5 kV and 25 μF with a resistance of 1 kΩ. Set the time constant to 23 ms (*see* **Note 13**).
7. Transfer all the cell suspensions after an electric shock into 1 mL LB medium and incubate at 30 °C and 200 rpm for 3–4 h.
8. Spread the culture on an LB agar plate (LBK) containing 50 mg/mL kanamycin.
9. Incubate at 30 °C for 3 days to carry on the first homologous recombination. Recombinant bacteria carrying antibiotic gene in its genome will survive on LBK agar plate (*see* **Note 14**).

10. Pick up single colonies and inoculate each one into a 3 mL LBT (with 0.05% Tween 80 in LB) containing 50 mg/mL kanamycin.
11. Incubate at 30 °C and 200 rpm for 2 days before performing the second homologous recombination. The primary recombinant bacteria are subjected to secondary homologous recombination in the medium without antibiotics.
12. Dilute the cell culture to 10^3~10^5-fold in LBT medium (*see* **Note 15**).
13. Spread 100 μL of the diluted cell suspension on a LBK agar plate containing 10% sucrose (LBS) and spread another 100 μL of the same diluted cell suspension on LBK agar plate as a control. Since the genome of the primary recombinant bacteria contains antibiotic and sucrose lethal genes, these bacteria cannot grow on either sucrose-containing LB plates or kanamycin-containing LB plates. Thus, the growing single colonies on LBS plates are the bacteria that perform secondary recombination (*see* **Note 16**).
14. Incubate both types of plates at 30 °C for 3 days (*see* **Note 17**).
15. Pick 5–10 colonies from the LBS plates for colony PCR verification (Fig. 2).
16. Transfer positive colonies to 5 mL of LBT medium and incubate at 30 °C and 200 rpm for 2 days.
17. Extract their genomic DNAs for PCR verification (*see* **Note 18**).
18. Add glycerol to cell culture to a final concentration of 15% and store at −80 °C.

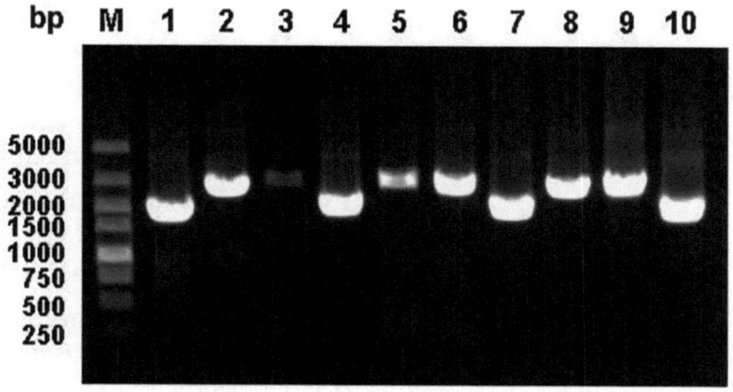

Fig. 2 Agarose gel electrophoresis imaging of colony PCR screening kstd-knockout mutant strains after the second homologous recombination. M—DNA marker; lanes 1, 4, 7, and 10—kstd-knockout mutants; and lanes 2, 3, 5, 6, 8, and 9—the wild-type strain. (Reproduced from Ref. [14] with permission from Springer Nature)

4 Notes

1. Make sure to select appropriate restriction sites and avoid duplicated restriction sites within the upstream or downstream homology arms.
2. The length of the upstream and downstream arms of the targeted gene is selected in the range of 700–2000 bp, which is beneficial to the efficiency of homologous recombination.
3. Purified DNA can be stored in short term (<1 week) at 4 °C or long term (>1 week) at −25 °C.
4. The GC content in the genome of *Mycobacteria* is generally higher than that in other bacteria such as *E. coli*. When conducting molecular cloning, the annealing temperature in PCR setup should be appropriately increased (≥65 °C).
5. If digestion or ligation is difficult, the reaction time can be appropriately extended.
6. The choice of buffer in the double digestion system is important, and TaKaRa recommends to use the double digestion reaction buffer system.
7. Restriction enzyme-digested upstream and downstream fragments and p2NIL–SacB vector can be linked in a single step with T4 ligase.
8. When preparing mycobacterial competent cells, strictly control cell density. In general, the $OD_{600\ nm}$ value of 0.4 gives the best condition for DNA electroporation; otherwise, the recombination efficiency will be significantly reduced.
9. Before the preparation of *Mycobacterium* competent cells, the washing solution containing glycerol and Tween 80 should be precooled on ice in advance, and the cells should be resuspended in the precooled washing solution. Furthermore, collection of competent cells should ensure that the cells are in a low-temperature and sterile environment, and the centrifugation speed should not be too high to damage competent cells.
10. For DNA electroporation, the plasmid concentration should be above 5 μg, while the volume of the plasmid and cell mixtures should be lower than 10 μL.
11. The concentrated plasmid needs to be dissolved with Milli-Q water, ensuring that the plasmid solution contains no metal ions to avoid electric sparking.
12. Electroporation cuvette should be cleaned with autoclaved water and ethanol and sterilized by ultraviolet light for 20 min in biosafety hood, and the outside of the cuvette should be kept dry to avoid electric sparking.

13. If the gene electroporation instrument has not been used for a long time, the instrument should be charged and discharged at low voltage first. The voltage should be gradually increased until the desired value is reached before the DNA electroporation can be performed.
14. Mycobacteria grow slowly and generally need 2–4 days. Thus, a blank medium should be used as a control. In addition, the plates after coating the bacterial solution need to be sealed with parafilm.
15. As the secondary homologous recombination is quite efficient, it needs to be properly diluted when spreading the plate, and it is recommended here that the number of spread colonies is more appropriate when diluted 10^5-folds.
16. Agarose plates containing sucrose (LBS) should not contain any antibiotics, and fresh plates should be used and cannot be stored for a long time to prevent contamination.
17. In general, the efficiency of the secondary recombination is much lower than that of the primary recombination, and the number of colonies on the sucrose-containing LB plate is significantly less than the number of single colonies on the kanamycin-containing LB plate.
18. To verify the correctness of the recombinant bacteria, we generally pick up few putatively positive colonies to culture and extract their genomes for PCR verification. If necessary, DNA sequencing is conducted for verification.

Acknowledgments

This work was supported by the Key Laboratory of Industrial Fermentation (Ministry of Education), Cooperative Innovation Center of Industrial Fermentation (Ministry of Education & Hubei Province), and Hubei Key Laboratory of Industrial Microbiology to Z.D.S.

References

1. Marques M, Carvalho F, Carvalho C, Cabral J, Fernandes P (2010) Steroid bioconversion: towards green processes. Food Bioprod Process 88:12–20
2. Donova MV, Gulevskaya SA, Dovbnya DV, Puntus IF (2005) *Mycobacterium* sp. mutant strain producing 9α-hydroxyandrostenedione from sitosterol. Appl Microbiol Biotechnol 67:671–678
3. Wei W, Fan S, Wang F, Wei D (2010) A new steroid-transforming strain of *Mycobacterium neoaurum* and cloning of 3-ketosteroid 9-α-hydroxylase in NwIB-01. Appl Microbiol Biotechnol 162:1446–1456
4. Donova MV, Egorova OV (2012) Microbial steroid transformations: current state and prospects. Appl Microbiol Biotechnol 94:1423–1447
5. Lorena FC, Beatriz G, García J (2018) New insights on steroid biotechnology. Front Microbiol 9:958

6. Gupta RS, Lo B, Son J (2019) Corrigendum: phylogenomics and comparative genomic studies robustly support division of the genus *Mycobacterium* into an emended genus *Mycobacterium* and four novel genera. Front Microbiol 10:714
7. Wang H, Song S, Peng F, Yang F, Chen T, Li X, Cheng X, He Y, Huang Y, Su Z (2020) Whole-genome and enzymatic analyses of an androstenedione-producing *Mycobacterium* strain with residual phytosterol-degrading pathways. Microb Cell Factories 19:187
8. Olivera ER, Luengo JM (2019) Steroids as environmental compounds recalcitrant to degradation: genetic mechanisms of bacterial biodegradation pathways. Genes 10:512
9. Garcia JL, Uhia I, Galan B (2012) Catabolism and biotechnological applications of cholesterol degrading bacteria. Microb Biotechnol 5:679–699
10. Capyk JK, Casabon I, Gruninger R, Strynadka NC, Eltis LD (2011) Activity of 3-ketosteroid 9alpha-hydroxylase (KshAB) indicates cholesterol side chain and ring degradation occur simultaneously in *Mycobacterium tuberculosis*. J Biol Chem 286:40717–40724
11. Bragin EY, Shtratnikova VY, Schelkunov MI, Dovbnya DV, Donova MV (2019) Genome-wide response on phytosterol in 9-hydroxyandrostenedione-producing strain of *Mycobacterium* sp. VKM Ac-1817D. BMC Biotechnol 19:39
12. Xu LQ, Liu YJ, Yao K, Liu HH, Tao XY, Wang FQ, Wei DZ (2016) Unraveling and engineering the production of 23,24-bisnorcholenic steroids in sterol metabolism. Sci Rep 6:21928
13. Thomas ST, VanderVen BC, Sherman DR, Russell DG, Sampson NS (2011) Pathway profiling in *Mycobacterium tuberculosis*: elucidation of cholesterol-derived catabolite and enzymes that catalyze its metabolism. J Biol Chem 286:43668–43678
14. Li X, Chen T, Peng F, Song S, Yu J, Sidoine DN, Cheng X, Huang Y, He Y, Su Z (2021) Efficient conversion of phytosterols into 4-androstene-3,17-dione and its C1,2-dehydrogenized and 9alpha-hydroxylated derivatives by engineered mycobacteria. Microb Cell Factories 20:158
15. Nesbitt NM, Yang X, Fontan P, Kolesnikova I, Smith I, Sampson NS, Dubnau E (2010) A thiolase of *Mycobacterium tuberculosis* is required for virulence and production of androstenedione and androstadienedione from cholesterol. Infect Immun 78:275–282
16. Thomas ST, Sampson NS (2013) *Mycobacterium tuberculosis* utilizes a unique heterotetrameric structure for dehydrogenation of the cholesterol side chain. Biochemistry 52:2895–2904
17. Yang M, Guja KE, Thomas ST, Garcia-Diaz M, Sampson NS (2014) A distinct MaoC-like enoyl-CoA hydratase architecture mediates cholesterol catabolism in *Mycobacterium tuberculosis*. ACS Chem Biol 9:2632–2645
18. Galan B, Uhia I, Garcia-Fernandez E, Martinez I, Bahillo E, de la Fuente JL, Barredo JL, Fernandez-Cabezon L, Garcia JL (2017) *Mycobacterium smegmatis* is a suitable cell factory for the production of steroidic synthons. Microb Biotechnol 10:138–150
19. Piddington DL, Fang FC, Laessig T, Cooper AM, Orme IM, Buchmeier NA (2001) Cu, Zn superoxide dismutase of *Mycobacterium tuberculosis* contributes to survival in activated macrophages that are generating an oxidative burst. Infect Immun 69:4980–4987
20. Fu Y, Foden JA, Khayter C, Maeder ML, Sander JD (2013) High-frequency off-target mutagenesis induced by CRISPR-Cas nucleases in human cells. Nat Biotechnol 31: 822–826
21. Link AJ, Phillips D, Church GM (1997) Methods for generating precise deletions and insertions in the genome of wild-type *Escherichia coli*: application to open reading frame characterization. J Bacteriol 179:6228–6237
22. Quandt J, Hynes MF (1993) Versatile suicide vectors which allow direct selection for gene replacement in gram-negative bacteria. Gene 127:15–21
23. Hoang TT, Karkhoff-Schweizer R, Kutchma AJ, Schweizer HP (1998) A broad-host-range Flp-FRT recombination system for site-specific excision of chromosomally-located DNA sequences: application for isolation of unmarked *Pseudomonas aeruginosa* mutants. Gene 212:77–86
24. Pelicic JM, Reyrat JM, Gicquel B (1996) Expression of the *Bacillus subtilis sacB* gene confers sucrose sensitivity on mycobacteria. J Bacteriol 178:1197–1199
25. Pelicic V, Reyrat JM, Gicquel B (2010) Generation of unmarked directed mutations in mycobacteria, using sucrose counter-selectable suicide vectors. Mol Microbiol 20:919–925

Part III

Genetic and Biochemical Analyses

Chapter 6

RNA Preparation and RNA-Seq Bioinformatics for Comparative Transcriptomics

Antonio Rodríguez-García, Alberto Sola-Landa, and Carlos Barreiro

Abstract

The principal transcriptome analysis is the determination of differentially expressed genes across experimental conditions. For this, the next-generation sequencing of RNA (RNA-seq) has several advantages over other techniques, such as the capability of detecting all the transcripts in one assay over RT-qPCR, such as its higher accuracy and broader dynamic range over microarrays or the ability to detect novel transcripts, including non-coding RNA molecules, at nucleotide-level resolution over both techniques. Despite these advantages, many microbiology laboratories have not yet applied RNA-seq analyses to their investigations. The high cost of the equipment for next-generation sequencing is no longer an issue since this intermediate part of the analysis can be provided by commercial or central services. Here, we detail a protocol for the first part of the analysis, the RNA extraction and an introductory protocol to the bioinformatics analysis of the sequencing data to generate the differential expression results.

Key words RNA extraction, Transcriptomics, RNA-seq, Bioconductor, Differential expression

1 Introduction

The high-throughput RNA sequencing or RNA-seq allows transcriptome analysis with unprecedented possibilities and has displaced the previous microarray technology. Although it is possible to sequence RNA directly [1], most RNA-seq studies are carried out with cDNA generated from the retro-transcription of the RNA. The preparation of the cDNA library is a complex and critical step for the analysis success that depends on the platform to be used and the type of RNA to be analyzed [2, 3].

Differential gene expression (DGE) analysis is the most common application of RNA-seq in microbiology. In addition, RNA-seq is an exceptional technique for the identification of non-coding RNAs (ncRNA) or small RNAs (sRNA), as reported in the first bacterial RNA-seq analysis [4] (*see* **Note 1**). Most ncRNAs exert their action as regulators at the post-transcriptional level, mainly

Carlos Barreiro and José-Luis Barredo (eds.), *Microbial Steroids: Methods and Protocols*, Methods in Molecular Biology, vol. 2704,
https://doi.org/10.1007/978-1-0716-3385-4_6,

negatively and in association with the RNA chaperone Hfq [5–7]. However, these ncRNAs differ considerably in length, sequence, and secondary structure, which results in a very complex bioinformatics prediction and identification. RNA-seq of Hfq-bound transcripts showed that sRNAs are not only expressed from intergenic regions but also that a considerable percentage derives from 3′ untranslated regions (UTR) of mRNA, expressed from its own promoter or after mRNA processing [8]. In the last years, other mechanisms of regulation mediated by sRNA have been described [6, 9–11], showing the importance of these RNAs, most of them very difficult to identify by techniques other than RNA-seq [12].

RNA-seq makes it possible to localize the transcription start sites (TSS) massively, as demonstrated by Sharma and co-workers by differential RNA-seq (dRNA-seq) in *Helicobacter pylori* for the first time [13] and widely employed with other bacteria [14]. It is also possible to determine the gene organization in operons [15] or even to identify RNA modifications in mRNA and ncRNA [16], among other applications [2–4, 17].

This chapter details a method routinely used in our lab for RNA extraction. This protocol can be applied for RNA-seq and other techniques needing high-quality RNA, such as qRT-PCR or micro-arrays. Next, we describe a simple bioinformatics analysis to obtain the results of differential expression from RNA-seq data, intending to be an introduction to this type of analysis.

2 Materials

The process presented here comprises several steps divided into phases: firstly, the bench procedures aimed to isolate high-quality RNA from culture samples; secondly, the bioinformatics analysis that employs the FASTQ files provided by the sequencing service system.

Working with RNA requires special care to avoid RNase degradation since these ubiquitous proteins are highly resistant and persistent. It is indispensable to wear globes throughout the procedure, changing as often as possible. All the solutions (including Milli-Q water) and consumable materials like tips and tubes should be certified for this use or autoclaved; small equipment like pipettors and grids can be cleaned with RNaseZap® (Ambion) or similar products. This solution is applied with a paper towel to the surfaces and rinsed with abundant water.

2.1 Buffers, Solutions, and Reagents

1. TE buffer: 10 mM Tris–HCl, pH 8.0, and 1 mM EDTA pH 8.0.
2. Lysozyme solution: lysozyme (15 mg/mL) in TE buffer (*see* **Note 2**).

3. Lysis buffer ML (Macherey-Nagel): lysis buffer containing guanidinium thiocyanate (*see* **Note 3**).
4. Acid phenol for molecular biology applications (e.g., AquaPhenol™ from MP Biomedicals).
5. Chloroform:isoamyl alcohol (24:1) solution (CIA).
6. Absolute ethanol and 75% aqueous ethanol solution (v/v).
7. 3 M NaCl solution.
8. RNAprotect™ Bacteria Reagent (QIAGEN).
9. TURBO DNA-free™ kit (Invitrogen). This kit includes DNase I, 10× DNase buffer, and inactivation reagent.

2.2 Laboratory Equipment for RNA Extraction

1. Microcentrifuge.
2. FastPrep-24 instrument and Lysing Matrix B 2 mL tubes (MP Biomedicals) for cell disruption and homogenization.
3. Phase Log Gel™ Heavy 2 mL tubes (5 PRIME GmbH), three tubes per sample (*see* **Note 4**).
4. Thermocycler.
5. NanoDrop spectrophotometer or equivalent.

2.3 Bioinformatics Hardware

The programs indicated here, fed with the data of a typical bacterial RNA-seq project, can be run on a desktop computer. A minimum of 8 GB of RAM (random access memory) and more than 100 GB of free disk space are recommended.

2.4 Software

Several bioinformatics programs are available for each step of the analysis; however, some steps are only available in the form of Linux command-line programs. For Windows and macOS users, running a Linux distribution as a virtual machine is a convenient solution (*see* **Note 5**). The programs selected for the differential expression analysis of the RNA-seq data are the following:

1. The short-read aligner BBMap is a Java program that can run on any platform and has no dependencies other than a Java Runtime Environment (JRE). It is available from https://sourceforge.net/projects/bbmap/.
2. Samtools [18–20] is a suite of Linux command-line programs widely used for working with high-throughput sequencing data. Available at http://www.htslib.org/.
3. FastQC serves for quality control of high-throughput sequence data. It is a Java-based program and requires an installed JRE. Available at http://www.bioinformatics.babraham.ac.uk/projects/fastqc/.
4. The "R" software for statistical computing is available from https://www.r-project.org/. There are versions for Windows, macOS, and Linux systems. Once the last version is installed,

the bioconductor packages that perform the analysis are downloaded and installed through the following typed commands:

```
source("https://bioconductor.org/biocLite.R")
biocLite()
biocLite("splineTimeR")
biocLite("GenomicFeatures")
biocLite("Rsamtools")
biocLite("GenomicAlignments")
```

2.5 Input Files

In this introductory protocol, we do not cover the process of high-throughput sequencing, which is usually outsourced. The most used platform for RNA-seq is Illumina. The service provider will require RNA samples of enough quality and concentration (*see* **Note 6**) and will conduct the steps of rRNA removal, library preparation, cDNA sequencing, demultiplexing, and quality trimming of the reads. A guide on the sequencing depth for transcriptomics has been published [21]. To illustrate the following steps, two files containing the sequencing reads in FASTQ format are used. These files are generated from a strand-specific library fed to an Illumina NextSeq 500 sequencer that provides reads at a length of 75 bp.

Genome data of the bacterial strain, both the sequence and the annotation, are required. A file containing the genome sequence in FASTA format is used for mapping the reads; a second file should be provided with the annotation, usually in GFF3 format (more information on https://www.ncbi.nlm.nih.gov/genbank/genomes_gff/). The genome annotation is used to obtain the table of read counts per gene.

3 Methods

Establishing a sound experimental design is the first and crucial step to ensuring a successful transcriptomics analysis. Commonly, the plan is guided by an interrogation to the biological system. In our research area of phytosterols bioconversion to steroid precursors, an interesting question is which genes are activated by phytosterols. Our studied strain is *Mycolicibacterium neoaurum* B-3805 (formerly *Mycobacterium neoaurum*), whose genome sequence and annotation are available through GenBank accession CP011022.1 [22]. We use a simple comparative experiment of two conditions, control vs. treatment, to reveal the genes induced by the presence in the medium of phytosterols. The control condition is the basal culture condition (defined medium with glycerol as the carbon source). Meanwhile, the treatment condition only differs in the medium composition: half of the glycerol content of the basal medium is replaced by phytosterols.

No biological replicates are included in the proposed experimental design. Biological replicates are required to ensure the results' reliability and to perform statistical tests that provide statistical significance in the form of *p*-values. However, replication increases the cost of the experiment. Moreover, it might be impossible to define a priori the best experimental conditions, e.g., which culture time is more suitable for the analysis. The knowledge gathered from an exploratory trial is the surest way to decide which conditions are more informative and suitable for replicating. Therefore, the proposed process is intended to perform this exploratory analysis on a short time series. We will analyze here three culture times of both medium composition, i.e., medium with or without phytosterols. The R package splineTimeR [23] is chosen here since this method is capable of estimating values without the use of replicates (*see* refs. 24 and 25 for reviews about differential expression tools). Once the experimental conditions are validated, the process with added replicates will be straightforward. The following sections cover the steps to obtain total RNA preparations of enough quantity and quality and obtain the differential expression analysis results.

3.1 Samples Harvest and RNA Stabilization

1. Collect one volume of culture sample (*see* **Note** 7) and immediately transfer it to a tube containing two volumes of RNAprotect™ Bacteria Reagent (*see* **Note 8**).
2. Mix the contents using a vortex mixer for 5 s.
3. Incubate for 5 min at room temperature (*see* **Note 9**).
4. Centrifuge at 5000 × *g* for 15 min (10 mL tubes) or 16,000 × *g* for 5 min (1.5–2.0 mL tubes).
5. Remove the supernatant and follow with RNA extraction or freeze the pellet until use (*see* **Note 10**).

3.2 RNA Extraction

In contrast with standard column-based kits, this purification protocol must effectively recover all sizes of RNA molecules, which is mandatory for analyzing ncRNAs.

1. Thaw the samples at room temperature for about 10 min.
2. Add 125 μL of lysozyme solution, suspend the cells by pipetting, and transfer them to a new tube.
3. Incubate for exactly 10 min at room temperature (*see* **Note 11**).
4. Add 450 μL of lysis buffer ML and mix by pipetting.
5. Transfer to a Lysing Matrix B tube, add 360 μL of acid phenol, and shake longitudinally. Keep on ice for a few minutes.
6. Select a speed of 6.5 m/s in the FastPrep instrument and agitate for 30 s.

7. Keep on ice for 1 min.
8. Repeat **steps 6** and 7.
9. Centrifuge at room temperature at 16,000 × *g* for 1 min to collect the matrix and cell debris.
10. Transfer the lysate to a Phase Log Gel (PLG) tube and add 0.5 volumes of CIA (*see* **Note 4**).
11. Shake the tubes vigorously longitudinally until they get a uniform whitish suspension. Keep shaking during 30–60 s.
12. Centrifuge at 16,000 × *g* for 5 min.
13. Collect the supernatant in a new PLG tube. Add 0.5 volumes of CIA and 0.5 volumes of acid phenol.
14. Repeat **steps 11** and **12**.
15. Collect the supernatant in a new PLG tube and add 1 volume of CIA.
16. Repeat **steps 11** and **12**.
17. Transfer the supernatant to a new (non-PLG) tube. Add two volumes of absolute ethanol and 0.1 volumes of 3 M NaCl.
18. Let the RNA precipitate overnight at −20 °C.
19. Centrifuge at 16,000 × *g* for 1 h at 4 °C. Discard the supernatant and wash the pellet with 200 μL of 75% ethanol.
20. Centrifuge at 16,000 × *g* for 10 min and discard the supernatant thoroughly with the help of a fine-tip pipette.
21. Let air-dry the pellet for a few minutes and dilute it in 50 μL of RNase-free water.

3.3 Removal of Contaminant DNA

1. Determine the RNA concentration by absorbance or fluorescence methods.
2. Take 10 μg of RNA in a 0.2 mL tube and prepare the following reaction mix (50 μL final volume) (*see* **Note 12**):

10 μg of RNA ($x \leq 44$)	x μL
Milli-Q water	(41.1-x) μL
10× DNase buffer	5.0 μL
DNase I	1.0 μL

3. Mix by repeatedly pipetting up and down, and incubate at 37 °C for 45 min (*see* **Note 13**).
4. Add 1 μL more of DNase I, mix well, and incubate for 45 min at 37 °C.

5. Add 10 μL of inactivation reagent. This reagent must be well resuspended before use by flicking or vortexing the tube.
6. Incubate at room temperature for 5 min. It is crucial to maintain the reagent suspended by flicking the tube three times during the incubation.
7. Centrifuge at 16,000 × *g* for 1.5 min and carefully transfer the supernatant to a new tube. Avoid the contact of the tip with the pellet to avoid reagent transfer.
8. Store the cleaned RNA solution frozen, preferably at −20 °C for weeks or at −80 °C for months or years.

3.4 Alignment of the Sequencing Reads to the Genome Sequence

Each bioinformatics step of analysis generally consists of the input file(s) that is processed using a program that creates output file(s). The output file(s) will be, in turn, the input file(s) of the next step (*see* Fig. 1 for an overview of the process). In the case of Linux command-line programs, the program is executed by a line of text typed or copy-pasted in a terminal window. To run some programs, the working directory may be anywhere in the directory tree, but for others, it must be in the same directory where the program file is. The following Linux commands indicate this requirement by "(*)" at the beginning of the line (*see* **Note 14**).

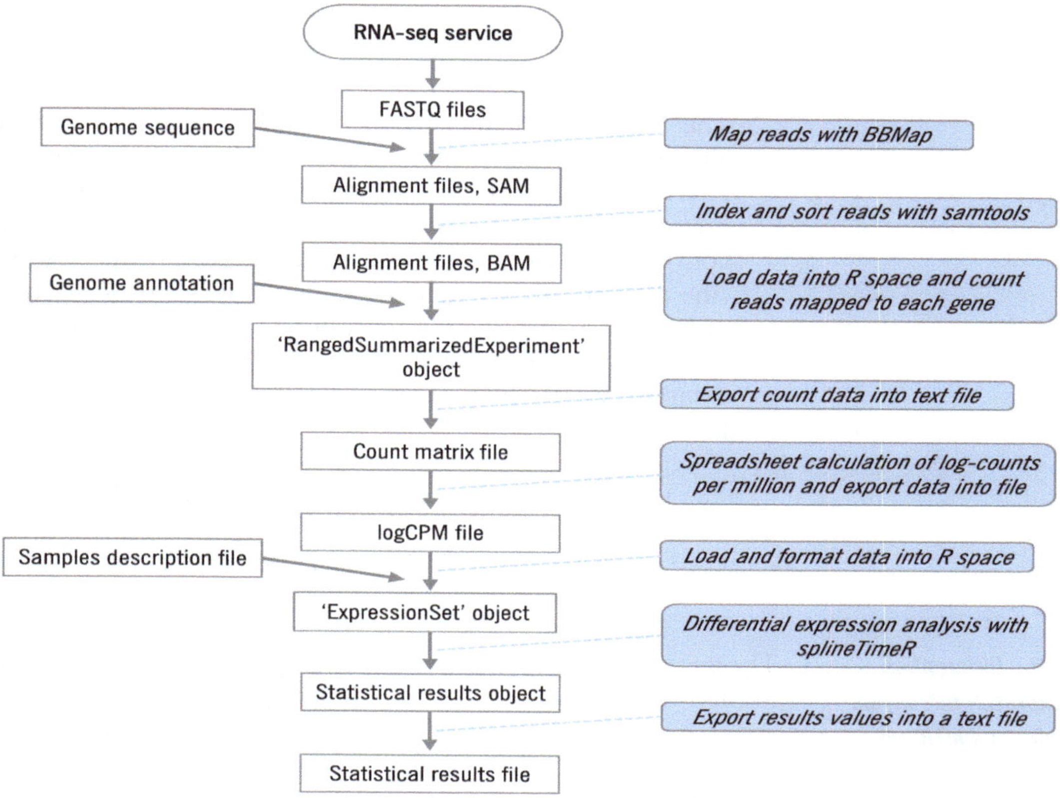

Fig. 1 Workflow through the bioinformatics analysis of RNA-seq data

1. The FASTA file of the strain genome sequence ("genome.fna") must be indicated to the mapper program. This step is done by typing, in a terminal window, the first line of the following commands. Then, next commands will produce the alignments of the sequencing libraries of reads (Control.T1.fastq, Control.T2.fastq, Control.T3.fastq, Phytosterol.T1.fastq, Phytosterol.T2.fastq, and Phytosterol.T3.fastq) obtained from cell samples from two media at three culture times, respectively. Please note that the character "\" in the following command serves to type a single command as separate lines. The input FASTQ files (*see* **Note 15**) can be either compressed (file extension ".fastq.gz") or not (file extension ".fastq"):

```
(*)./bbmap.sh ref=genome.fna
(*)./bbmap.sh ./bbmap.sh in=Control.T1.fastq \
outm=Control.T1.sam slow=t ambiguous=random
(*)./bbmap.sh ./bbmap.sh in=Control.T2.fastq \
outm=Control.T2.sam slow=t ambiguous=random
(*)./bbmap.sh ./bbmap.sh in=Control.T3.fastq \
outm=Control.T3.sam slow=t ambiguous=random
(*)./bbmap.sh ./bbmap.sh in=Phytosterol.T1.fastq \
outm=Phytosterol.T1.sam slow=t ambiguous=random
(*)./bbmap.sh ./bbmap.sh in=Phytosterol.T2.fastq \
outm=Phytosterol.T2.sam slow=t ambiguous=random
(*)./bbmap.sh ./bbmap.sh in=Phytosterol.T3.fastq \
outm=Phytosterol.T3.sam slow=t ambiguous=random
```

2. The Sequence Alignment/Map (SAM) format is a generic alignment format for storing short-read alignments against reference sequences. The BAM format is the binary, compressed version of a SAM file to save disk space. Firstly, it is convenient to inspect the alignment SAM files with FastQC (the program manual guides the interpretation of the output graphs); secondly, the samtools program is used to convert, index, and order the read alignment files with the following commands (only shows the set of commands for one alignment file):

```
samtools sort Control.T1.sam -o Control.T1_Or-
dered.bam
samtools index Control.T1_Ordered.bam
```

3.5 Differential Expression Analysis with splineTimeR

The differential expression analysis is conducted with packages run in the R environment. A working directory is created to copy the above BAM files as input of the analysis, as well as the genome annotation. For Windows users, R can be installed and run directly in this system. In our example, the working directory is located and named as "C:/Transcriptomics/Exp1" (Windows system).

1. Open an R session and inside the console, run the following commands (comments are preceded by "#" and can be copy-pasted in the terminal as well):

```
library("splineTimeR") #Loading the required
packages.
library("Rsamtools")
library("GenomicFeatures")
library("GenomicAlignments")
library(limma)
wdir <- "C:/Transcriptomics/Exp1"
list.files(wdir) #To check the files of the working
directory.
filenames <- file.path(wdir, list.files(wdir,pat-
tern="Ordered.bam$"))
bamfiles <- BamFileList(filenames, yield-
Size=2000000)
```

2. Load to the R space the genome annotation with the following commands:

```
Annotfile <- file.path(wdir,"Annotation.gff")
(txdb <- makeTxDbFromGFF(Annotfile, for-
mat="gff3"))
(genes <- transcriptsBy(txdb, "gene"))
```

3. Before performing the statistical analysis, read count data are computed from the alignment files. Count data is a matrix of integers: for each gene (in rows), the integers tell the total of reads that map on that gen in each sample (in columns). Store the output as an object of class *RangedSummarizedExperiment* that is named "se" in the following commands:

```
se <- summarizeOverlaps(features=genes, reads=bam-
files,
  mode="Union",
  singleEnd=TRUE,
  ignore.strand=FALSE,
  fragments=FALSE)
se #To see the structure of the object "se".
head(assay(se)) #This shows the first rows of the
read counts table.
colSums(assay(se)) #This shows the counts totals.
summary(assay(se)) #This shows the counts statis-
tics.
```

4. The object "se" contains the count values for each gene in each sample. In order to prepare these data for export, a new object is created, the names of its columns are abbreviated, and finally, the data are exported into a tab-delimited file ("Counts.txt"):

```
counts <- assay(se)
head(counts) #This shows the first lines of the object.
colnames(counts) <- sub("_Ordered.bam", "",colnames
(counts))
head(counts) #Compare column names before and after the
shortening.
setwd(wdir) #To change the working directory of R.
write.table(counts,"Counts.txt", row.names=TRUE,
sep="\t")
```

5. Now, count data can be imported into a common spreadsheet program, like Microsoft Excel. Before performing the differential expression analysis with splineTimeR, counts should be normalized by library size and log-transformed. Both steps are easily done in a spreadsheet with the formula used by Law et al. [26] that yield log-counts per million (logCPM). Once calculated, export the values into a tab-separated file named "logCPM.txt," with the data structure and column headings as follows:

Gene	Control. T1	Control. T2	Control. T3	Phytosterol. T1	Phytosterol. T2	Phytosterol. T3

6. Create also a tab-delimited text file ("samplesDescription.txt") with the content of the next table; this is the description of the samples. Please note that the cultivation time is now included. For this analysis, the time points should be identical between compared treatment groups.

SampleName	Treatment	Time
Control.T1	Control	24
Control.T2	Control	36
Control.T3	Control	48
Phytosterol.T1	Phytosterol	24
Phytosterol.T2	Phytosterol	36
Phytosterol.T3	Phytosterol	48

7. Load the logCPM data and also the description of samples into the previously open R session:

```
logCPM <- as.matrix(read.csv("logCPM.txt",
sep="\t",row.names="Gene"))
dim(logCPM) #To check the full table has been
loaded.
logCPM.eset <- ExpressionSet(logCPM) #To format the
data into the required object type.
samplesDescription <- read.csv("samplesDescrip-
tion.txt",sep="\t")
pData(logCPM.eset) <- samplesDescription) #To in-
corporate the sample information into the input
object for splineTimeR.
```

8. Then, the statistical results of the treatment comparison (phytosterol culture vs. control culture) are obtained with these commands:

```
DE.Phytosterol_vs_Control <- splineDiffExprs(eSe-
tObject = logCPM.eset[,c(4:6,1:3)], df = 1, refer-
ence = "Control", intercept = TRUE)
```

9. The easiest way to work with the differential expression results is to export the values to a text file and import them into a spreadsheet. To create the file, type the command:

```
write.table(as.data.frame(DE.Phytosterol_vs_Con-
trol),file="Results_SplineTimeR.txt",
sep="\t", quote=FALSE, row.names=FALSE)
```

4 Notes

1. First RNA-seq analyses were done with the yeasts *Saccharomyces cerevisiae* and *Schizosaccharomyces pombe*, published in May 2008 [27, 28]. The first bacterial RNA-seq analysis was published in March 2009 [4]. In this work, the authors identified 13 ncRNA analyzing the differential expression of intergenic regions. All these 13 ncRNAs were not included among the 213 putative ncRNA computationally predicted in previous work. Recently, RNA-seq also served to identify very small ncRNA (<16 nt) in bacteria [29].
2. The spatula used to take the lysozyme, thoroughly cleaned, must be flamed before use to eliminate the possible presence of RNases. The solution is aliquoted and stored at −20 °C. No more than two freeze–thaw cycles are recommended.

3. This lysis buffer was chosen because it is sold independently and not only as part of the NucleoSpin® miRNA kit.
4. PLG tubes must be previously pre-packed by centrifugation (30 s at maximum speed). The pre-package can be done during Subheading 3.2, **step 9**.
5. We recommend the free VirtualBox software (https://www.virtualbox.org/) to build the virtual machine and the simple Xubuntu distribution (http://xubuntu.org/). Once the guest operating system is installed inside a virtual machine, it is recommended to install the so-called Guest Additions to create a shared folder as the most convenient way of sharing files between the Windows/macOS host and the Linux guest systems.
6. Usually, a minimum of 5 μg of total RNA is required at 100 ng/μL or higher concentration. The quality of the RNA preparation depends on its purity and the integrity of the RNA molecules (absence of RNase-dependent degradation). The purity can be checked easily with a spectrophotometer since a pure preparation shows an A_{260}/A_{280} ratio of 2.1. The standard for integrity control is the RIN (RNA integrity number) provided by the Agilent 2100 Bioanalyzer instrument. The usual requirement for the RNA samples being sequenced is a RIN ≥ 8.
7. The culture sample volume will depend on the growth phase and culture density. In our experience, an average of 3.5 μg of RNA is extracted from 1 OD_{600} of cells. *See* **Note 6** for the requirements of the RNA-seq procedure.
8. In bacteria, RNA biosynthesis and degradation are continuous processes. To minimize the RNA degradation after taking the samples, it is necessary to flash-freeze the samples or add a reagent that inhibits the RNase activity, such as the RNAprotect™ reagent. To reduce as much as possible the time the samples are untreated, it is preferable—disposing of an array of tubes containing the necessary amount of the reagent—to add the culture sample to the reagent tubes instead of adding the reagent to the sample tubes.
9. A temperature higher than 15 °C is necessary to avoid the precipitation of the RNAprotect components.
10. A maximum of 2 weeks at −20 °C or 4 weeks at −80 °C is recommended by QIAGEN. However, we have obtained RNA preparations of the same high quality, compared with 1-month storage, from frozen samples (−20 °C) for 1 year.
11. This step is critical since more incubation time might result in appreciable degradation, while less time implies a lower yield.

12. TURBO DNA-free™ kit is our preferred DNase treatment. Although 0.5 μL of DNase I and 30 min are the standard conditions recommended by the manufacturer, the indicated conditions ensure the complete removal of DNA, at least for RNA-seq applications.
13. Because of the small reaction volume, 0.2 mL tubes and a thermocycler are used for the incubation.
14. To conduct the indicated steps efficiently, it is highly recommended to follow a Linux command-line tutorial. The steps carried out in the R environment are also executed with command lines but with a different language and the input and output data are not saved as files on the disk, but as "objects" inside the software. An excellent and free introductory guide to R is found at https://cran.r-project.org/doc/manuals/R-intro.pdf.
15. The FASTQ format is widely used to store sequencing reads. It is a text file that contains for each read four lines. The first line always starts with "@" and specifies the name of the read (the software of the Illumina sequencer generates a systematic identifier); the second line contains the sequence letters, and the third line begins with "+", which is optionally followed by the same sequence identifier of the first line; the fourth line encodes the quality values (the Phred score) for the sequence of the second line: both lines must contain the same number of characters.

Acknowledgments

This work was supported by a grant of the European Union program ERA-IB [MySterI (EIB.12.010)] through the APCIN call of the Spanish Ministry of Economy and Competitiveness (MINECO, Spain) (PCIN-2013-024-C02-01). The authors thank the European Union program ERA-IB; the Spanish Ministry of Economy and Competitiveness (MINECO, Spain); the MySter I Consortium (INBIOTEC, Pharmins Ltd., University of York, SINTEF, Technische Universität Dortmund, and Gadea Biopharma S. L.), and the Syntheroids project (ERA-CoBioTech 1st call; founded through the APCIN call of the Spanish Ministry of Science, Innovation and Universities; ID: PCI2018-093066). They also thank J. Merino, B. Martín and A. Casenave for their excellent technical assistance and the degree students of the group A. Morales and P. González.

References

1. Garalde DR, Snell EA, Jachimowicz D et al (2018) Highly parallel direct RNA sequencing on an array of nanopores. Nat Methods 15: 201–206. https://doi.org/10.1038/nmeth.4577
2. Hrdlickova R, Toloue M, Tian B (2017) RNA-seq methods for transcriptome analysis. Wiley Interdiscip Rev RNA 8:1364. https://doi.org/10.1002/wrna.1364
3. Shi H, Zhou Y, Jia E et al (2021) Bias in RNA-seq library preparation: current challenges and solutions. Biomed Res Int 2021: 6647597. https://doi.org/10.1155/2021/6647597
4. Yoder-Himes DR, Chain PSG, Zhu Y et al (2009) Mapping the *Burkholderia cenocepacia* niche response via high-throughput sequencing. Proc Natl Acad Sci U S A 106:3976–3981. https://doi.org/10.1073/pnas.0813403106
5. Valentin-Hansen P, Eriksen M, Udesen C (2004) The bacterial Sm-like protein Hfq: a key player in RNA transactions. Mol Microbiol 51:1525–1533. https://doi.org/10.1111/j.1365-2958.2003.03935.x
6. De Lay N, Schu DJ, Gottesman S (2013) Bacterial small RNA-based negative regulation: Hfq and its accomplices. J Biol Chem 288: 7996–8003. https://doi.org/10.1074/jbc.R112.441386
7. Miyakoshi M, Chao Y, Vogel J (2015) Regulatory small RNAs from the 3′ regions of bacterial mRNAs. Curr Opin Microbiol 24:132–139. https://doi.org/10.1016/j.mib.2015.01.013
8. Chao Y, Papenfort K, Reinhardt R et al (2012) An atlas of Hfq-bound transcripts reveals 3′ UTRs as a genomic reservoir of regulatory small RNAs. EMBO J 31:4005–4019. https://doi.org/10.1038/emboj.2012.229
9. Miyakoshi M, Chao Y, Vogel J (2015) Cross talk between ABC transporter mRNAs via a target mRNA-derived sponge of the GcvB small RNA. EMBO J 34:1478–1492. https://doi.org/10.15252/embj.201490546
10. Feng L, Rutherford ST, Papenfort K et al (2015) A qrr non-coding RNA deploys four different regulatory mechanisms to optimize quorum-sensing dynamics. Cell 160:228–240. https://doi.org/10.1016/j.cell.2014.11.051
11. Papenfort K, Sun Y, Miyakoshi M et al (2013) Small RNA-mediated activation of sugar phosphatase mRNA regulates glucose homeostasis. Cell 153:426–437. https://doi.org/10.1016/j.cell.2013.03.003
12. Stiens J, Arnvig KB, Kendall SL et al (2022) Challenges in defining the functional, non-coding, expressed genome of members of the *Mycobacterium tuberculosis* complex. Mol Microbiol 117:20–31. https://doi.org/10.1111/mmi.14862
13. Sharma CM, Hoffmann S, Darfeuille F et al (2010) The primary transcriptome of the major human pathogen *Helicobacter pylori*. Nature 464:250–255. https://doi.org/10.1038/nature08756
14. Sharma CM, Vogel J (2014) Differential RNA-seq: the approach behind and the biological insight gained. Curr Opin Microbiol 19:97–105. https://doi.org/10.1016/j.mib.2014.06.010
15. Conway T, Creecy JP, Maddox SM et al (2014) Unprecedented high-resolution view of bacterial operon architecture revealed by RNA sequencing. mBio 5:e01442-14. https://doi.org/10.1128/mBio.01442-14
16. Marbaniang CN, Vogel J (2016) Emerging roles of RNA modifications in bacteria. Curr Opin Microbiol 30:50–57. https://doi.org/10.1016/j.mib.2016.01.001
17. Stark R, Grzelak M, Hadfield J (2019) RNA sequencing: the teenage years. Nat Rev Genet 20:631–656. https://doi.org/10.1038/s41576-019-0150-2
18. Li H, Handsaker B, Wysoker A et al (2009) The sequence alignment/map format and SAMtools. Bioinformatics 25:2078–2079. https://doi.org/10.1093/bioinformatics/btp352
19. Li H (2011) A statistical framework for SNP calling, mutation discovery, association mapping and population genetical parameter estimation from sequencing data. Bioinformatics 27:2987–2993. https://doi.org/10.1093/bioinformatics/btr509
20. Li H (2011) Improving SNP discovery by base alignment quality. Bioinformatics 27:1157–1158. https://doi.org/10.1093/bioinformatics/btr076
21. Lei R, Ye K, Gu Z et al (2015) Diminishing returns in next-generation sequencing (NGS) transcriptome data. Gene 557:82–87. https://doi.org/10.1016/j.gene.2014.12.013
22. Rodríguez-García A, Fernández-Alegre E, Morales A et al (2016) Complete genome sequence of "*Mycobacterium neoaurum*" NRRL B-3805, an androstenedione (AD) producer for industrial biotransformation of sterols. J Biotechnol 224:64–65.

https://doi.org/10.1016/j.jbiotec.2016.03.021

23. Michna A (2022) splineTimeR: time-course differential gene expression data analysis using spline regression models followed by gene association network reconstruction. R package version 1.24.0. https://doi.org/10.18129/B9.bioc.splineTimeR
24. Seyednasrollah F, Laiho A, Elo LL (2015) Comparison of software packages for detecting differential expression in RNA-seq studies. Brief Bioinform 16:59–70. https://doi.org/10.1093/bib/bbt086
25. Chowdhury HA, Bhattacharyya DK, Kalita JK (2020) Differential expression analysis of RNA-seq reads: overview, taxonomy and tools. IEEE/ACM Trans Comput Biol Bioinform 17:566–586. https://doi.org/10.1109/TCBB.2018.2873010
26. Law CW, Chen Y, Shi W et al (2014) voom: precision weights unlock linear model analysis tools for RNA-seq read counts. Genome Biol 15:R29. https://doi.org/10.1186/gb-2014-15-2-r29
27. Nagalakshmi U, Wang Z, Waern K et al (2008) The transcriptional landscape of the yeast genome defined by RNA sequencing. Science 320:1344–1349. https://doi.org/10.1126/science.1158441
28. Wilhelm BT, Marguerat S, Watt S et al (2008) Dynamic repertoire of a eukaryotic transcriptome surveyed at single-nucleotide resolution. Nature 453:1239–1243. https://doi.org/10.1038/nature07002
29. Diallo I, Ho J, Lalaouna D et al (2022) RNA sequencing unveils very small RNAs with potential regulatory functions in bacteria. Front Mol Biosci 9:914991. https://doi.org/10.3389/fmolb.2022.914991

Chapter 7

Bidimensional Analyses of the Intra- and Extracellular Proteomes of Steroid Producer Mycobacteria

Carlos Barreiro and Ana M. Ibáñez

Abstract

The importance of the pathogenic mycobacteria has mainly focused the omic analyses on different aspects of their clinical significance. However, those industrially relevant mycobacteria have received less attention, even though the steroid market sales in 2021 were estimated in $56.45 billion.

The extracellular proteome, due to its relevance in the sterol processing and uptake, and the intracellular proteome, because of its role in steroids bioconversion, are the core of the present chapter. Both, mono-dimensional gels, as preparatory analysis, and bidimensional gels as proteome analysis are described. As a proof of concept, the protein extraction methods for both sub-proteomes of *Mycobacterium* are described. Thus, procedures and relevant key points of these proteome analyses are fully detailed.

Key words Proteome, Secretome, Steroid, 2D-DIGE, *Mycobacterium*, B-3805, Sterol, Phytosterols, Cholesterol

1 Introduction

Mycobacterium genus can be divided, as its close-related genus *Corynebacterium*, under an effective, but informal, classification into (a) "bad guys," these with clinical significance (*Mycobacterium tuberculosis*, *Mycobacterium leprae*) and (b) "good guys," which are involved in biotechnological applications such as steroid production (*Mycobacterium neoaurum*) [1–3]. Additionally, other classifications based on growth rate, tuberculous or nontuberculous mycobacteria, etc. can also be applied [4].

These actinobacteria included in the recently created new order of Corynebacteriales are located in the family Mycobacteriaceae. Nowadays, the *Mycobacterium* genus contains 260 species and 24 subspecies as describes Jean P. Euzéby on the website "List of prokaryotic names with standing in nomenclature (LPSN)" (http://www.bacterio.net/mycobacterium.html) [5, 6]. This fact

Carlos Barreiro and José-Luis Barredo (eds.), *Microbial Steroids: Methods and Protocols*, Methods in Molecular Biology, vol. 2704, https://doi.org/10.1007/978-1-0716-3385-4_7,

presents the genus as a wide taxonomic group. Its first genome sequence was released in 1998 [7] and currently more than 173 genome entries appear in the NCBI database (http://www.ncbi.nlm.nih.gov/genome/). This opened the so-called post-genomic era, mainly focused on omic analysis as proteomics [8] or transcriptomics (microarrays [9], RNAseq [10, 11]) of different *Mycobacterium* species.

The World Health Organization (WHO) estimated 1.5 million of tuberculosis death in 2020 and 10 million people fell ill with tuberculosis (https://www.who.int/news-room/fact-sheets/detail/tuberculosis), which contrasts with the successful story of leprosy control program in the last 20 years [12, 13]. Thus, due to the relevance of these ancient diseases, the proteome analyses in *Mycobacterium* genus have been largely focused on the pathogen species. The proteomics applied to mycobacterial diseases have achieved different goals, such as (a) mycobacterial strain characterization [14], (b) pathogenic strain characterization [15–18], (c) immunogenic target definition [19, 20], (d) immune response attenuation [12], (e) extracellular proteome (secretome) analysis [21–23], (f) 2D PAGE database generation [24], and also (g) nonpathogenic model strains being analyzed (e.g., *Mycobacterium smegmatis*) [25].

In contrast, the omic analysis based on industrial-relevant mycobacteria has been less tackled, even though the steroid production by *Mycobacterium* strains is an old fashion from the late 1960s [26]. Nowadays, according to the market research report of the data provider company Data Bridge (databridgemarketresearch.com, accessed on August 2022), in 2021 the total revenue from steroid sales was estimated at $56.45 billion and is expected to reach $131.98 billion by 2029 [27–29]. In the case of the steroid production, three are the main concerns standing in the way of the industrial process that also difficult the omics approach: (1) low solubility of sterols as phytosterols (~0.1 mM), which are the raw materials for steroid production; (2) steroid toxicity for the producer bacteria (end-product inhibition), as well as (3) the development of microorganisms for the production a la carte of high-added-value steroids [30]. These facts somehow affect the protein sample collection due to the traditional use of, for example, two-phase cultures [aqueous and oil (soya, rape, sunflower) phases], which is critical for secretome analysis. Anyway, some interesting studies applied to the steroid producer mycobacteria have been published [10, 31], in addition to the increased number of available genome sequences, which are the key step for the subsequent protein identification [3, 32].

Three are the sub-proteomes typically analyzed from one cell: intracellular, membrane and extracellular (secretome). The last one is complex in terms of protein purification, due to the presence of phytosterols or oil derivatives disturbing the protein extraction. The analysis of the extracellular proteome is based on some preliminary concepts: (a) use of defined media, if possible, to avoid two-phase culture broth, which make extremely the process complex; (b) the lowest amount of sterol precursors (e.g., cholesterol, phytosterols) compatible with steroid production and detection (e.g., steroid production proteins should be active); and (c) proper growth phase (e.g., lag phase) to avoid culture lysis, which could enrich the intracellular proteins present in the secretome (degradome), to avoid their identification as false moonlighting proteins [33, 34]. These moonlighting proteins comprise an interesting group of multifunctional proteins, which presents two functions in a single polypeptide chain [35, 36]. Several examples of moonlighting proteins are described at the MoonProt database (http://www.moonlightingproteins.org/).

In contrast, the intracellular environment presents a high protein concentration. It allows, if needed, the sample precipitation by using larger volumes of precipitants as acetone-based solutions, which decreases the number of low soluble proteins precipitated when TCA (trichloroacetic acid) is used. TCA complicates the subsequent dissolution in the urea-based buffers. In a similar way as described above, the lower the amount of sterol precursors, the better for protein purification. An excess of phytosterols in the culture media results in white clumps that precipitate in parallel to the mycobacterial cells when they are collected. Thus, the use of defined media and the lowest amount of sterol precursors compatible with steroid production and detection is convenient.

The extracellular proteome, due to its relevance in the sterol processing and uptake, and intracellular proteome, because of its role in steroid bioconversion, are the core of the present chapter.

As a proof of concept, both sub-proteomes of *Mycobacterium neoaurum* NRRL B-3805, a relevant industrial strain involved in steroid production, have been used to develop the described protein extraction methods. Besides, the intracellular proteome was used as an example of DIGE (differential in gel electrophoresis) labeling. The target strain, *M. neoaurum* NRRL B-3805 (CECT3018; DSM 2967; NCIMB 11678), was initially obtained after two rounds of ultraviolet irradiation from soil-isolated mycobacteria as part of a mutagenesis program for efficient strains in sterol bioconversion [37] (patent no. US3759791). This strain is a suitable AD (4-androstene-3,17-dione) producer and a low producer of the by-product ADD (androsta-1,4-diene-3,17-dione). Its genome sequence has been published, which eases these omic analyses [32].

2 Materials

2.1 Bacterial Growth Media

1. SINTEF's defined cultivation medium for shake flasks [38]: 10.0 g of glycerol (99.5%), 4.0 g of PIPES [piperazine-N, N′-bis(2-ethanesulfonic acid)], 2.5 g of NH_4Cl, 0.50 g of $MgSO_4{\cdot}7H_2O$, 0.50 g of K_2HPO_4, 0.30 g of NaH_2PO_4, 0.10 g of aspartic acid, and 4.0 mL of SINTEF's trace mineral solution, pH 7.0–7.2 (see below). As a control condition, this defined medium with glycerol as a carbon source (5.7 g/L, 87% purity) is selected. However, sterol biotransformation analyses are carried out by the addition of 7.0 g/L of cholesterol or phytosterols, respectively, to the control medium, which also includes glycerol (5.7 g/L, 87% purity) as a carbon source.
2. SINTEF's trace mineral solution [38] (per liter): 3.0 g of $FeCl_3$, 500 mg of $Na_2EDTA{\cdot}2H_2O$, 400 mg of $MnCl_2{\cdot}4H_2O$, 300 mg of $ZnCl_2$, 250 mg of $CuCl_2{\cdot}2H_2O$, 20 mg of $CoCl_2{\cdot}6H_2O$, 10 mg of $Na_2MoO_4{\cdot}2H_2O$, 25 mL of conc. HCl (37%), and 975 mL of distilled water (or similar quality).

2.2 Bacterial Strain

1. *Mycobacterium neoaurum* NRRL B-3805 (CECT3018; DSM 2967; NCIMB 11678): a mutant strain derived from *Mycobacterium* sp. NRRL B-3683 by ultraviolet irradiation, which lacked the ability to 1-dehydrogenate steroids. The ancestor of *Mycobacterium* sp. NRRL B-3683 was a soil-isolated bacterium able to bioconvert stigmast-4-en-3-one to androsta-1,4-diene-3,17-dione also irradiated with ultraviolet light [32, 37].

2.3 Biological and Chemical Materials

1. 2D Clean-Up Kit (Cytiva/Merck).
2. 2-Propanol.
3. Acetic acid.
4. Acetone.
5. Acrylamide/bis-acrylamide solution (37.5:1) (ProtoGel (30%); National Diagnostics).
6. Ammonium persulfate.
7. Ammonium sulfate.
8. Benzonase (DNase and RNase; Merck).
9. Bradford reagent (Bio-Rad Protein Assay Dye Reagent Concentrate; Bio-Rad).
10. Bromophenol Blue.
11. BSA (bovine serum albumin).
12. CHAPS (3-[(3-cholamidopropyl)dimethylammonio]-1-propanesulfonate).

13. cOmplete™ Protease Inhibitor Cocktail (Roche).
14. Coomassie Brilliant Blue G-250.
15. DIGE Fluor Minimal Labeling Kit (Cytiva/Merck).
16. DTT (dithiothreitol).
17. Ethanol.
18. DryStrips Cover Fluid (Cytiva/Merck).
19. Glycerol (87%).
20. Glycine.
21. Hydrochloric acid.
22. Iodocetamide.
23. IPG buffer 3–11 NL (Cytiva/Merck).
24. Lysine.
25. Lysing Matrix B (MPbio).
26. Methanol.
27. NaCl.
28. NaOH.
29. N,N-dimethylformamide (99% anhydrous).
30. Phosphoric acid.
31. Precision Plus Protein Standards (All Blue; Bio-Rad).
32. SDS (sodium dodecyl sulfate).
33. TCA (trichloroacetic acid).
34. TEMED (tetramethylethylenediamine).
35. Tris-base [Tris(hydroxymethyl)aminomethane].
36. Urea.
37. Cellulose chromatography papers 3MM Chr sheets (Whatman).

2.4 Buffers and Solutions

As a general prevention, the proteomics solutions should be filtered through 0.45 μm filter to avoid insoluble contaminants and interfering particles.

1. BSA master solution: 0.25 mg/mL bovine serum albumin dissolved in distilled water.
2. Breaking buffer: It is the same with washing buffer (50 mM Tris–HCl, pH 7.2) but supplemented with protease inhibitor (e.g., cOmplete, Roche).
3. Colloidal Coomassie solution (blue silver): 0.12% dye (Coomassie Brilliant Blue G-250), 10% ammonium sulfate, 10% phosphoric acid, and 20% methanol in distilled water [39].
4. Denaturation buffer: 350 mM Tris–HCl (pH 6.8), 30% (v/v) glycerol, 10% (v/v) SDS, 0.6 M DTT, and 0.02% bromophenol blue.

5. DIGE pH buffering stock solution: 250 mM NaOH.
6. DIGE pH buffering working solution: 100 mM NaOH.
7. Equilibration buffer: 2% (w/v) SDS, 50 mM Tris–HCl pH 8.8, 6 M urea, 30% (v/v) glycerol, and 0.002% bromophenol blue.
8. Fixation solution: 7% methanol and 7% acetic acid in distilled water.
9. 5× lower electrophoresis buffer: 15.2 g/L Tris, 72 g/L glycine, and 5 g/L SDS.
10. NaCl 0.9%: sodium chloride in distilled water.
11. RB (rehydration) buffer: 8 M urea and 2% CHAPS in water. Filter through 0.45 μm. Store at −20 °C. Prior to use supplement with 1% IPG buffer 3–11 NL (nonlinear) and 60 mM DTT.
12. Running solution (10× Laemmli SDS electrophoresis buffer): 250 mM Tris, 2 M glycine, and 1% (w/v) SDS.
13. SDS solution: 10% SDS in distilled water.
14. TCA precipitation solution: 20% TCA and 0.14% DTT in acetone.
15. Washing buffer: 50 mM Tris–HCl, pH 7.2.
16. Washing solution: 0.07% DTT in acetone.
17. Resuspension buffer (RB2): 8 M urea and 4% (w/v) CHAPS (3-[(3-cholamidopropyl) dimethylammonio]-1-propanesulfonate). Filter through 0.45 μM membrane and store at −20 °C.
18. Stop solution: Lysine 10 mM in water. Store at −20 °C.
19. Upper electrophoresis buffer 5×: 8.5 g/L Tris, 40.3 g/L glycine, and 2.8 g/L SDS.

2.5 Acrylamide/Bis-acrylamide Gels

1. Monodimensional SDS-PAGE running section of two small gels (Mini-PROTEAN format; 10 mL). Three acrylamide concentrations (7.5%, 12%, and 15%) are shown according to the molecular masses of the proteins to be separated (7.5% for 250–50 kDa; 12% for 100–20 kDa; 15% for 60–10 kDa).

Reagent	7.5%	12.0%	15.0%
Acrylamide/bis-acrylamide 30% (*see* **Note 1**)	2.5 mL	4.0 mL	5.0 mL
Tris–HCl 1.5 M, pH 8.8	2.5 mL	2.5 mL	2.5 mL
SDS 20% (w/v)	50 μL	50 μL	50 μL
Ammonium persulfate 10% (w/v)	50 μL	50 μL	50 μL
TEMED (*see* **Note 2**)	5 μL	5 μL	5 μL
Distilled water	4.9 mL	3.4 mL	2.4 mL

2. Stacking section of two 1D gels (10 mL).

Reagent	Volume
Acrylamide/bis-acrylamide 30%	1.3 mL
Tris–HCl 0.5 M, pH 6.8	2.5 mL
SDS 20% (w/v)	50 μL
Ammonium persulfate 10% (w/v)	50 μL
TEMED	10 μL
Distilled water	6.09 mL

3. SDS-PAGE mixture for six gels (21 × 26 cm each gel; 400 mL) of the second-dimensional analysis of DIGE.

Reagent	Volume
Acrylamide/bis-acrylamide 30%	166.8 mL
Tris–HCl 1.5 M, pH 8.8	100.0 mL
SDS 20% (w/v)	4.0 mL
Ammonium persulfate 10% (w/v)	2.0 mL
TEMED	132 μL
Distilled water	127.2 mL

2.6 Equipment

1. 0.45 μm filter membrane and SteriCup device (0.22 μm filter) (Millipore).
2. DALTsix Gel Caster (GE Healthcare).
3. Ettan DALTsix Electrophoresis Unit (GE Healthcare).
4. FastPrep®-24 Instrument (MPbio).
5. IPG strip holder (Cytiva/Merck).
6. Micropipette (PIPETMAN P2, P20, P200, P1000).
7. Mini-PROTEAN® 3 Cell (Bio-Rad).
8. Power supply.
9. Shaking platform.
10. Spectrophotometer.
11. Thermomixer.
12. Tubes (10 and 50 mL tubes) and microtubes (1.5 mL tubes).
13. Vortex shaker.
14. Kitasato flask.

3 Methods

3.1 Protein Collection: Secretome

The secretome or extracellular proteome (less commonly named as exoproteome) is the rich group of molecules secreted from living cells, which in a broader sense also includes molecules shed from the living cell surface and the cellular machinery involved in the protein secretion [40–42]. The protocol for mycobacterial extracellular protein obtaining is as follows (Fig. 1):

1. Grow *M. neoaurum* NRRL B-3805 in 100 mL of (preferable defined) liquid medium, nearly 48 h to ensure the exponential phase (if needed).
2. Harvest 25 mL of culture in a 50 mL tube by centrifugation for 10 min at 4 °C and 5000 × *g*.
3. Filter the supernatant (SteriCup 0.22 μm) to avoid any mycobacterial cell and sterol particle.
4. Add 25 mL of TCA precipitation solution per centrifuge tube with 25 mL of supernatant. The final TCA percentage is 10% (*see* **Note 3**).
5. Keep at −20 °C, overnight (*see* **Note 4**).
6. Centrifuge for 10 min at 4500 × *g* and discard the supernatant. A brilliant white (to yellowish) pellet should be observed.
7. Wash the pellet with 10 mL of washing solution. Resuspend the pellet by vortexing (*see* **Note 5**).
8. Centrifuge for 10 min at 4500 × *g*. Discard the supernatant.
9. Repeat the washing step.
10. Wash with 10 mL of 80% acetone in distilled water (*see* **Note 6**).
11. Centrifuge for 10 min at 4500 × *g*. Discard the supernatant.
12. Centrifuge for 1 min at 4500 × *g*. Carefully remove all the remaining liquid with a micropipette. Let air-dry up to 2 min (*see* **Note 7**).
13. Resuspend the protein pellet in 100–200 μL (according to the obtained pellet) of RB buffer (*see* **Note 8**).
14. Shake the samples for at least 1 h at 20–25 °C (thermomixer) (*see* **Note 9**). Transfer the protein solution to 1.5 mL microtube and centrifuge for 1 min at top speed to discard insoluble particles.
15. Transfer the supernatant to a new 1.5 mL microtube.
16. Preserve the samples at −20 °C.

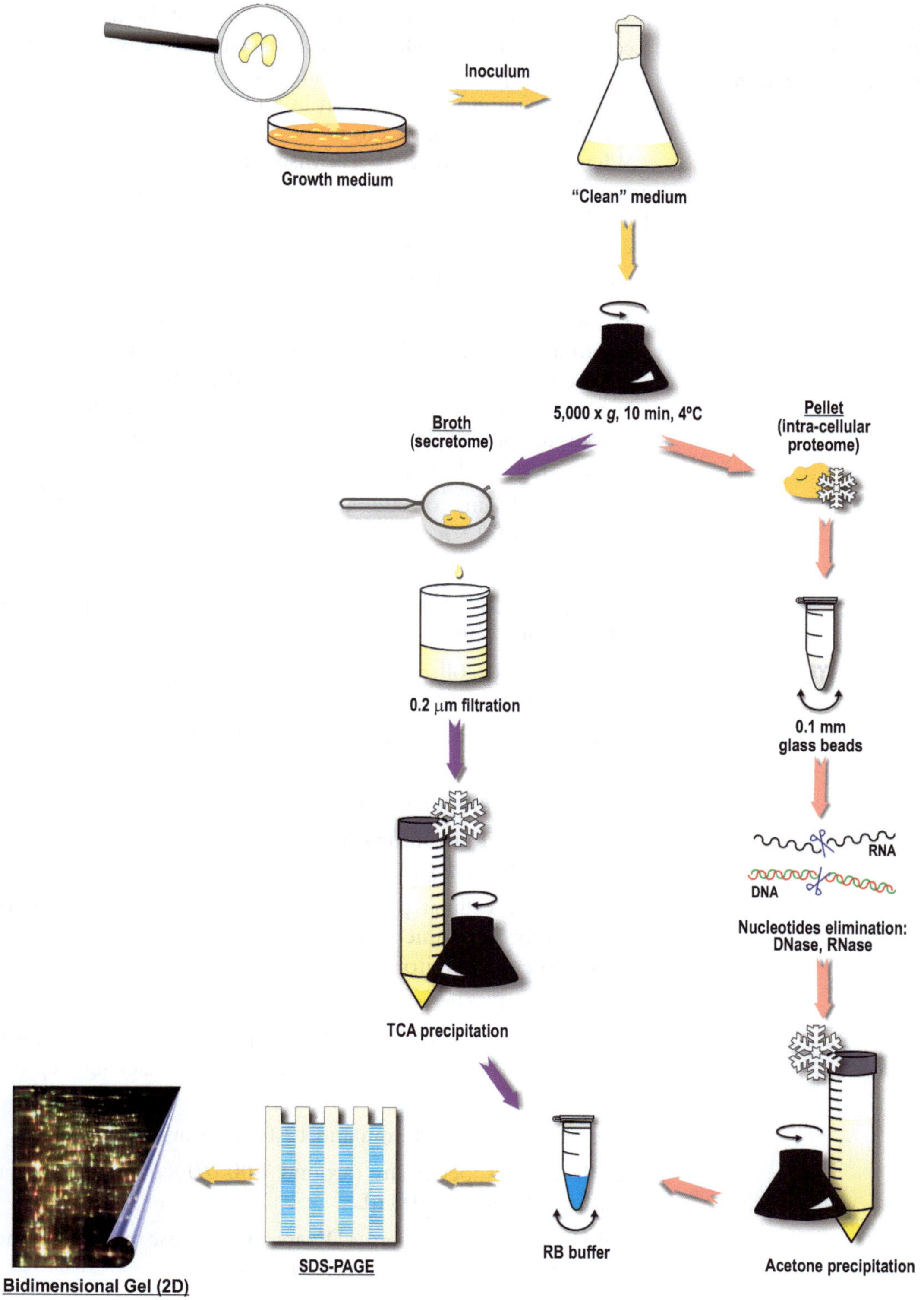

Fig. 1 General overview of the protein extraction methods applied to the analysis of the extracellular (purple arrows) and intracellular (pink arrows) proteomes of *M. neoaurum* NRRL B-3805

3.2 Protein Collection: Intracellular Proteome

The intracellular proteome presents a snapshot of the microorganism cellular metabolism. Due to the initial harvest steps the amount of the obtained proteins, usually, is higher than the secreted ones; then, the precipitation process can be carried out by using higher precipitant volumes (e.g., acetone).

When the proteome analysis consists in the comparison of the nutrient effects (e.g., glycerol vs. sterols), it is recommended to collect the samples in the lag or early stationary phase to avoid the analysis of proteins from lysis phase. The protocol for mycobacterial intracellular protein extraction is as follows (Fig. 1):

1. Grow *M. neoaurum* NRRL B-3805 in 100 mL of (defined) liquid medium up to the desired optical density (nearly 48 h to ensure the exponential phase, if needed). If the sterols supplemented to the media interfere the growth measurements by the traditional optical density method, the methylene blue procedure can be used [43].
2. Collect 5–10 mL of culture by centrifugation for 10 min at 4 °C and 5000 × *g*.
3. Discard the supernatant.
4. Resuspend (e.g., vortexing) and wash the cell with pre-cooled (4 °C) NaCl 0.9% (*see* **Note 10**).
5. Centrifuge to collect the cellular pellet. Keep at −20 °C or continue the protocol (*see* **Notes 11** and **12**).
6. Resuspend and wash the cellular pellet with pre-cooled (4 °C) 0.9% NaCl.
7. Centrifuge 5000 × *g* to collect the cellular pellet.
8. Resuspend and wash the cells with pre-cooled (4 °C) washing buffer (*see* **Note 13**).
9. Harvest the cellular pellet and resuspend in 0.8–1.0 mL of breaking buffer, same with washing buffer but supplemented with protease inhibitor (e.g., cOmplete) (*see* **Note 14**).
10. Transfer to FastPrep tubes (lysing matrix B) (*see* **Note 15**).
11. Shake 3 times for 30 s in the FastPrep instrument at top speed (6.5 m/s). Place 1 min on ice among every shacking step (*see* **Note 16**).
12. Centrifuge for 15–30 min and 12,000 × *g* at 4 °C.
13. Collect the supernatant (around 600–700 μL if the initial buffer volume was 800 μL).
14. Add 2 μL of Benzonase (DNase and RNase mixture), and incubate for 30 min at 37 °C (*see* **Note 17**).
15. Centrifuge for 15–30 min and 12,000 × *g* at 4 °C.

16. Collect the supernatant. Do not disturb cell debris. If any cell debris is accidentally collected, a new spin should be done (5 min, 12,000 × *g*, 4 °C). Store at −20 °C or precipitate (*see* **Note 18**).
17. If precipitate, add 4–9 volumes of cold acetone (−20 °C). Gently mix. White clouds of precipitating proteins should be observed, which slowly fall down when the movement stops. Keep at −20 °C overnight (*see* **Note 19**).
18. Centrifuge for 10 min and 12,000 × *g* at 4 °C. A brilliant white protein pellet should be observed. An optional washing step with 80% acetone (5 mL) to discard salts can be done.
19. Discard the supernatant. Air-dry 1 h in a fume hood (acetone smell must disappear).
20. Resuspend in 100–500 μL (conditioned by the observed pellet) of RB buffer.
21. Shake the samples for at least 1 h at room temperature (20–25 °C) by using a thermomixer at top speed.
22. Centrifuge for 1 min at 12,000 × *g* (*see* **Note 20**).
23. Transfer the supernatant to a new 1.5 mL microtube.
24. Conserve the protein samples at −20 °C. Small aliquots to avoid freeze and thaw cycles are recommended (optional).

3.3 Protein Quantitation by Bradford Method

Different methods and improvements for protein quantitation have been described based on (a) dyes, as Coomassie (Bradford) [44] or (b) copper, depending on the well-known "biuret reaction" (Lowry [45]; BCA (bicinchoninic acid) [46]). These are the base for the commercial kit currently available from different suppliers: (a) Pierce 660 nm protein assay (Thermo), (b) Bradford protein assay (Bio-Rad), and (c) 2D Quant Kit (GE Healthcare) (https://tools.thermofisher.com/content/sfs/brochures/1602063-Protein-Assay-Handbook.pdf). The protocol for protein quantitation presented in this chapter is the Bradford method [44], which is carried out as follows:

1. Prepare the BSA master solution (0.25 μg/μL).
2. Following Table 1, distribute the master solution in 1.5 mL microtubes and add distilled water (up to 800 μL) in duplicate.
3. In parallel, prepare the protein samples as follows: (a) dilution 1 (1:100), 5 μL of proteins:495 μL of distilled water and (b) dilution 2 (1:2000), 40 μL of dilution 1:760 μL distilled water (*see* **Note 21**).
4. Add 200 μL of Bradford reagent. Mix and keep 5 min in the dark.
5. Mix again and measure spectrophotometrically (DO_{595}) the standards to obtain the standard curve and sample absorbance.

Table 1
Volumes to reach the standard amounts required to perform the Bradford calibration curve

Protein (μg)	0.0	0.5	1.25	2.5	3.75	5.0	6.25	7.5	8.75	10.0
Master sol. (μL)	0	2	5	10	15	20	25	30	35	40
Distilled water (μL)	800	798	795	790	785	780	775	770	765	760
Bradford reagent	200	200	200	200	200	200	200	200	200	200

3.4 Sample Integrity Validation

Some actinobacteria as *Streptomyces* species can produce colored antibiotics (e.g., actinorhodin, prodigiosins), which can remain along the whole protein extraction process. As a result, these and other metabolites can affect protein quantitation (under- or over-estimation). Besides, the protein integrity should be evaluated to detect possible protein degradation or heterogeneity among biological replicates, which should be highly similar. Thus, protein visualization in a monodimensional (1D) and subsequently in a small 2D gel [47, 48] is strongly recommended to check these parameters and validate the extraction procedures. It is done as follows:

1. Clean Mini-PROTEAN® 3 Cell including glass plates, spacers, gel caster, and equipment very thoroughly (*see* **Note 22**). Rinse with distilled water and absolute ethanol. If fluorescence labeling is considered, replace ethanol by a few drops of 10% SDS solution (*see* **Note 23**).
2. Place the glass plates (spacers also if these are not joined to the glasses) in the gel caster following the supplier instructions (*see* **Note 24**). A 0.75 mm thickness gel is enough for loading and visualization of 20 μg of proteins per well. When small 2D gels are used, 1.5 mm is the recommended thickness.
3. Mix the reagents of the running section solution (*see* Subheading 2.5), with the exception of ammonium persulfate (*see* **Note 25**). Mixture degasification by means of a Kitasato flask and a vacuum pump is always recommended. It is mandatory for large gels as the Ettan Dalt format (21 × 26 cm) (*see* **Note 26**).
4. Immediately before pouring the polyacrylamide solution, add the ammonium persulfate (*see* **Note 27**). Shake briefly and pour the solution to the gels. Keep space for the stacking solution and the comb if a 1D gel is being performed. Save a few milliliters as polymerization reporter.
5. Immediately after pouring the solution overlay each gel with a layer of water-saturated 1-butanol or 2-propanol to minimize gel exposure to oxygen and to create a flat gel surface (*see* **Note 28**). For 1D small gels 30–60 min of polymerization is enough. When bigger gels are used, overnight polymerization is recommended (*see* **Note 29**).

6. Prepare the solution of the stacking section (*see* Subheading 2.5), if needed [the strip of the isoelectrofocusing (IEF, first dimension) acts as staking gel when 2D gels are used].
7. After 30 min to 1 h [in function of gel size, room temperature, and the solidification of the reaming solution out of the gel (reporter)], the gel should be polymerized. Discard the covering solution (butanol or isopropanol) and add the stacking gel (for monodimensional gels). Place the comb for 1D gels over the running gel and fill the gaps with stacking gel solution. Avoid bubble formation by continuous addition of the solution until the solidification begins. Full polymerization usually takes 15–20 min.
8. If a 2D gel is done instead of the stacking gel, directly add running solution (10× Laemmli SDS electrophoresis buffer) and save at 4 °C.
9. Taking 20 μg of each protein sample, add denaturation buffer and boil for 5 min.
10. Place on ice, spin down the samples, and load into the gel wells located inside the running chamber (covered by running buffer).
11. When 2D gels are used, place the IEF strip laying over the running gel carrying 80–100 μg.
12. Apply 100–150 V continuous current along 1 h for small gels (Mini-PROTEAN format). When Ettan Dalt format gels are used, 3 W/gel, 45 min, and 20 W/gel during 3.5 h are recommended.
13. Run is completed when the bromophenol blue line (from denaturation buffer) reaches the bottom of the gel. Remove the gels from the running chamber, disassemble the glass plates, and proceed to stain. 20 μg of total proteins is easily visualized by using colloidal Coomassie staining [49, 50] (*see* **Note 30**).
14. Fix the proteins (*see* **Note 31**) with fixation solution for 30 min in a shaking platform.
15. Discard the fixation solution. Wash three times for 10 min with distilled water.
16. Cover the gels with colloidal Coomassie [39]. Keep overnight (*see* **Note 32**).
17. Save colloidal Coomassie (up to 6 months in the dark). Wash the gel with distilled water until background becomes transparent (1–2 h) (*see* **Note 33**). 0.75 mm thickness gels are easily destained (1–2 h).

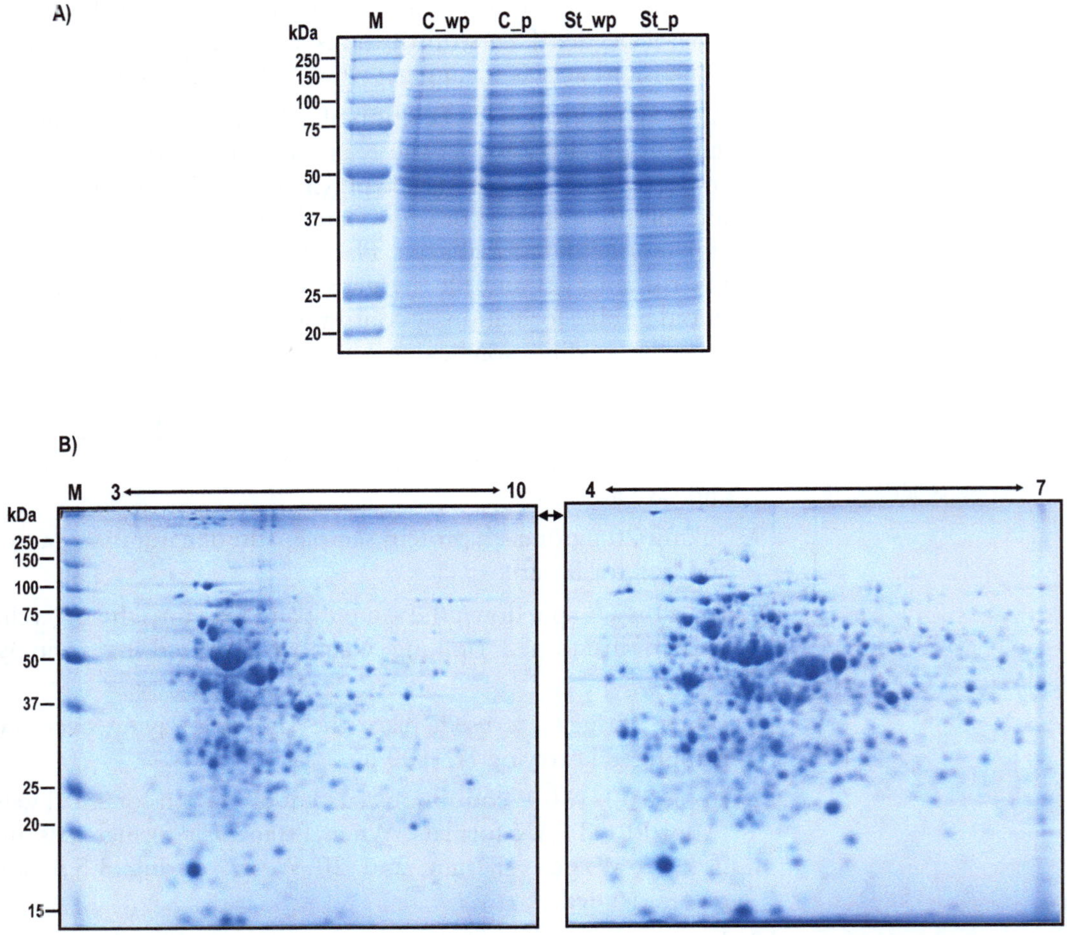

Fig. 2 Polyacrylamide gels for protein extraction method validation of intracellular proteins. (**a**) Monodimensional gel (1D) comparing samples from control (C) versus sterol (St)-supplemented conditions. Besides, the effect of the acetone precipitation was also tested (wp, without precipitation; p, precipitated). (**b**) Bidimensional (2D) gels using a pI range 3–10 NL (non-lineal) (left) and 4–7 (right). The loaded sample was the same without (left) and with (right) acetone precipitation. Note the big-size proteins are lost (double arrowhead middle upper part) when precipitation is carried out, as well as the dirtiest aspect of the proteins without precipitation in the central section (75–25 kDa) of the gel (left). M: Precision Plus Protein Standards (Barreiro et al. 2022)

18. Test the protein integrity by visual inspection, verify the similarities among biological replicates of the same condition, and validate the Bradford quantitation.

Once these protein extraction steps and the subsequent quantification and method validation have been carried out (Fig. 2), the protein samples are ready for bidimensional analyses (e.g., 2D-DIGE).

3.5 2D Clean-Up

Protein extraction and sample clean-up are the main steps to ensure an optimal resolution of 2D gels. This last step is designed to prepare the samples for an optimal labeling since salts, detergents, denaturants, and/or organic solvents should be removed. Thus, the 2D Clean-Up Kit can be used as follows, which yields clean samples and allows the subsequent pH adjustment prior to the DIGE labeling.

1. Transfer 200 μg (to guarantee the final of amount protein needed, i.e., at least 75 μg) of protein in a maximum volume of 100 μL to a new 1.5 mL microtube and add 300 μL of precipitant solution. Mix by vortexing and incubate on ice for 15 min.
2. Add 300 μL of co-precipitant and mix by vortexing again. Centrifuge at 12,000 × *g* and 4 °C for 5 min.
3. Remove the supernatant by pipetting, assuring to eliminate any remaining liquid to the bottom of the tubes.
4. Without disturbing the pellets, add 40 μL of co-precipitant on the top of the pellet and incubate on ice for 5 min.
5. Centrifuge at 12,000 × *g* and 4 °C for 5 min.
6. Remove the liquid assuring to eliminate any remaining liquid from the bottom of the tubes.
7. Resuspend the pellet in 100 μL of distilled water by pipetting or vortexing. It seems a milky solution not fully dissolved.
8. Add 1 mL of prechilled wash buffer (−20 °C) and 5 μL of wash additive and mix by vortexing until the pellet is fully dispersed, but not dissolved. Incubate at −20 °C for at least 30 min. Shake every 10 min if the incubation time is 30 min. Overnight incubation can also be considered. Centrifuge at 12,000 × *g* and 4 °C for 5 min and remove completely the supernatant by pipetting. Centrifuge the tubes to bring any remaining liquid to the bottom of the tube and ensure the supernatant elimination avoiding precipitation solutions (carrying TCA), which results in very acidic protein solutions.
9. Resuspend the pellet in 20 μL of resuspension buffer (RB2) supplemented with 30 mM Tris and centrifuge at 12,000 × *g* for 30 s to remove any insoluble particle and reduce any foam. Carefully adjust the pH to 8.0–8.5 with 100 mM NaOH. Measure the final pH by means of pH indicator strips (±1 μL per measurement).
10. Quantify the proteins samples by Bradford method (*see* Subheading 3.3).
11. Final solution should range between 5 and 10 μg/μL and pH 8.5, preferably.

3.6 2D-DIGE Analysis

DIGE (differential in gel electrophoresis) [51, 52] methodology is based on three spectrally resolvable fluorescent dyes (Cy2, Cy3, and Cy5), which contain an N-hydroxysuccinimidyl ester reactive group to form a covalent bond with amino groups. This property is used to label the lateral amino group of lysine residues of proteins.

In order to ensure minimal labeling (CyDye DIGE Fluor minimal dyes), the concentration of dye present in a protein labeling is limited, aimed to stain 1–2% of lysine residues [53]. The protocol is developed as follows:

1. Keep the dyes at RT for at least 10 min before using them to avoid water condensation.
2. Reconstitute CyDye DIGE Fluor minimal dyes with 99% anhydrous N,N-dimethylformamide (DMF) to a final concentration of 1 nmol/μL (*see* **Note 34**). Dilute the solution to 0.4 nmol/μL with DMF (*see* **Note 35**). If you selected CyDye DIGE Fluor Minimal Labeling Kit (5 nmol), which allows to label samples from 12 gels, the final volume should be 12.5 μL of DMF.
3. Add 1 μL of the dilute solution of Cy3 or Cy5 to a sample containing 50 μg of protein, one tube per sample, and replicate (*see* **Note 36**). For internal standard dye, add 25 μg of each analyzed replicate to a new microtube, all in the same tube, and add X μL of Cy2 (where X is the number of total analyzed gels). Mix by vortexing and centrifuge at 12,000 × *g* for 30 s.
4. Incubate all the tubes for 30 min on ice in the dark (labeling process). Stop the labeling reaction by adding 1 μL of 10 mM lysine, and incubate for 10 min on ice in the dark (*see* **Note 37**).
5. Use each gel to run two samples (replicates) of different condition (*see* **Note 38**). Mix the pair of samples labeled with Cy3 and Cy5 and proportionally add the internal standard. Add RB buffer supplemented with 1% IPG buffer and 80 mM DTT to a final volume of 350 μL. Centrifuge at 12,000 × *g* for 30 s to remove any possible insoluble particle and reduce any foam.
6. Place the mix into an 18 cm IPG strip holder. Place the strips in the holder, over the sample mix, with the acrylamide contacting the sample. Cover the strip with 2.0–2.5 mL of DryStrips Cover Fluid.
7. Run the first dimension at 100 μA/strip and 20 °C following this program: 1 h 0 V; 12 h 30 V (rehydration); 1 h 60 V; 1 h 500 V; 1 h 1000 V; 30 min gradient until 8000 V; and 8000 V until the samples reach 50,000 kVh (~6–7 h). If samples are properly obtained and labeled (e.g., salt-free), the run takes 23.5 h, approximately. Store the strips at −80 °C until use.

8. Clean glass plates, spacers, gel caster, and equipment very thoroughly (*see* **Note 22**). Rinse with distilled water and 10% SDS solution (*see* **Note 23**).
9. Place the low-fluorescence 24 cm glass plates in the gel caster (*see* **Note 39**).
10. Mix the reagents of the running solution (*see* Subheading 2.5), with the exception of ammonium persulfate (*see* **Note 25**). Mixture degasification, by means of a Kitasato flask and a vacuum pump, is highly recommended (*see* **Note 26**).
11. Immediately before pouring the polyacrylamide solution, add the ammonium persulfate (*see* **Note 27**). Shake briefly and pour the solution to the gel caster. Carefully, slightly hit the gel caster to ease the release of any air bubble before the final polyacrylamide polymerization, which could disturb the subsequent SDS-PAGE run.
12. Immediately after pouring the solution, overlay each gel with a layer of water-saturated 1-butanol or 2-propanol to minimize gel exposure to oxygen and to create a flat gel surface (*see* **Note 28**). Overnight polymerization is recommended (*see* **Note 29**).
13. Equilibrate each IEF strip with 10 mL of equilibration buffer supplemented with 1% DTT for 15 min under constant agitation. For the second equilibration step, replace DTT by 4% (w/v) iodoacetamide, and incubate again for 15 min under constant agitation. Wash the IEF strips with distilled water after the equilibration steps.
14. Insert the gels into the electrophoresis chamber and add 5 L of 1× lower electrophoresis buffer in the lower chamber. Create the upper buffer chamber of the Ettan Dalt System and fill in with 1.4 L of 1× upper electrophoresis buffer in the upper chamber. Place the strips on the top of each gel (*see* **Note 40**).
15. Run the electrophoresis at 3 W/gel for 45 min and 1.3 W/gel until 100 W (~12 h).
16. Run is completed when the bromophenol blue line reaches the bottom of the gel. Remove the gels from the running chamber, carry out the image acquisition, disassemble the glass plates, do a visible stain as control (e.g., Colloidal Coomassie), and proceed to image analysis (Fig. 3).

4 Notes

1. Bis-acrylamide (N,N′-methylene-bisacrylamide) allows the polymerization process (co-polymerization of acrylamide and bis-acrylamide) mediated by ammonium persulfate and TEMED. It is the responsible of the random crosslink of the

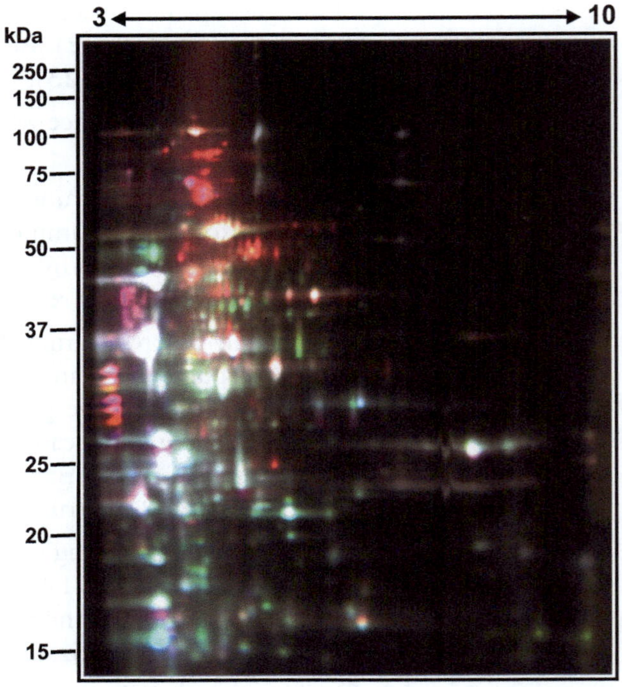

Fig. 3 2D-DIGE comparison of the *M. neoaurum* NRRL B-3805 secretome grown under control (glycerol) and induced (sterol) conditions. IEF was carried out by means of 3–10 NL (non-lineal) strips and the second dimension (SDS-PAGE) was 12.5% of polyacrylamide (Barreiro et al. 2022)

elongating polymer chains. This results in the characteristic gel porosity, which depends on the monomer concentrations and polymerization conditions.

2. TEMED is subject to oxidation, which changes its color from transparent to yellowish, decreasing the catalytic activity. The practical result is greater amounts required for an adequate polymerization [*see* Tech note no. 1156 from Bio-Rad: Acrylamide Polymerization—A Practical Approach (http://www.bio-rad.com/webroot/web/pdf/lsr/literature/Bulletin_1156.pdf)]. Thus, the renovation of the TEMED once per year is strongly suggested.
3. TCA (trichloroacetic acid) is easily dissolved in acetone at room temperature. The solution should be kept at −20 °C before use. Traditionally acetone precipitation is recommended in a ratio 4:9 per sample volume, which is unapproachable when secretome are analyzed due to the large volumes used in such analyses. Thus, TCA precipitation in a (1:1 vol/vol) ratio is used.

4. If the microorganism presents high secretion capability protein precipitation can be detected almost immediately and precipitation time can be decreased. On the contrary, if white lumps are not easily observed after the precipitant addition, keep at − 20 °C overnight.
5. TCA is an extremely acidic compound, and washing steps with acetone and subsequently with 80% acetone improve TCA trace elimination. It is important to avoid protein acidification, but it is mandatory if DIGE labeling is considered. DIGE optimal labeling conditions are pH 8.5 and 5–10 μg/μL of protein when minimal labeling is carried out. However, if TCA is present in the sample, the amount of NaOH needed to reach the optimal pH will dilute the sample behind the recommended concentration. Thus, a careful removal of the washing solutions also helps in the later pH equilibration.
6. Protein resuspension along the washing steps is not a "real resuspension." It means to detach the pellet from the bottom of the tube, but the proteins continue insoluble in acetone. Then, this "resuspension" is a kind of milky (preferable) to yellowish lumps floating in the solution. If proteins became totally or suddenly soluble in cold acetone solutions, something is going wrong (e.g., water solution instead of acetone was used).
7. When acetone is used, long drying period (up to 1 h, R/T) can be accepted without disturbing the protein integrity and solubilization. In contrast, when TCA is used as the precipitant agent, the over-dry is a real risk, which deeply affects the subsequent protein solubilization. TCA is a more aggressive precipitant than acetone, which results in the precipitation of some protein or stuffs hardly soluble. If additionally the protein pellet is over-dried, the final number of insoluble proteins could be increased.
8. Concentrated chaotropes as urea (or thiourea and urea combinations) in addition to one (or more) non-ionic or zwitterionic detergents are used to enhance the protein solubility. On the one hand, urea and thiourea denature (solubilize and unfold) most proteins to their fully random conformation. Thus, their ionizable groups are exposed to solution. On the other hand, zwitterionic detergents ensure complete sample solubilization, preventing protein aggregation through hydrophobic interactions (*see* "2-D Electrophoresis: Principles and Methods" [Ref. No. 80-6429-60AC (2005) or 80-6429-60 AD (2010); GE Healthcare]).
9. Urea solutions (high concentration), if rise above 30–37 °C, begin to form cyanates (isocyanate), which are detrimental for the proteins by carbamylation, resulting in artifactual "charge

trains" [*see* "Urea Solution 8 M; Cat. No: U4883" (SIGMA) and "2-D Electrophoresis: Principles and Methods" (GE Healthcare)]. Thus, protein solubilization should be done by means of shaking periods without any kind of temperature support.

10. Washing steps are capital for intracellular proteome analysis in order to avoid growth media interferences. The protein cleanliness influences in the subsequent isoelectrofocusing (IEF) process. For example, high salt concentrations (>50–100 mM) disturb the first dimension.
11. When several samples from different conditions, which delay in the growth speed, are collected, it is usual to develop the collection process at different times along the day. Thus, one washing step and a fast-frozen process are recommended. Later, when all the samples have been saved at −20 °C, a second washing step helps to thaw the samples all together and start the disruption process at the same time.
12. Frozen proteins can be stored at −20 °C for long periods of time. Up to 1 year could be possible. Acetone precipitation is also a good preservation solution for long periods of time. However, TCA could affect the protein solution if long periods of precipitation are used.
13. Washing and breaking buffers present the same composition to adapt (e.g., osmotic conditions) the cells to the breaking buffer in the last washing step. Note the low salt content (50 mM) used to avoid the interference in the IEF. In addition, the pH of the buffer (7.2–7.4) is similar to the typically described as intracellular.
14. Proteases may be liberated or activated upon cell disruption, which induce proteolysis when contacting their target protein. This fact greatly complicates the analysis of the 2D resulting gels. Protein manipulation at low temperatures decreases the protease activity. In addition, the presence of Tris, sodium carbonate, or basic carrier ampholyte mixtures in the breaking buffers can often inhibit the proteolysis. Anyway, the use of cocktails of protease inhibitors is highly recommended to inhibit those recalcitrant proteases. Several inhibitors or cocktails of inhibitors are commercially available. Along this work, cOmplete™ Protease Inhibitor Cocktail was used.
15. Matrix lysis is composed of glass beads (silica spheres). A 0.1 mm diameter is quite convenient for the protein extraction in actinobacteria. However, the appropriate bead diameter should be optimized before each extraction process.
16. Due to the violent movement of the FastPrep apparatus, it is highly recommended to keep the samples on ice among breaking steps. It avoids protein degradation by system overheating.

17. Nucleic acids interfere the 2D gel development because (a) they cause smear, (b) they increase sample viscosity, (c) they stick to the proteins disturbing focusing, (d) if silver staining is used, the nucleic acids are also detected (for a detailed description, see the manual from GE Healthcare: "2-D Electrophoresis. Principles and Methods"). Thus, one step of nucleotide degradation is highly recommended. Benzonase, a cocktail of DNase plus RNase, allows an effective viscosity reduction and removal of nucleic acids without disturbing the final protein analyses.
18. If possible, it is better to avoid protein precipitation since every additional step means a protein lost. Thus, it is highly recommended to test both options (with and without precipitation) when a new procedure is updated. Sometimes the number of proteins lost is recompensed by the better resolution of the 2D gels. Usually, these lost proteins correspond to high molecular mass spots. The validation of the protocol by 1D gel to test protein integrity and later by mini 2D gels [7 cm strips (IEF) in a Mini-PROTEAN format gel (SDS PAGE, 8 × 6 cm)] is enough to detect protein lost and resolution variations.
19. In contrast to the secretome (*see* **Note 1**), the intracellular proteome can be precipitated by the addition of acetone. The cells are concentrated in a small volume after centrifugation and the final collected product after the breaking process is also small. Acetone precipitation allows an easy protein pellet resuspension in contrast to the TCA, which always presents insoluble particles that should be discarded after the centrifugation.
20. After protein resuspension a short centrifuge spin is recommended to collect, if any, those insoluble particles. Usually, no particles are visible when acetone precipitation is used.
21. The yield of the described methods for protein extraction ranges from 2 to 15 μg/μL. Thus, the suggested dilutions allow a proper measurement in the linear section of a standard calibration curve. Based on our experience, concentrations below 1 μg/μL do not generate fully trustable measures by Bradford method.
22. It is recommended to work with SDS-PAGE using the cautions described for mass spectrometry, even though the final application of the gel is just for visualization. Also, for 1D gel only high-quality chemicals must be used. Every insignificant mistake in an 8 × 6 cm gel is strongly magnified in a 2D(-DIGE) gel (21 × 26 cm). In addition, current mass spectrometers are extremely sensitive, showing every contamination, which affects the final result and subsequent protein/modification identifications. Keratin stemming from hair, pet hair, wool clothes, and skin (typical nude finger pointing to the protein

band in the gel along the destaining process) is the most frequently contaminant source. Thus, it is a recommended practice to filter each buffer solution through a membrane filter (0.45 μm pore) before use. Powder-free gloves and paper tissue are both mandatory.

23. When the room temperature is higher than 20–25 °C, the gel polymerization, as well as the previous mixture, should be done using pre-cooled reagents (e.g., water, acrylamide, Tris buffers except the SDS, which became solid at low temperatures) in a cold room or a refrigerator. In contrast to the agarose gels, the polymerization of acrylamide runs slow at low temperatures, which allows better polymerization. High room, reagent, or glass plate temperatures cause extremely fast and irregular acrylamide polymerization.
24. When multiple gel handling is required, the use of glass plates carrying the spacers is a good option.
25. Ammonium persulfate is recommended to be prepared immediately prior to use (*see* 2D Electrophoresis. Principles and Methods), but some laboratory traditions support the stock conserved at −20 °C for long periods of time.
26. Oxygen can cause incomplete gel polymerization, and oxygen removal from the gel environment for 5–10 min of degas using a Kitasato flask and a vacuum pump is highly recommended.
27. Polyacrylamide polymerization process starts with ammonium persulfate as a catalyst, whereas TEMED is the tertiary amino group provider to release the radicals. Once the gel reagent mixture is completed, the shaking is done, and the degasification process (if considered) is finished, the ammonium persulfate can be added. Then, just around 10 min is available before the polymerization process shows its effects [54]. Proceed as fast as possible in a safe way.
28. Westermeier and co-workers [54] supported the 0.1% SDS spraying as the best option to reach straight gel surfaces. They explain it is due to "very small detergent solution droplets glide down on the glass surface very quickly and will not mix with the monomer solution, because they are very small and have a low density" in contrast to pipetting water-saturated butanol or isopropanol, since they have a high affinity to the vinyl spacer material and can cause bent surfaces. Anyway, the alcoholic solutions are used widely with successful results.
29. The polyacrylamide gel polymerization has an initially fast solidification step, which is well-known, but a slow silent polymerization also occurs along several hours to complete the final matrix formation. Incomplete gel polymerization causes high background in the mass spectrometer. Thus, the gels should be used not more than 2 h of polymerization or saved in the

refrigerator overnight (4 °C) [54]. When gels are stored overnight, a practical way is to store them at 4 °C for 1 h covered with isopropanol and replace it with 1× lower electrophoresis buffer to avoid isopropanol evaporation, which could cause over-dried gel surface and later will disturb the protein electrophoresis.

30. From the discovery of Coomassie as a wool dye in textile industry to its current use in staining protein, different adaptations have been carried out to reach the colloidal Coomassie dye employed nowadays (for a complete method review on Coomassie Blue staining of polyacrylamide gels, *see* [55]).
31. Protein denaturalization and precipitation in the gel matrix is the aim of the fixation. It prevents protein diffusion, which enhances the band resolution, as well as removes buffer components as SDS (for additional details see https://www.nationaldiagnostics.com/electrophoresis/article/protein-fixation-gels). In order to diminish a ester bond formation in the fixation buffer, by means of well-known chemical reactions of acid plus alcohol that yields an ester plus water, both acetic acid and methanol at 7% can be used, which decreases the final ester production.
32. Silver Blue is an endpoint staining [39]. It is an effective, low reagent consuming and fast staining protocol. Water destaining saves reagents because of it does not need the traditional ethanol and acetic acid in this destaining solution. Besides, thin gels (e.g., 0.75 mm) after 10 min of staining can be observed to evaluate how the proteins are. Moreover, the gel background is totally transparent after water destaining.
33. The higher the number of washing steps, the faster the gel destains.
34. Avoid direct sunlight on the dyes. After DMF addition the solution is stable for 2 months at −20 °C.
35. This solution is stable for 2 weeks at −20 °C.
36. Prepare each sample in an independent 1.5 mL microtube. Add the volume corresponding to 50 μg of the protein for each sample (5–10 μL of the pH-adjusted protein sample, which should range from 5 to 10 μg/μL). Label half of the tubes of one condition with Cy3 dye and the other half with Cy5 (dye swap). To ensure the same label for every sample, open the tubes and add the same volume of the dye inside the cap of each tube. When the correspondent dye is added to every sample in the same quantity, close the tubes and centrifuge them. If the volume is not enough for all the samples (e.g., poor micropipette calibration), collect the dye again from all the tube caps and start adding a lesser volume. Label replicates of the same condition with different dyes (dye swap).

37. The lysine overdose stops the labeling process since it depleted any dye excess that could continue. Thus, a putative cross-labeling is avoided when the different samples are mixed in the analysis steps (IEF, SDS-PAGE).
38. Two replicates from different condition run in each gel in addition to the proportional volume of the internal standard (Cy2 labeled control). So, three dyes (two replicates plus the standard) can be found in each first dimension and, subsequently, in the second dimension. Aimed to avoid mistakes, first make a loading table (e.g., Excel file) describing the samples (replicates) loaded in each strip/gel. Following the table locate the Cy3 labeled samples over the Cy5 labeled ones. Mix and spin down. Then, add the correspondent volume of Cy2 labeled standard. Again, to ensure the same amount for each sample, open all the tubes and add the same volume of the Cy2 standard inside the cap of each microtube. If the volume is not enough for all the samples (e.g., poor micropipette calibration), collect the Cy2 labeled standard again from all the microtube caps and start over, adding a lesser volume. If the Cy2 standard is enough, then close the tubes and harvest them by centrifugation.
39. Store the assembly glasses at 4 °C prior to use to ensure a low temperature, which guarantees the correct acrylamide polymerization.
40. Use 5 μL of Precision Plus Protein Standards as molecular weight marker. Since no comb is used, a 0.3 × 1.5 cm piece of Whatman 3 MM paper can be used to dispense the standard over the gel.

Acknowledgments

Special thanks to (a) the ESTELLA project ("DESign of bio-based Thermoset polymer with rEcycLing capabiLity by dynAmic bonds for bio-composite manufacturing") (Project no. 101058371) funded by the European Union through the Horizon Europe Framework Programme (call: HORIZON-CL4-2021-RESILIENCE-01-11) and (b) the BioPac project (*Development of bioactive and lifespan-controlled bioplastics*) (Ref. no. TED2021-131864B-C21) funded by the MCIN (*Ministerio de Ciencia e Innovación*)/AEI (*Agencia Estatal de Investigación*)/ 10.13039/501100011033 (Digital Object Identifier) and the European Union "NextGenerationEU"/PRTR (Recovery, Transformation and Resilience Plan).

Ana Ibañez is supported by a "Margarita Salas" modality postdoctoral grant (Reference no. UP2021-025) through the University of

León awarded by the Spanish Ministry of Universities within the Recovery, Transformation and Resilience Plan (Modernization and Digitalization of the Educational System), the funding of which comes from the European Recovery Instrument European Union-NextGenerationEU.

References

1. Bott M (2007) Corynebacteria: the good guys and the bad guys. Microbiol Today 34:74–77
2. Barreiro C, Martín JF (2015) *Corynebacterium* genus, a dual group of clinical and industrial relevant bacteria. In: Barreiro C (ed) New trends in *Corynebacterium glutamicum*: beyond the amino acids. Nova Science Publishers, Inc, New York
3. Shtratnikova VY, Bragin EY, Dovbnya DV et al (2014) Complete genome sequence of steroltransforming *Mycobacterium neoaurum* strain VKM Ac-1815D. Genome Announc 2:12–13
4. Tortoli E (2014) Microbiological features and clinical relevance of new species of the genus *Mycobacterium*. Clin Microbiol Rev 27:727–752
5. Euzéby JP List of Prokaryotic names with standing in nomenclature (https://www.bacterio.net/)
6. Goodfellow M, Kämpfer P, Busse H-J et al (2012) Bergey's manual of systematic bacteriology, The Actinobacteria, part B, vol 5. Springer
7. Cole ST, Brosch R, Parkhill J et al (1998) Deciphering the biology of *Mycobacterium tuberculosis* from the complete genome sequence. Nature 393:537–544
8. Mattow J, Siejak F, Hagens K et al (2009) Two-dimensional gel electrophoresis-based proteomics of mycobacteria. Methods Mol Biol 465:111–142
9. Uhía I, Galán B, Kendall SL et al (2012) Cholesterol metabolism in *Mycobacterium smegmatis*. Environ Microbiol Rep 4:168–182
10. Liu M, Zhu Z-T, Tao X-Y et al (2016) RNA-Seq analysis uncovers non-coding small RNA system of *Mycobacterium neoaurum* in the metabolism of sterols to accumulate steroid intermediates. Microb Cell Fact 15:64
11. Cornejo-Granados F, López-Leal G, Mata-Espinosa DA et al (2021) Targeted RNA-Seq reveals the *M. tuberculosis* transcriptome from an *in vivo* infection model. Biology 10:848
12. Korb VC, Chuturgoon AA, Moodley D (2016) *Mycobacterium tuberculosis*: manipulator of protective immunity. Int J Mol Sci 17:131
13. Kumar B (2015) World Leprosy Day 2015: renewing commitment for a leprosy free world! Indian J Med Res 141(1):1–4
14. Bisht D, Singhal N, Sharma P et al (2006) Analysis of mycobacterial strains by two-dimensional gel electrophoresis. J Commun Dis 38:255–262
15. Betts JC, Dodson P, Quan S et al (2000) Comparison of the proteome of *Mycobacterium tuberculosis* strain H37Rv with clinical isolate CDC 1551. Microbiology 146:3205–3216
16. Jungblut PR, Schaible UE, Mollenkopf HJ et al (1999) Comparative proteome analysis of *Mycobacterium tuberculosis* and *Mycobacterium bovis* BCG strains: towards functional genomics of microbial pathogens. Mol Microbiol 33: 1103–1117
17. Kruh NA, Troudt J, Izzo A et al (2010) Portrait of a pathogen: the *Mycobacterium tuberculosis* proteome *in vivo*. PLoS One 5:e13938
18. Schmidt F, Donahoe S, Hagens K et al (2004) Complementary analysis of the *Mycobacterium tuberculosis* proteome by two-dimensional electrophoresis and isotope-coded affinity tag technology. Mol Cell Proteomics 3:24–42
19. Gupta MK, Subramanian V, Yadav JS (2009) Immunoproteomic identification of secretory & subcellular protein antigens & functional evaluation of the secretome fraction of *Mycobacterium* immunogenum a newly recognized species of the *Mycobacterium chelonae-Mycobacterium abscessus*. J Proteome Res 8:2319–2330
20. Kruh-Garcia NA, Murray M, Prucha JG et al (2014) Antigen 85 variation across lineages of *Mycobacterium tuberculosis*-implications for vaccine and biomarker success. J Proteome 97:141–150
21. Mattow J, Schaible UE, Schmidt F et al (2003) Comparative proteome analysis of culture supernatant proteins from virulent *Mycobacterium tuberculosis* H37Rv and attenuated M. bovis BCG Copenhagen. Electrophoresis 24:3405–3420
22. Ranganathan S, Garg G (2009) Secretome: clues into pathogen infection and clinical applications. Genome Med 1:113

23. Yadav JS, Gupta M (2012) Secretome differences between the taxonomically related but clinically differing mycobacterial species *M. abscessus* & *M. chelonae*. J Integr OMICS 2:64–79
24. Mollenkopf HJ, Jungblut PR, Raupach B et al (1999) A dynamic two-dimensional polyacrylamide gel electrophoresis database: the mycobacterial proteome via Internet. Electrophoresis 20:2172–2180
25. He Z, De Buck J (2010) Cell wall proteome analysis of *Mycobacterium smegmatis* strain MC2 155. BMC Microbiol 10:121
26. Sih CJ, Wang KC (1965) A new route to estrone from sterol. J Am Chem Soc 87: 1387–1388
27. Al Jasem Y, Khan M, Taha A et al (2014) Preparation of steroidal hormones with an emphasis on transformations of phytosterols and cholesterol – a review. Mediterr J Chem 3:796–830
28. Donova MV, Egorova OV (2012) Microbial steroid transformations: current state and prospects. Appl Microbiol Biotechnol 94:1423–1447
29. Fernández-Cabezón L, Galán B, García JL (2018) New insights on steroid biotechnology. Front Microbiol 9:958
30. Wang F, Yao K, Wei D (2011) From soybean phytosterols to steroid hormones. In: El-Shemy H (ed) Soybean and health. InTech, Rijeka, Croatia, pp 231–252
31. Bragin EY, Shtratnikova VY, Dovbnya DV et al (2013) Comparative analysis of genes encoding key steroid core oxidation enzymes in fast-growing *Mycobacterium* spp. strains. J Steroid Biochem Mol Biol 138:41–53
32. Rodríguez-García A, Fernández-Alegre E, Morales A et al (2016) Complete genome sequence of "*Mycobacterium neoaurum*" NRRL B-3805, an androstenedione (AD) producer for industrial biotransformation of sterols. J Biotechnol 224:64–65
33. Barreiro C (2015) Methods in proteomics applied to *Corynebacterium glutamicum*. In: Barreiro C (ed) New trends in *Corynebacterium glutamicum*: beyond the amino acid. Nova Science Publishers, Inc, New York
34. Barreiro C, Martín JF, García-Estrada C (2012) Proteomics methodology applied to the analysis of filamentous fungi – new trends for an impressive diverse group of organisms. In: Prasain JK (ed) Tandem mass spectrometry – applications and principles. InTech, Rijeka, Croatia
35. Jeffery CJ (2014) An introduction to protein moonlighting. Biochem Soc Trans 42:1679–1683
36. Jeffery CJ (2009) Moonlighting proteins-an update. Mol BioSyst 5:345–350
37. Marsheck WJ, Kraychy S, Muir RD (1972) Microbial degradation of sterols. Appl Microbiol 23:72–77
38. Josefsen KD, Nordborg A, Sletta H (2017) Bioconversion of phytosterols into androstenedione by *Mycobacterium*. In: Barredo JL, Herráiz I (eds) Microbial steroids: methods and protocols. Humana Press, New York, pp 177–197
39. Candiano G, Bruschi M, Musante L et al (2004) Blue silver: a very sensitive colloidal Coomassie G-250 staining for proteome analysis. Electrophoresis 25:1327–1333
40. Greenbaum D, Luscombe NM, Jansen R et al (2001) Interrelating different types of genomic data, from proteome to secretome: 'oming in on function. Genome Res 11:1463–1468
41. Makridakis M, Roubelakis MG, Vlahou A (2013) Stem cells: insights into the secretome. Biochim Biophys Acta 1834:2380–2384
42. Tjalsma H, Bolhuis A, Jongbloed JD et al (2000) Signal peptide-dependent protein transport in *Bacillus subtilis*: a genome-based survey of the secretome. Microbiol Mol Biol Rev 64:515–547
43. Fischer M, Sawers RG (2013) A universally applicable and rapid method for measuring the growth of *Streptomyces* and other filamentous microorganisms by methylene blue adsorption-desorption. Appl Environ Microbiol 79:4499–4502
44. Bradford MM (1976) A rapid and sensitive method for the quantitation of microgram quantities of protein utilizing the principle of protein-dye binding. Anal Biochem 72:248–254
45. Lowry OH, Rosebrough NJ, Farr AL et al (1951) Protein measurement with the Folin phenol reagent. J Biol Chem 193:265–275
46. Smith PK, Krohn RI, Hermanson GT et al (1985) Measurement of protein using bicinchoninic acid. Anal Biochem 150:76–85
47. Laemmli UK (1970) Cleavage of structural proteins during the assembly of the head of bacteriophage T4. Nature 227:680–685
48. O'Farrell PH (1975) High resolution two-dimensional electrophoresis of proteins. J Biol Chem 250:4007–4021

49. Fazekas de St Groth S, Webster RG, Datyner A (1963) Two new staining procedures for quantitative estimation of proteins on electrophoretic strips. Biochim Biophys Acta 71:377–391
50. Meyer TS, Lamberts BL (1965) Use of coomassie brilliant blue R250 for the electrophoresis of microgram quantities of parotid saliva proteins on acrylamide-gel strips. Biochim Biophys Acta 107:144–145
51. Unlü M, Morgan ME, Minden JS (1997) Difference gel electrophoresis: a single gel method for detecting changes in protein extracts. Electrophoresis 18:2071–2077
52. Timms JF, Cramer R (2008) Difference gel electrophoresis. Proteomics 8:4886–4897
53. Gorg A (2004) 2-D electrophoresis. Principles and methods. GE Healthcare
54. Westermeier R, Naven T, Hopker H-R (2008) Proteomics in practice: a guide to successful experimental design, 2nd edn, completely revised edition. WILEY-VCH Verlag GmbH & Co. KGaA, Weinheim
55. Westermeier R (2006) Sensitive, quantitative, and fast modifications for Coomassie Blue staining of polyacrylamide gels. Proteomics 6 Suppl 2:61–64

Chapter 8

Fungal Sterol Analyses by Gas Chromatography–Mass Spectrometry Using Different Derivatives

Aurélie Mossion, Isabelle Ourliac-Garnier, and Gaetane Wielgosz-Collin

Abstract

Sterols are the main components of the fungal membrane. Their study can be used to describe the chemical biodiversity among the strains and species or to work on antifungal treatment. Those molecules can be analyzed by gas chromatography coupled with mass spectrometry (GC–MS) as free molecules or after derivation as acetate or trimethylsilyl ether (TMSi). Here we describe how to extract sterols from fungal biomass according to its physiological form and the culture conditions (liquid and solid media). The different methodologies that can be used to obtain free sterols, acetate, and/or TMSi derivatives are presented. Identification keys using the fragmentation at 70 eV are also described.

Key words Fungal sterols, Extraction, Saponification, Acetate, TMS derivatives, GC–MS, Identification

1 Introduction

Sterols are a class of hydrophobic ringed lipid molecules that plays a crucial role in the organization and functioning of all eukaryotic organisms since they determine the permeability and fluidity of the membrane [1, 2]. Although sterols are present in all organisms, a great diversity can be observed depending on the kingdom studied.

Fungal sterol analysis can be achieved with two main different goals. The first one is to study the huge diversity of molecules which can be expressed by fungi when grown on different media using the one strain many compounds approach (OSMAC). For example, this strategy has been used to study fungal sterols and allowed to isolate eburicol and evaluate its biological activities [3].

The second goal is to help understanding why some fungal pathogenic strains may develop resistance against azole drugs. In fact, azole drugs are one of the five classes of compounds in the

Carlos Barreiro and José-Luis Barredo (eds.), *Microbial Steroids: Methods and Protocols*, Methods in Molecular Biology, vol. 2704, https://doi.org/10.1007/978-1-0716-3385-4_8,

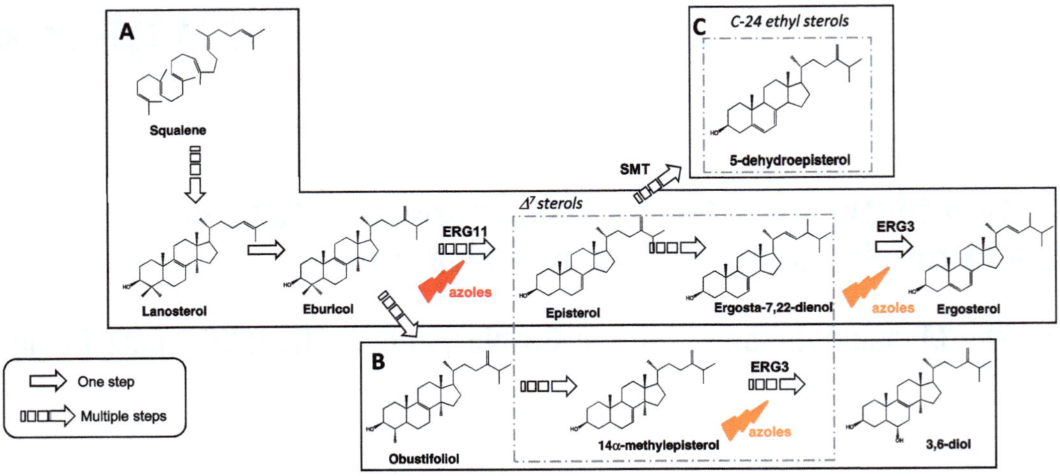

Fig. 1 Schematic representation of ergosterol biosynthesis pathways in fungal pathogens. (A) Sterol main intermediates from squalene to ergosterol. Sterol intermediates accumulate with azole inhibition of ERG11, 14α-lanosterol demethylase, and (B) lead to 3,6 diol that is toxic for the cell. ERG3, sterol $\Delta^{5,6}$ desaturase, is a secondary target of azole compounds which inhibition leads to accumulation of Δ^{7}-sterols and can (C) drive to an increased Δ^{24}-sterol methyltransferase activity (SMT)

antifungal arsenal [4]. Fluconazole, itraconazole, voriconazole, posaconazole, and isavuconazole are the first-line prevention and treatment of *Aspergillus* infections [5] and an alternative initial therapy for *Candida* infections [6] that affect millions of people around the world and are associated with a high mortality rate, particularly in immunocompromised patient. Azole derivatives are also used in agriculture [7]. They act by inhibiting the activity of the enzyme 14α-lanosterol demethylase [8], which plays a key role in ergosterol biosynthesis (Fig. 1). Overuse, long treatment courses and environmental exposure of azoles in the past decade have resulted in drug resistance [9] but also in the selection and emergence of pathogens that have innate resistance.

There is a true need for the development of new antifungals because (i) the antifungal therapeutic arsenal is limited and the compounds are highly toxic [10] and (ii) either intrinsic or acquired antifungal resistance has raised in the past decades [11]. Antifungal drug discovery is a real challenge because fungal pathogens use much of the same machinery as human. Being selective against fungal pathogens is one of the challenges faced when targeting ergosterol biosynthesis since human membrane is composed of cholesterol that shares only part of the biosynthesis pathway with ergosterol.

Elucidating sterol composition in azole-susceptible and azole-resistant strains is of interest in basal conditions and under antifungal treatments to understand (i) innate or acquired mechanism of resistance [12] and (ii) mechanism of action of new drugs targeting ergosterol biosynthesis [13, 14].

To determine the sterol composition, the first step is to extract the sterols from the biomass using saponification. Then, gas chromatography coupled with mass spectrometry is the most commonly analytical method used to identify the molecules as free sterols and/or after derivation.

2 Materials

Prepare all the reagents using ultrapure water (18 MΩ cm at 25 °C) and analytical grade reagents. Prepare and store in glass vessels only at room temperature all the solutions (unless indicated otherwise). Waste materials must be treated following local rules.

2.1 Reagents

2.1.1 Fungal Strains

1. *Aspergillus fumigatus* (ATCC204305 or any strain identified as *Aspergillus fumigatus*).
2. *Candida* spp. (ATCC22019 or any strain identified as *Candida* sp.) (*see* **Note 1**).

2.1.2 Media

1. Sabouraud agar: Weigh 40 g of dextrose, 10 g of peptone, and 15 g of agar and add 1 L of distilled water; adjust pH at 5.6 (at 25 °C). Autoclave to sterilize. Cool to 45/50 °C and pour into Petri dishes. Store at 4 °C (*see* **Note 2**).
2. YPD 20% glucose broth: Weigh 5 g of yeast extract and 10 g of peptone and dissolve in 450 mL of distilled water. Autoclave to sterilize and add 50 mL of sterile glucose 10× (20%) just before the first use. Store at 4 °C.
3. Glucose 10× (20%): Dissolve 200 g of D-glucose in 1 L of distilled water. Autoclave or filter to sterilize. Store at 4 °C for no more than 1 year.
4. PBS 10×: Phosphate-buffered saline solution concentrated 10× contains 1.37 M NaCl, 27 mM KCl, 100 mM Na_2HPO_4, and 18 mM KH_2PO_4 with a pH of 6.8. Weigh 17.8 g of Na_2HPO_4, $2H_2O$, 2.4 g of KH_2PO_4, 80 g of NaCl, and 2 g of KCl. Adjust final volume to 1 L and pH to 6.8. Filter to sterilize and store at 4 °C.
5. PBS 1×: Take 100 mL of PBS 10× and adjust the volume at 1 L with distilled water. Verify that pH is 7.4. Filter to sterilize and store at 4 °C.
6. Tween 20 (0.1%): Add 10 μL of Tween 20 to 10 mL of sterile distilled water. Homogenize. Keep the solution at 4 °C.

2.1.3 Saponification

1. KOH in ethanol 3.8 M: To prepare extemporaneously 50 mL of ethanolic potassium hydroxide solution (3.8 M), weigh 12.5 g of KOH (86%), add 18 mL of distilled water, and make up to 50 mL with absolute ethanol. Don't keep it for the next experiment (*see* **Note 3**).

2. Hexane.
3. Methylene chloride (CH_2Cl_2).
4. Anhydrous Na_2SO_4.

2.1.4 Acetylation

1. Acetic anhydride.
2. Pyridine.
3. Methylene chloride (CH_2Cl_2).
4. HCl 2 M: Prepare 83.5 mL of water in a graduated cylinder. Put around 50 mL of water in a glass bottle. Carefully add 16.5 mL of HCl 37%. Homogenize and add the rest of water.
5. Saturated $NaHCO_3$ solution: Prepare 10 mL of water. Add NaHCO3(s) until crystals remain in the solution.
6. Anhydrous Na_2SO_4.

2.1.5 Trimethylsilylation

1. *N*-methyl-*N*-(trimethylsilyl)trifluoroacetamide (CAS 24589-78-4).
2. Cholesterol 10 μg/mL in hexane: Weigh 100 μg of cholesterol (>99%, CAS 57-88-5) into a glass vial with a cork and add 10 mL of hexane. Homogenize and keep at 4 °C.
3. Methylene chloride (CH_2Cl_2) GC grade.

2.1.6 GC–MS

1. Hexane.
2. Methylene chloride (CH_2Cl_2) GC grade.

2.2 Common Equipments

2.2.1 Inoculation

1. Sterile glass Pasteur pipets.
2. Sterile oses.
3. Sterile Erlenmeyers.
4. Petri dishes.

2.2.2 Biomass Obtention

1. 50 mL Falcon test tube.
2. Büchner.
3. Whatman filter paper N°1.
4. Scalpel.

2.2.3 Saponification

1. Pyrex tube.
2. Heating block.
3. Centrifuge.
4. Glass Pasteur pipet.
5. Nitrogen flux (or at least compressed air).

2.3 Lab Equipment

2.3.1 Extraction and Derivation

1. Horizontal laminar flow hood.
2. Chemical fume hood.
3. Cell culture incubator (for culture on solid medium).
4. Orbital shaking incubator (for culture in liquid medium).
5. Refrigerated centrifuge.
6. Oven.

2.3.2 GC–MS

1. Gas chromatograph coupled with mass spectrometer with specific software.
2. Carrier gas: Helium.
3. Column: SLB-5™ (60 m × 0.25 mm × 0.25 μm) or equivalent.
4. Detector: 280 °C/70 eV.
5. Vials.

3 Methods

3.1 Cell Culture

Under horizontal laminar flow hood.

3.1.1 On Solid Medium

1. Prepare a conidial suspension with 1.10^6 cells/mL in sterile water.
2. Inoculate a 10 cm Petri dish with 500 μL of the cell suspension and incubate 10 days at 27 °C.
3. Harvest the biomass by scrapping it off using a scalpel or a scrapper.
4. Dry until stable weight at 65 °C.

3.1.2 Aspergillus fumigatus *in Liquid Medium*

1. From a 24 h culture on Sabouraud agar plate at 37 °C, prepare a 4.10^7 spores/mL suspension in sterile water with Tween 20 (0.1%).
2. Inoculate 200 μL of the suspension to 25 mL of YPD 20% glucose broth and incubate 48 h at 37 °C under agitation (130 rpm) (*see* **Note 4** for antifungal treatment).
3. Harvest the mycelia obtained by filtration on a Büchner with two layers of Whatman filter paper (n°1), wash two times with distilled water, dry, transfer into a Pyrex test tube with a cap (20 mL) and weigh the biomass obtained (*see* **Note 5**).

3.1.3 Candida *spp. in Liquid Medium*

1. Make a 24 h pre-culture on Sabouraud agar plate at 35 °C.
2. Inoculate one colony into 25 mL of YPD 20% glucose broth and incubate 24 h at 30 °C under agitation (130 rpm) (*see* **Note 4**).

3. Transfer the culture medium into a 50 mL Falcon tube and centrifuge at 1236 × *g* for 5 min at room temperature.
4. Wash two times with 5 mL PBS 1×.
5. Transfer the biomass into a Pyrex test tube with a cap (20 mL) and weigh the biomass obtained (*see* **Notes 5** and **6**).

3.2 Saponification

1. Under the chemical hood, add 5 mL of KOH in 3.8 M ethanol, into the Pyrex test tube containing the fungal biomass. Close the tube and heat at 80 °C for 90 min.
2. After cooling, add 1 mL of water and 2.5 mL of hexane in the test tube. Mix gently to extract sterols in hexane. Centrifuge at 1236 × *g* for 5 min (*see* **Note 7**).
3. Transfer the hexane phase (upper one) to another test tube using a glass Pasteur pipet. Add enough anhydrous Na_2SO_4 to dry the organic phase and mix gently. Centrifuge at 1236 × *g* for 5 min. Transfer the supernatant in a pre-weighted test tube and measure the volume (*see* **Note 7**).
4. To analyze free sterols: Dry under a stream of nitrogen. Weigh. Prepare a solution of 1 mg/mL in CH_2Cl_2. Use 1 mL to analyze in GC–MS and keep the remaining solution at −20 °C before undergoing derivation.

3.3 Derivation

Different derivatives can be used depending on the objectives of the study. Work under the chemical hood.

3.3.1 Acetylation

1. If the response is correct for the free sterols, dry the remaining solution (*see* Subheading 3.2, **step 4**). Add 500 μL of anhydride acetic and 2 drops of pyridine.
2. Leave 1 night at room temperature or heat for 90 min at 80 °C.
3. Add 500 μL of water and 3 mL of CH_2Cl_2. Centrifuge at 1236 × *g* for 5 min.
4. Discard the aqueous phase (upper one). Add 1 mL of HCl 2 M, mix, and centrifuge at 1236 × *g* for 5 min.
5. Discard the aqueous phase (upper one). Add 1 mL of saturated $NaHCO_3$, mix, and centrifuge at 1236 × *g* for 5 min.
6. Discard the aqueous phase (upper one) and add enough anhydrous Na_2SO_4 to dry the organic phase. Centrifuge at 1236 × *g* for 5 min.
7. Transfer the supernatant in a pre-weighed test tube, dry under a stream of nitrogen, and weigh.
8. Prepare a solution at 1 mg/mL in CH_2Cl_2 to analyze the acetylated sterols by GC–MS.

3.3.2 Trimethylsilylation

1. Collect a volume of the solution from Subheading 3.2, **step 3** equivalent to 100 mg of the initial biomass. Add 2.5 μg cholesterol per sample as an internal standard, i.e., 250 μL of cholesterol solution at 10 μg/mL. Dry by heating at 80 °C.
2. After cooling, incubate with 100 μL of *N*-methyl-*N*-trimethylsilyltrifluoroacetamide (TMSi) during 30 min at room temperature.
3. Evaporate the solvent at 80 °C under air flux.
4. After cooling, add 200 μL CH_2Cl_2 to solubilize dried TMSi derivatives, transfer to a glass chromatography vial, and store at −20 °C until analysis.

3.4 GC–MS Analyses

3.4.1 General Parameters

Injector: 250 °C.

Injection: 1 μL in splitless mode.

Detector: 280 °C/70 eV.

3.4.2 Free and Acetylated Sterols

Gaz: He at 1 mL/min.

Oven: 200 °C and then 3 °C/min until 310 °C.

3.4.3 TMSi Ether Derivatives

Gaz: He at 1.2 mL/min.

Oven: 150 °C then until 290 °C at 50 °C/min and from 290 °C to 305 °C at 2 °C/min, hold for 7 min, and finally 10 °C/min to 315 °C and hold for 16 min.

The protocol is resumed in Fig. 2.

4 Interpretation Keys

General sterol analysis using GC–MS has deeply been described [15] and also for TMSi derivatives in pathogenic fungi [16, 17]. Here are presented some of the main interpretation keys used for the analysis of fungal sterols. The main sterols that can be identified in pathogenic or nonpathogenic fungi are summarized in Table 1, and their chemical structures are detailed in Fig. 3.

When analyzing sterol mass spectrum, the first important thing to identify is the molecular ion, named $[M]^+$, which allows the total number of carbons and unsaturations to be deduced. Then, the fragmentation of the side chain and rings gives the position of the unsaturations. The base peak (100% of intensity) also helps to identify the sterol under study. It is noteworthy that squalene, sterol precursor (Fig. 1) is observed in *Candida glabrata* and *Candida auris* samples, for example, with m/z value of $[M]^+$ 410.

The main fragmentation ions encountered for free sterols, acetates, and TMSi are explained in Fig. 4.

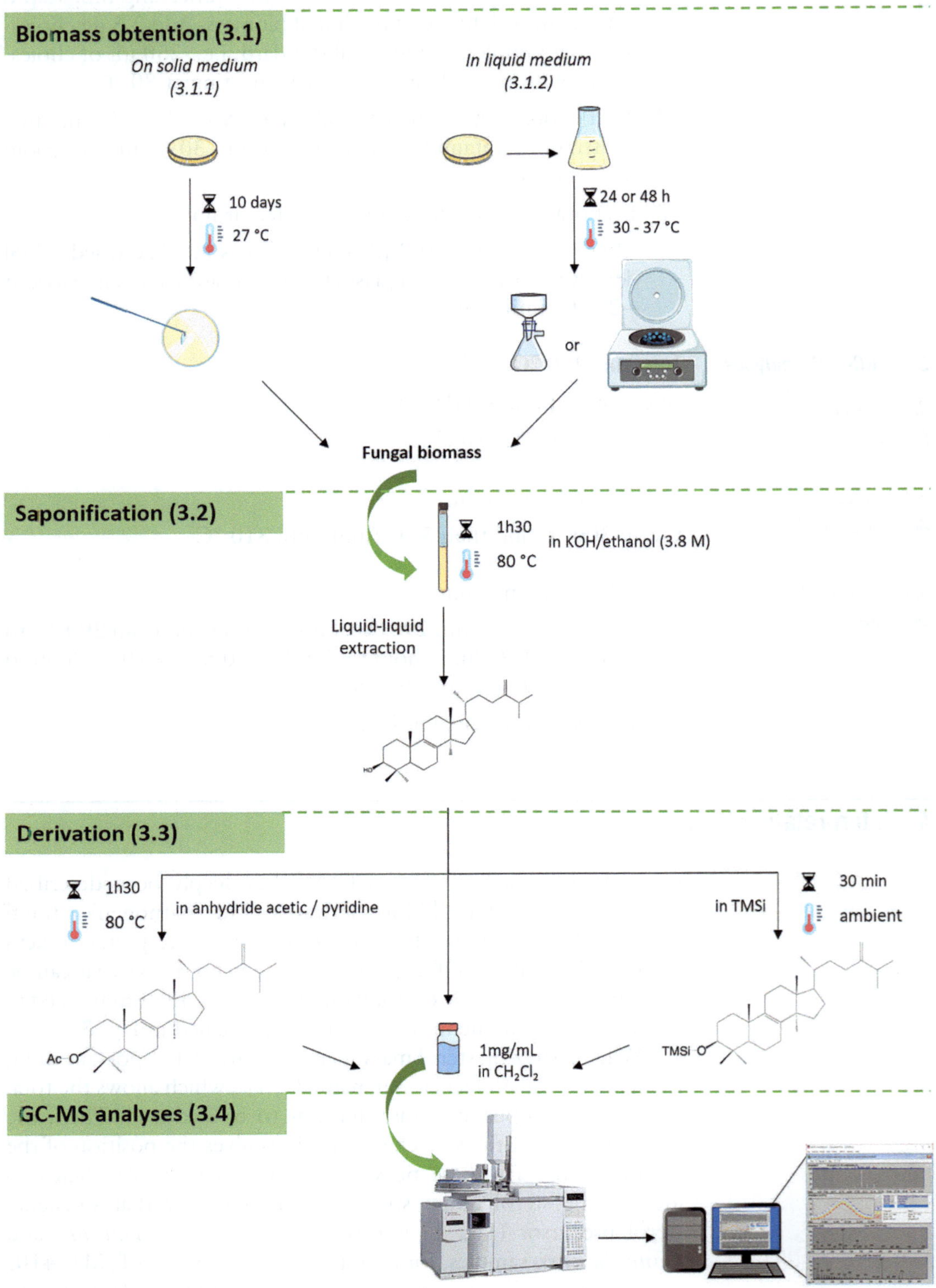

Fig. 2 Brief protocol to analyze fungal sterols

Table 1
Main sterols described in pathogenic or nonpathogenic fungi with their total number of carbons and unsaturations, the characteristic ions for free molecules and acetates and TMSi derivatives encountered when analyzed by GC–MS

Total number of carbon and unsaturations	Common name	IUPAC name	Structure (Fig. 3)	Free sterols $[M^{+}]$	Acetates $[M\text{-}OAc]^{+}$ or $[M\text{-}60]^{+*}$	TMSi $[M\text{-}TMSi]^{+}$
27:2	Zymosterol	Cholesta-8,24(28)-dien-3β-ol	6a	384	426	456
27:4	Neoergosterol	19-Nor-24-methylcholest-5,7,9,22-tetraen-3β-ol	4e	380	362*	452
28:1	Fungisterol	24-methylcholesta-7-en-3β-ol or Ergosta-7-en-3β-ol	2d	400	442	472
28:2	Episterol	Ergosta-7,24(28)-dien-3β-ol	2b	398	440	470
28:2	Stellasterol or 5-dihydroergosterol	24-Methylcholesta-7,22-dien-3β-ol or Ergosta-7,22-dien-3β-ol	2e	398	440	470
28:2	Fecosterol	Ergosta-8,24(28)-dien-3β-ol	6b	398	440	470
28:2	4-Methylzymosterol	4α-Methyl-cholesta-8,24(28)-dien-3β-ol	9a	398	440	470
28:3	5-Dehydroepisterol	24-Methylcholesta-5,7,24(28)-trien-3β-ol Ergosta-5,7,24(28)-trien-3β-ol	3b	396	378*	468
28:3	Ergosterol[a]	24-Methylcholesta-5,7,22-trien-3β-ol or Ergosta-5,7,22-trien-3β-ol	3e	396	378*	468
28:3		24-Methylcholesta-8,22,24(28)-trien-3β-ol Ergosta-8,22,24(28)-trien-3β-ol	6f	396	438	468
28:4	9-Dehydroergosterol	24-Methylcholesta-5,7,9(11),22-tetraen-3β-ol or Ergosta-5,7,9(11),22-tetraen-3β-ol	5e	394	376*	466
29:2	14α-Methylepisterol	14α,24-Dimethylcholesta-7,24(28)-dien-3β-ol or 14-methylergosta-7,24(28)-dien-3β-ol	7b	412	454	484
29:2	4-Methylfecosterol	4α,24-Dimethylcholesta-8,24(28)-dien-3β-ol or 4α-methylergosta-8,24(28)-dien-3β-ol	9b	412	454	484

(continued)

Table 1
(continued)

Total number of carbon and unsaturations	Common name	IUPAC name	Structure (Fig. 3)	Free sterols $[M^+]$	Acetates $[M\text{-}OAc]^+$ or $[M\text{-}60]^{+*}$	TMSi $[M\text{-}TMSi]^+$
29:2	14α-Methylfecosterol	14α,24-Dimethylcholesta-8,24(28)-dien-3β-ol or 14α-Methylergosta -8,24(28)-dien-3β-ol	10b	412	454	484
29:2	T-MAS	4,4-Dimethylcholesta-8,24(28)-dien-3β-ol	11b	412	454	484
29:2	3,6-Diol	14α,24-Dimethylcholesta-8,24(28)-dien-3β,6α-diol 14α-Methylergosta -8,24(28)-dien-3β,6α-diol	12b	428	512	572
29:3	7-Dehydrostigmasterol	24-Ethyl-cholesta-5,7,22-trien-3β-ol	3c	410		482
29:3	FF-MAS	4,4-Dimethylcholesta-8,14,24(28)-trien-3β-ol	13b	410		482
30:2	Lanosterol	4α,4α,14α-Trimethylcholesta-8,24-dien-3β-ol	1a	426		498
30:2	Obtusifoliol[b]	4α,14α,24-Trimethylcholesta-8,24(28)-dien-3β-ol or 4α,14α-dimethylergosta-8,24(28)-dien-3β-ol	8b	426		498
31:2	Eburicol	4α,4ß,14α,24-Tetramethylcholesta-8,24(28)-dien-3β-ol or 4α,4ß,14α-trimethylergosta-8,24(28)-dien-3β-ol	1b	440		512

[a]Ergosterol can be detected as isomer Z or E with retention times close but different
[b]4α,14α,24-Trimethylcholesta-8,24(28)-dien-3ß-one (obtusifolione) can be identified with $[M]^+$ 424

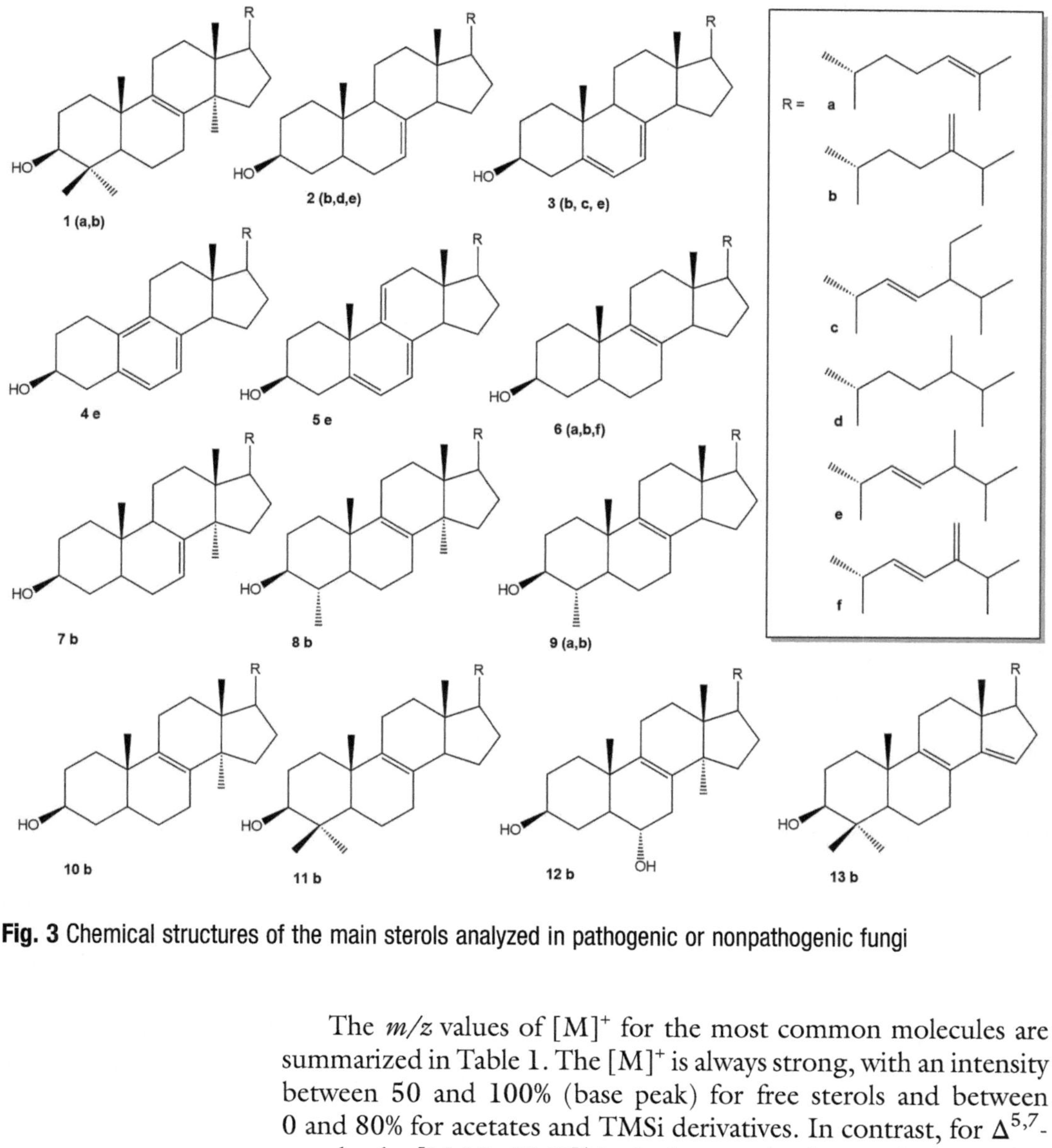

Fig. 3 Chemical structures of the main sterols analyzed in pathogenic or nonpathogenic fungi

The *m/z* values of $[M]^+$ for the most common molecules are summarized in Table 1. The $[M]^+$ is always strong, with an intensity between 50 and 100% (base peak) for free sterols and between 0 and 80% for acetates and TMSi derivatives. In contrast, for $\Delta^{5,7}$-sterols, the $[M\text{-}Me\text{-}H_2O]^+$ is the base peak when analyzing as free molecules. Moreover, $\Delta^{5,7}$- and Δ^5-steryl acetates undergo an easy loss of acetic acid (60 u.m.a) due to the proximity of the double bound, the $[M]^+$ is not significant, and then the $[M\text{-}60]^+$ ion becomes the base peak. This particular fragmentation is very useful in differentiating $\Delta^{5,7}$- and Δ^5-sterols from the others and justifying the acetate derivation. Δ^5-sterol TMSi ethers have characteristic ions at *m/z* 129 and $[M\text{-}129]^+$ for the loss of the TMSiOH group together with C-1, C-2, and C-3 of the sterol A-ring (129 u.m.a). For sterol with two unsaturations such as $\Delta^{5,7}$-sterol TMSi ethers, characteristic fragments are at *m/z* 131 and $[M\text{-}131]^+$.

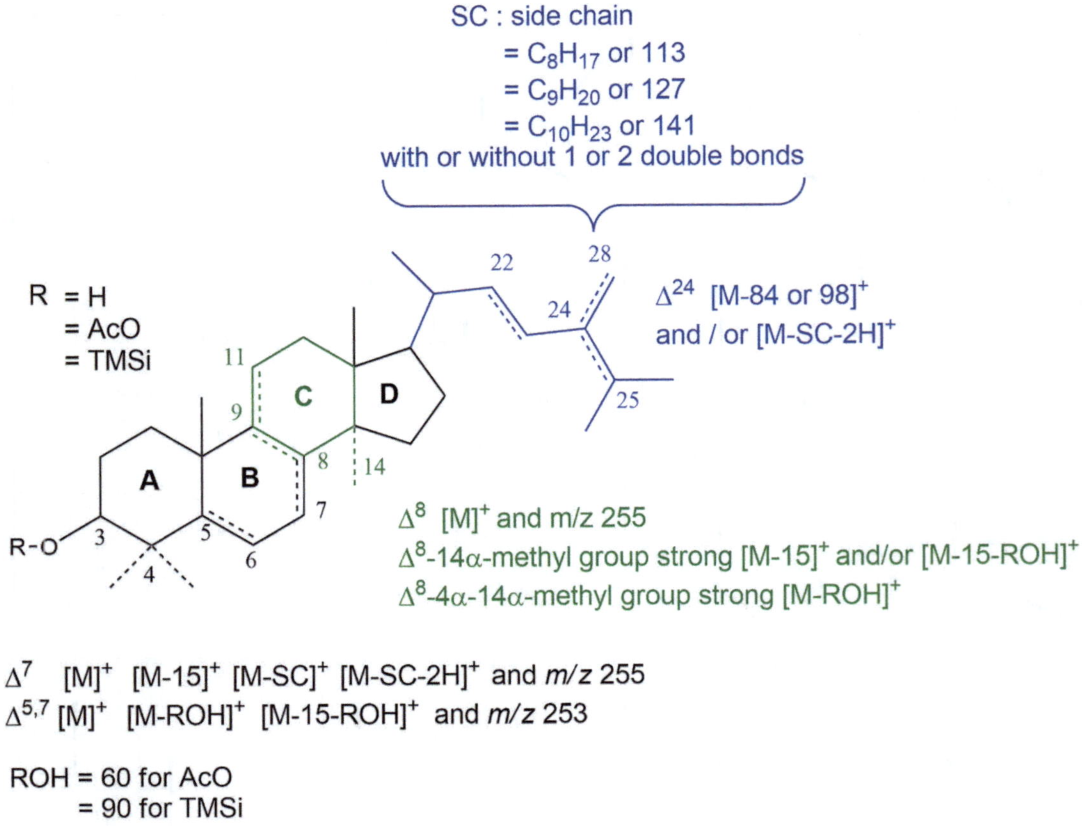

Fig. 4 Main fragmentation pattern of sterols as free molecules, acetates, and TMSi derivatives

Concerning the fragmentation of the side chain (SC), the most easily obtained ion results from a loss of the 3β-hydroxyl function together with the loss of ring D (42 a.m.u.): $[M\text{-}SC\text{-}42\text{-}ROH]^+$, where R = H for free sterol, R = AcO for steryl acetate, and R = TMSi for sterol TMSi ether. The *m/z* value of the $[M\text{-}SC\text{-}42\text{-}ROH]^+$ ion gives the number of double bonds in the B or C rings. Thus, with a sterol monoene ($\Delta^{5,}$ $\Delta^{7,}$ $\Delta^{8,}$ $\Delta^{9(11)}$), there is an abundant ion at *m/z* 213; with a sterol diene ($\Delta^{5,7}$), it is at *m/z* 211; and with a sterol triene ($\Delta^{5,7,9(11)}$), it is at *m/z* 209. Interestingly, 4α-methyl-Δ^7-sterols show an ion 14 a.m.u. higher at *m/z* 227 that allow to locate the methyl group at position C-4 rather than at C-14 or in the side chain.

The $[M\text{-}SC\text{-}ROH]^+$ ion at *m/z* 255 is usually observed for all derivatives.

With TMSi ethers, a characteristic fragmentation ion observed is at *m/z* $[M\text{-}90]^+$ which corresponds to the loss of trimethylsilanol (TMSiOH). Another common feature is the loss of a methyl group leading (often located at C-14) to ions at *m/z* such as $[M\text{-}15]^+$, $[M\text{-}90\text{-}15]^+$.

The mass spectra of eburicol analyzed as free sterol, acetate, and TMSi derivatives are given as an example in Fig. 5.

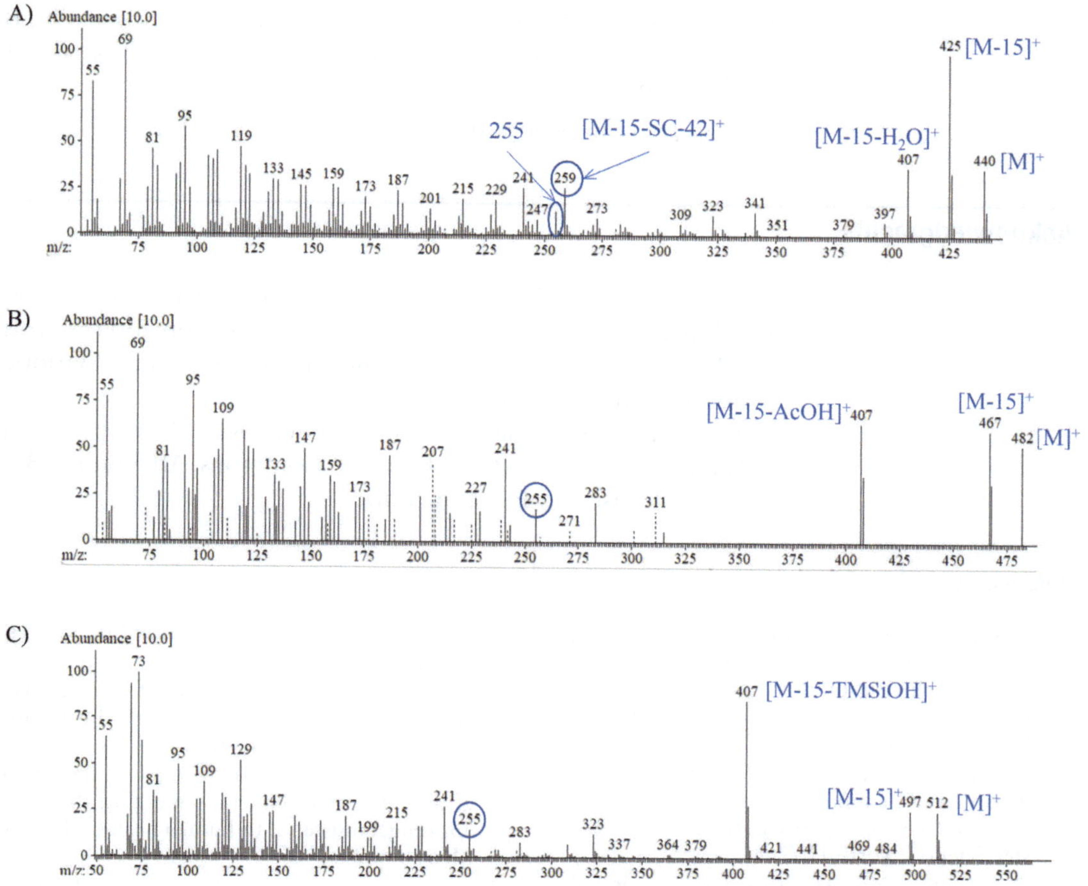

Fig. 5 Mass spectra of eburicol as (**a**) free sterol, (**b**) acetate, and (**c**) TMSi with main fragmentation ions

5 Notes

1. *Candida* spp. and *Aspergillus fumigatus* strains from the biocollection of IICiMed lab were studied following this protocol [12–14] as well as marine fungal strains from ISOMer lab collection [3].
2. Different culture media can be used to undergo one strain many compounds approach (OSMAC) [3].
3. The solution can be stored in the fridge if prepared as follows: Prepare 100 mL of ethanol 96° in a graduated cylinder. Weigh 11.2 g of KOH and transfer to a brown-glass bottle. Add ethanol. Mix using magnetic stirrer until complete dissolution (at least 2 h). Store at 4 °C. When the solution turns colored, throw it and prepare a fresh one.
4. If any antifungal treatment is planned, add it at **step 2**, to reach the desired concentration. If the antifungal is diluted in DMSO, add the same volume of DMSO into untreated sample.
5. The optimal fungal biomass for saponification is comprised between 100 mg and 1.5 g.

6. To be able to pipet the biomass, a volume of 500 μL of PBS is necessary. It would be of interest to dry the biomass at that step.
7. This step can be repeated twice to improve extraction.

Acknowledgments

The authors acknowledge Fabrice Pagniez to initiate the work in IICiMed lab and to be always available for discussion about sterols and Thibaut Robiou du Pont, Vony Rabesaotra, and Marjorie Albassier for their technical help.
Figure 2 was partly generated using Servier Medical Art, provided by Servier, licensed under a Creative Commons Attribution 3.0 unported license and Flaticon.com.

References

1. Kerr RG, Kerr SL, Pettit GR et al (1991) Sterols of marine invertebrates. 63. Isolation and structure elucidation of sutinasterol, the major sterol of the marine sponge *Xestospongia* sp. J Org Chem 56:58–62
2. Lipowsky R, Sackmann E (1995) Structure and dynamics of membranes: I. From cells to vesicles/ II. Generic and specific interactions. Elsevier Science, North Holland
3. Dos Santos Dias AC, Couzinet-Mossion A, Ruiz N et al (2019) Steroids from marine-derived fungi: evaluation of antiproliferative and antimicrobial activities of eburicol. Mar Drugs 17:372
4. Mota Fernandes C, Dasilva D, Haranahalli K et al (2021) The future of antifungal drug therapy: novel compounds and targets. Antimicrob Agents Chemother 65:e01719
5. Ullmann AJ, Aguado JM, Arikan-Akdagli S et al (2018) Diagnosis and management of Aspergillus diseases: executive summary of the 2017 ESCMID-ECMM-ERS guideline. Clin Microbiol Infect 24:e1–e38
6. Pappas PG, Lionakis MS (2018) Invasive candidiasis. Nat Rev Dis Primer 4:18026
7. Jørgensen LN, Heick TM (2021) Azole use in agriculture, horticulture, and wood preservation – is it indispensable? Front Cell Infect Microbiol 11:730297
8. Monk BC, Sagatova AA, Hosseini P et al (2020) Fungal Lanosterol 14α-demethylase: a target for next-generation antifungal design. Biochim Biophys Acta BBA - Proteins Proteomics 1868:140206
9. Whaley SG, Berkow EL, Rybak JM et al (2017) Azole antifungal resistance in Candida albicans and emerging non-albicans Candida species. Front Microbiol 7:2173
10. Wiederhold NP (2018) The antifungal arsenal: alternative drugs and future targets. Int J Antimicrob Agents 51:333–339
11. Fisher MC, Alastruey-Izquierdo A, Berman J et al (2022) Tackling the emerging threat of antifungal resistance to human health. Nat Rev Microbiol 20:1–15
12. Lavergne R-A, Albassier M, Hardouin J-B et al (2022) Impact of TR34/L98H, TR46/Y121F/T289A and TR53 alterations in azole-resistant Aspergillus fumigatus on sterol composition and modifications after in vitro exposure to Itraconazole and Voriconazole. Microorganisms 10:104
13. Montoir D, Guillon R, Gazzola S et al (2020) New azole antifungals with a fused triazinone scaffold. Eur J Med Chem 189:112082
14. Pagniez F, Lebouvier N, Na YM et al (2020) Biological exploration of a novel 1,2,4-triazole-indole hybrid molecule as antifungal agent. J Enzyme Inhib Med Chem 35:398–403
15. Goad JL, Akihisa T (1997) Analysis of sterols. Chapman & Hall, London
16. Müller C, Neugebauer T, Zill P et al (2018) Sterol composition of clinically relevant Mucorales and changes resulting from Posaconazole treatment. Mol Basel Switz 23:1218
17. Müller C, Binder U, Bracher F et al (2017) Antifungal drug testing by combining minimal inhibitory concentration testing with target identification by gas chromatography–mass spectrometry. Nat Protoc 12:947–963

Chapter 9

Cholesterol Assay Based on Recombinant Cholesterol Oxidase, ABTS, and Horseradish Peroxidase

Alexey V. Sviridov, Mikhail V. Karpov, Victoria V. Fokina, and Marina V. Donova

Abstract

Cholesterol determination by cholesterol oxidase reaction is a fast, convenient, and highly specific approach with widespread use in clinical diagnostics. Routinely, endpoint measurements with 4-aminophenazone or 4-aminoantipyrine as chromogens and sodium cholate, surfactants, or alcohols as solubilizing agents are used. Here we describe a novel kinetic method to determine cholesterol in 0.05–0.75 mM range in neutral or acidic buffers by use of recombinant cholesterol oxidase from *Nocardioides simplex* in a coupled reaction with horseradish peroxidase, ABTS as a chromogen, and methyl-β-cyclodextrin as a solubilizing agent.

Key words Cholesterol, Cholesterol oxidase, Enzymatic measurement, ABTS, Horseradish peroxidase, Methyl-β-cyclodextrin

1 Introduction

Cholesterol is one of the most important lipids with multiple functions in animal cells. It is an essential component of cell membranes, taking part in various signaling processes and serving as a precursor in the biosynthesis of steroid hormones and vitamins [1]. Cholesterol measurements in biological samples, especially in blood serum, are among the basic clinical procedures routinely used for the clinical diagnosis and prognosis of atherosclerosis, coronary heart disease, and other lipid disorders [2, 3]. Assessing of cholesterol content is also essential in academic field for various cell membrane studies [1].

Cholesterol oxidase (COX, EC 1.1.3.6) is a flavoenzyme that catalyzes the oxidation of 3β-hydroxysteroids with oxygen and the isomerization of the intermediate, $\Delta^{5(6)}$-ene- 3-ketosteroid (cholest-5-en-3-one), to produce $\Delta^{3(4)}$-ene-3-ketosteroid (cholest-4-en-3-one) and hydrogen peroxide [4]. An application of COX for

Carlos Barreiro and José-Luis Barredo (eds.), *Microbial Steroids: Methods and Protocols*, Methods in Molecular Biology, vol. 2704, https://doi.org/10.1007/978-1-0716-3385-4_9,

quantitative cholesterol analysis by color reaction of COX-produced H_2O_2 with titanium dioxide and xylenol orange was proposed in 1973 as a promising end-point assay with stable reagents, which does not require protein precipitation and is not subject to interference from hemoglobin or bilirubin [5]. A two-component kinetic method of cholesterol measurement in blood serum with cholesterol esterase (EC 3.1.1.13) and COX was developed later [6] and used acetylacetone as a chromogen. Now, a coupled reaction with horseradish peroxidase (HRP) is routinely used, thus enabling esterase/oxidase cholesterol assay to achieve adequate convenience, reproducibility, and specificity without prior hydrolysis or extraction [2]. Enzymatic cholesterol determination assumes either endpoint or kinetic approach. The latter offers advantages over the endpoint assay, such as short analysis time, less interfering substances, and no requirement for sample blank measurement, but requires detailed data on kinetic properties (K_M, V_{max}, k_{cat}) of specific COX used for analysis [3].

The known COX/HRP cholesterol assay protocols make use of 4-aminophenazone [3, 7] or 4-aminoantipyrine [8–12] as chromogens and Triton X-100, phenol, isopropanol, or sodium cholate to solubilize cholesterol which is poorly soluble in water-based media.

The compound 2,2′-azino-bis (3-ethylbenz-thiazoline-6-sulfonic acid) (ABTS) is a well-known substrate for peroxidases and readily participates in a number of other redox reactions [13]. Its ability to undergo single-electron oxidation yielding green-colored cation radical $ABTS^{•+}$ with characteristic absorption peak at 414 nm makes it highly suitable to visualize and quantify highly oxidizing intermediates in metal porphyrin system or hydrogen peroxide in glucose determinations [14]. For peroxidase assays, ABTS was long known a chromogen of choice as it is specific for peroxidase, does not inhibit the enzyme, gives a stable product with a well-defined absorption spectrum, allows easy quantification of the initial rates in the enzymatic reaction, and is non-toxic [15, 16].

Interestingly, no method of enzymatic cholesterol or COX measurement published up to date used ABTS as a chromogen. This could be attributed to the fact that optimum for ABTS oxidation lies in acidic range (pH 3.0–3.5) [13, 17], while COX and HRP demonstrate maximum activity within pH range from 6.0 to 8.0 [12, 18]. On the other hand, methods of measuring HRP activity per se with ABTS as a chromogen in the abovementioned pH range are available [15, 19, 20].

Complete cholesterol solubilization and avoiding of its precipitation during measurement in water-based samples is another task which is essential to achieve reliable and reproducible results. Available protocols generally use surfactants (Triton X-100) with phenol and isopropanol [8, 12] or phenol and sodium cholate [11] as solubilizing agents. Recently, cyclic glucopyranose oligomers known as α-, β-, or γ-cyclodextrins with hydrophilic exterior and hydrophobic central cavity which can incorporate different otherwise insoluble compounds in reversible manner were introduced as promising agents for drug delivery and cellular membrane lipids extraction [21]. Methyl-β-cyclodextrin (MCD) is most suitable for cholesterol molecule incorporation and it is used in experiments for membrane cholesterol depletion [1] and solubilization of the steroid in water/methanol systems prior to HPLC analysis [22].

Here we describe our findings on using ABTS as a chromogen and MCD as a solubilizing agent to develop simple, time-saving, and sensitive kinetic protocol of COX/HRP-based cholesterol assay.

2 Materials

Use deionized water (electrical resistance ≥18 MΩ cm^{-1}) for preparation of all solutions and analytical grade reagents.

Utilize Bradford reagent for measurements all of protein concentrations in liquid samples. One unit (U) of enzyme activity is defined as the amount of enzyme which transforms 1 μmol of substrate in 1 min. Specific activity is defined as a ratio of enzyme activity (U) to enzyme concentration (mg protein/mL).

2.1 Strains and Culture Media

1. *Nocardioides simplex* VKM Ac-2033D (All-Russian Collection of Microorganisms, VKM).
2. *E. coli* BL21(DE3) (Thermo Fisher Scientific).
3. LB medium: 10 g/L of tryptone, 5 g/L of yeast extract and 10 g/L of NaCl. Adjust pH to 7.0 with 40% NaOH prior to sterilization.
4. NS medium: 4 g/L of dipotassium phosphate trihydrate, 2 g/L of monopotassium phosphate, 6 g/L of yeast extract, 6 g of soy peptone, 15 g/L of glucose, 0.2 g/L of magnesium sulfate heptahydrate, 0.005 g/L of iron sulfate heptahydrate, and 0.002 g/L of zinc sulfate heptahydrate. Adjust pH to 7.0 prior to sterilization.
5. Kanamycin stock solution: 100 mg/mL kanamycin in water. Filter through 0.2 mM GD/X sterile filter (GE Healthcare), and store at −18 °C.
6. IPTG stock solution: 1 M IPTG in water. Filter through 0.2 mM GD/X sterile filter (GE Healthcare), and store at −18 °C.
7. Multitron thermostated rotary shaker (Infors HT, Switzerland) or equivalent

2.2 Molecular Cloning

1. AllPrep Bacterial DNA kit (Qiagen).
2. Q5 High-fidelity DNA polymerase (NEB).
3. Primer ChO-f: 5′-CAGTACATATGgcgagcatcggcaacggcg-3′.
4. Primer ChO-r: 5′-TCATCAAGCTTctagatgtcgttggccaggatg-3′.
5. pET-28a vector (Merck).
6. T4 DNA Ligase (Thermo Fisher Scientific).
7. QIAquick Gel Extraction Kit (Qiagen).
8. FastDigest *NdeI* and *HindIII* restriction enzymes with appropriate buffer (Thermo Fisher Scientific).
9. Mastercycler pro S thermal cycler (Eppendorf) or equivalent.
10. Ethidium bromide (Sigma).
11. Horizontal electrophoresis equipment with power supply and 365 nm transilluminator.

2.3 Recombinant COX_{Ns} Purification

1. Disruption buffer: 50 mM Tris–HCl, pH 7.5.
2. Binding buffer: 50 mM Tris–HCl, 50 mM NaCl, and 20 mM imidazole (PanReac Applichem), pH 7.5.
3. Elution buffer: 50 mM Tris–HCl, 50 mM NaCl, and 500 mM imidazole, pH 7.5.
4. Bio-Scale Mini Profinity 5 mL IMAC column (Bio-Rad).
5. DNAse solution: 400 Kunitz of DNAse I per mL (Sigma) in 50 mM Tris–HCl, pH 7.5.
6. J6-MI and L5-50 centrifuges (Beckman) or equivalent.
7. MiniSart 0.45 μm SFCA syringe filters (Sartorius).
8. Mini-Protean cell for vertical electrophoresis (Bio-Rad).
9. Q500 ultrasonic disintegrator (QSonica) or equivalent.
10. Membra-cel MC18 dialysis tubing (Viskase) or equivalent with cutoff range of 20 kDa or less.

2.4 Recombinant COX Activity Measurements

1. Recombinant COX from *Streptomyces hygroscopicus* (COX_{Sh}) (Sigma).
2. Recombinant COX from *Nocardioides simplex* (COX_{Ns}), obtained as described below.
3. Cholate assay buffer: 21.59 mM monosodium phosphate, 28.41 mM disodium phosphate, 64 mM sodium cholate (Sigma), 0.89 mM cholesterol (Steraloids), 0.4% Triton X-100 (w/v), 21 mM phenol, and 1.4 mM ABTS (Sigma), pH 7.0. Add cholesterol after all other components are

dissolved. Stir the mixture at 37 °C for 1 h until it becomes clear. Check the pH value and adjust it to 7.0 with 10% NaOH (w/v) if needed.

4. Phosphate assay buffer: monosodium phosphate, disodium phosphate, 5 mM MCD (Wacker Chemie), and 0.05–1 mM cholesterol, pH 6.2–8.0. Adjust pH to desired value by changing proportions of mono- and disodium phosphates according to standard buffer preparation manuals. To prepare 1 mM cholesterol stock add the lipid to 50 mM sodium phosphate buffer with 5 mM MCD and sonicate the mixture in ultrasonic bath at 50% amplitude 5 times for 1 min with 30 s off time until no visible solid particles remain and opalescent colloid pseudo-solution forms. Make phosphate assay buffer with desired concentration of cholesterol (0.05–1 mM) by mixing freshly prepared (*see* **Note 1**) 1 mM cholesterol stock with 50 mM sodium phosphate buffer with 5 mM MCD in corresponding proportions. Perform all operations with warm buffers to avoid precipitation of cholesterol. Buffers with 0.5 mM cholesterol or less should appear clear.
5. Britton–Robinson assay buffer: 50 mM boric acid, 50 mM phosphoric acid, 50 mM acetic acid (glacial) [23], 5 mM MCD, and 0.5 mM cholesterol, pH 5.0–6.2. Dissolve all components except cholesterol in water at 30 °C with stirring. Adjust pH to desired value with 40% NaOH (w/v). Prepare stock buffer with 1 mM cholesterol and, consequently, a series of buffers with desired cholesterol concentrations as described in Subheading 2.4, **item 4**.
6. ABTS 50× stock solution: 70 mM ABTS in water. Keep at −18 °C between uses.
7. Horseradish peroxidase stock solution: 500 U/mL HRP (Dia-M) in 50 mM sodium phosphate, pH 7.0. Keep at −18 °C between uses.
8. Hydrogen peroxide stock solution: 880 mM (3% v/v) H_2O_2 in water. Dilute stock solution of hydrogen peroxide to desired concentration with water directly before measurements.
9. Cholesterol oxidase (COX_{Ns} or COX_{Sh}) 4 U/mL solutions in 50 mM sodium phosphate buffer, pH 7.0.
10. Spectrophotometer (Shimadzu UV-1800 or equivalent) with a set of 1.5 mL glass cuvettes and TW-2.02 water thermostat (Elmi).
11. Elmasonic S40H ultrasonic bath (Elma Schmidbauer) or equivalent.

3 Methods

3.1 Cultivation of N. simplex

1. Cultivate *N. simplex* cells aerobically in NS medium in a rotary shaker (30 °C at 200 rpm) for 48 h to OD_{600} ~4.5.
2. Separate cells by centrifugation (20 min, 8000 × *g*, 4 °C).

3.2 Cloning of Cholesterol Oxidase Gene from N. simplex

Perform all operations involving commercial enzymes, competent cells, and kits according to the manufacturer's recommendations.

1. Isolate genomic DNA using an AllPrep Bacterial DNA kit.
2. Amplify the target DNA sequence encoding a truncated mature form of *N. simplex* COX_{Ns} (GenBank AIY19828.2, UniProt: A0A0A1DTF1, amino acid residues 42–541) by PCR with *N. simplex* DNA as a template, using Q5 DNA polymerase and the oligonucleotide primers ChO-f and ChO-r with *Nde*I and *Hind*III restriction endonuclease sites, respectively.
3. Digest resulting PCR mix with *NdeI* and *HindIII* endonucleases, and purify product by agarose gel electrophoresis with subsequent DNA extraction with QIAquick Gel Extraction Kit.
4. Clone resulting sequence into pET-28a vector with T4 DNA ligase.
5. Use ligase mixture to transform *E. coli* BL21(DE3) competent cells by heat shock.

3.3 Cultivation of Recombinant E. coli BL21(DE3)

1. Add kanamycin stock solution into sterile LB medium to obtain 50 mg/L concentration of antibiotic.
2. Seed recombinant *E. coli* BL21(DE3) in LB–kanamycin medium and incubate in a rotary shaker (37 °C at 200 rpm) overnight to OD_{600} ~5.0–6.0.
3. Use resulting culture to inoculate fresh LB–kanamycin medium so that initial OD_{600} is 0.05, then grow in a rotary shaker (37 °C at 200 rpm) until OD_{600} is 0.4–0.6.
4. Add IPTG stock solution up to 0.5 mM concentration, and cultivate in a rotary shaker (25 °C at 200 rpm) for 6 h.
5. Harvest cells at OD_{600} ~4.5 by centrifugation (20 min, 8000 × *g*, 4 °C).

3.4 Purification of Recombinant Cholesterol Oxidase

1. Wash recombinant *E. coli* BL21(DE3) cells by stirring cell pellet in 100 mL of disruption buffer. Centrifuge 20 min, 5000 × *g*, 4 °C. Resuspend the biomass in disruption buffer (~8 mL of buffer per 1 g of cell pellet).
2. Sonicate the cell suspension in an ultrasonic disintegrator at 50% amplitude 10 times for 30 s with 1 min off time, on ice.

Add 100 μL of freshly made DNAse solution into the homogenate, precipitate the debris by centrifugation (120 min, 25,000 × *g*, 4 °C), and then filter the supernatant through a 0.45 μm filter.

3. Use affinity chromatography for recombinant COX_{Ns} purification. Equilibrate the column with binding buffer, load the sample, then wash with six column volumes (CV) of the binding buffer. Elute the protein with ten CVs of an elution buffer. Pool the fractions with COX_{Ns} activity, dialyze resulting protein sample against disruption buffer and store at +4 °C (*see* **Note 2**).
4. Monitor the presence and purity of COX_{Ns} on all stages by SDS-PAGE in 12% gel.

3.5 Spectrophotometric Measurement of COX Activity and Cholesterol Concentration Assay

Calculate initial rates of COX-catalyzed reactions according to formula: $U = (K\ \Delta A_{420}\ V_{tot})/(\tau\ V_{prot})$, where $K = 0.02(7)$ is a coefficient derived from molar extinction of $ABTS^{\bullet +}$ (36,000 M^{-1} cm^{-1} [19]); ΔA_{420} is an increase in optical absorption at 420 nm in 1 min; V_{tot} is total volume of the sample (mL); τ is time (min); and V_{prot} is volume of protein added to the sample (mL). Calculate kinetic properties of COX_{Ns} and COX_{Sh} from corresponding saturation curves using Lineweaver–Burk and Hanes plots. Perform all measurements in triplicates at 30 °C.

3.5.1 COX Activity Detection in MCD-Based Buffers

1. Add 0.02 U (5 μL) of COX_{Ns} or COX_{Sh}, 10 or 20 μL (5 or 10 U) of HRP (*see* below), 20 μL of 50× ABTS stock to phosphate assay buffer or Britton–Robinson assay buffer up to final volume of 1 mL.
2. Conduct measurements in double beam spectrophotometer at 420 nm for 30 s; use similar samples without COX as references. Consider necessary adjustments to actual substrate concentrations due to dilution after addition of all components into the sample.
3. Determine the initial rate of enzymatic reactions by absorption change within a period of 0–10 s from the start of measurement. Use Britton–Robinson assay buffer in case measurements are to be made at pH below 6.2; otherwise phosphate assay buffer is recommended. Absorption curves must have almost perfect linearity at low substrate concentrations both in phosphate and Britton–Robinson assay buffers (Fig. 1) (*see* **Note 3**).
4. While measuring at high concentrations of cholesterol (0.75–1 mM), make sure that sufficient linearity of the curve is retained within 0–10 s time range (coefficient of determination of simple linear regression applied to the abovementioned

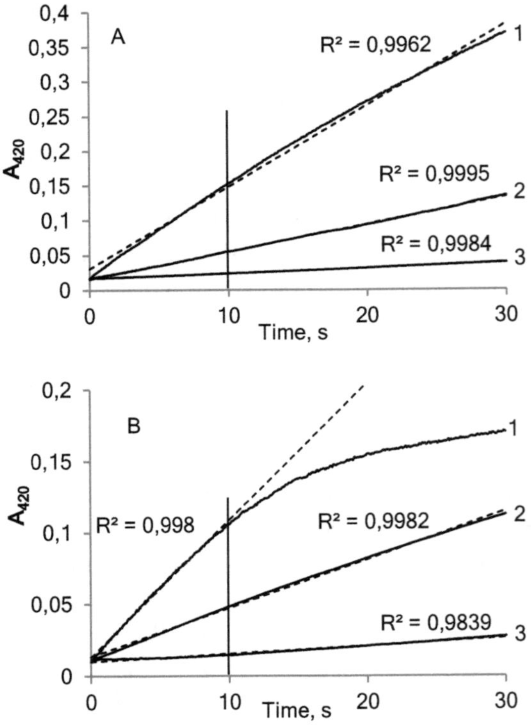

Fig. 1 Spectrophotometric measurements of ABTS oxidation by hydrogen peroxide produced by COX_{Ns} (**a**) and COX_{Sh} (**b**) in phosphate assay buffer with 5 mM MCD and 1 mM (1), 0.5 mM (2) or 0.1 mM (3) cholesterol. Dashed lines depict linear trends with respective coefficients of determination (R^2). Vertical lines depict the part of the curves used for reaction rate calculations

area of experimental curve must be ≥0.95). Note that linearity deterioration at high concentrations of cholesterol would negatively affect end-point or prolonged kinetic assays.

5. Use a series of standard buffer solutions (either phosphate or Britton–Robinson) with 5 mM MCD and different cholesterol concentrations to measure COX activity as described above and obtain saturation curve which should adhere to Michaelis–Menten kinetics (*see* **Note 4**) (Fig. 2). Recommended cholesterol concentrations for calibration with minimal errors are 0.1, 0.2, 0.3, 0.4, and 0.5 mM.
6. Plot Lineweaver–Burk diagram (Fig. 3) and use linear regression to calculate unknown cholesterol concentrations by measuring COX activity in experimental samples (*see* **Note 5**). To do so, mix aliquots with unknown concentration of cholesterol 1:1 with 2× assay buffer (e.g., 100 mM sodium phosphate and 10 mM MCD for phosphate assay buffer), sonicate as described above, and use for measurements. Twofold dilution of initial sample should be considered during calculations.

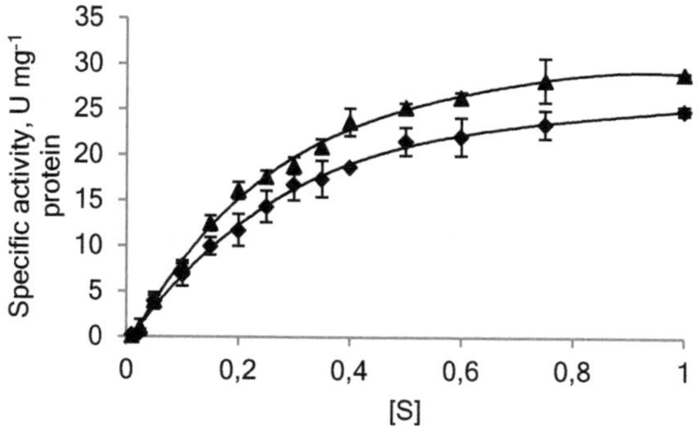

Fig. 2 Saturation curves of COX_{Ns} (filled diamonds) and COX_{Sh} (filled triangles) in sodium phosphate assay buffer pH 7.0. [S]—concentration of cholesterol, mM. Solid lines represent fourth power equation plots with best fit to experimental data

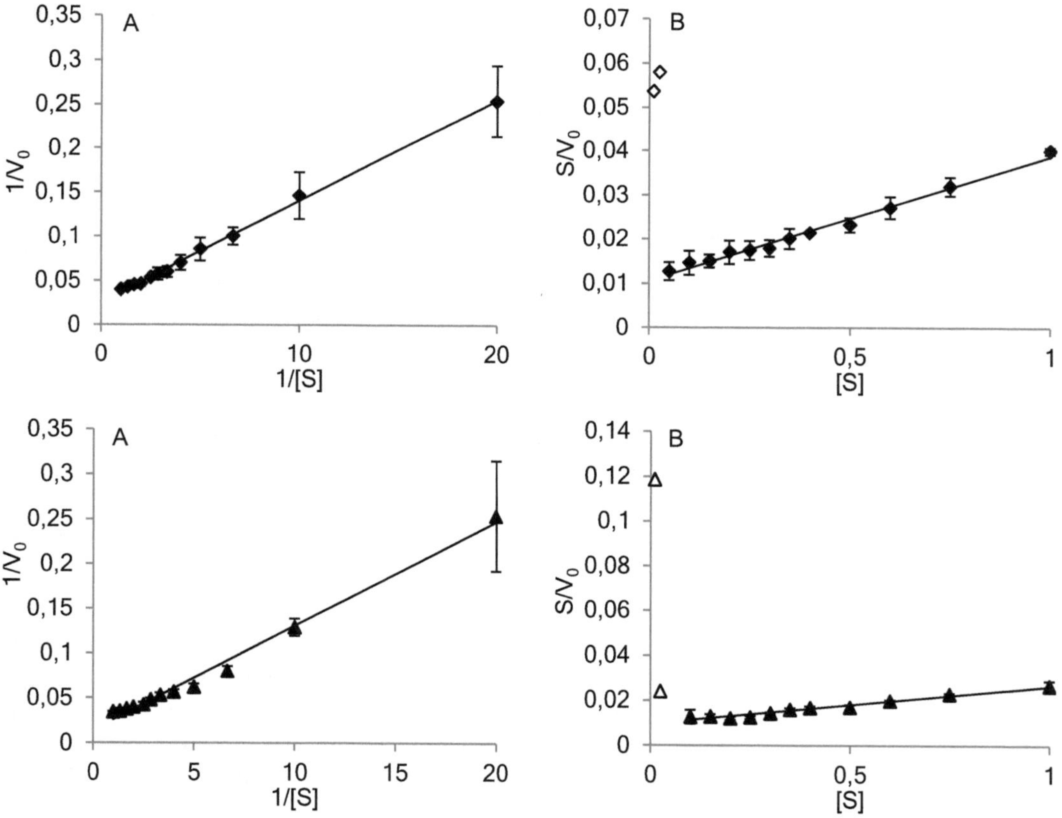

Fig. 3 Linearization of kinetic curves for COX_{Ns} (filled diamonds) and COX_{Sh} (filled triangles) in Lineweaver–Burk (**a**) coordinates. [S]—concentration of cholesterol, mM; V_0—initial rate of reaction. Hanes coordinates (**b**) are provided for consistency and to depict inapplicability of described method at low [S] values (0.01 and 0.025 mM), where experimental data (empty diamonds and triangles) do not fit the linear regression. The corresponding elements are not shown in the Lineweaver–Burk plots due to scale limitations

3.5.2 COX Activity Detection in Cholate Assay Buffer

1. Add 5–50 μL (0.02–0.2 U) of COX_{Ns} or COX_{Sh} solution and 10 μL (5 U) of HRP to cholate assay buffer up to a final volume of 1 mL and measure in double beam spectrophotometer at 420 nm for 5 min; use similar sample without COX as a reference.
2. For end-point measurements, incubate 1 mL samples containing 0.2 U of COX, 5 U of HRP, and cholate assay buffer for 6 h at 30 °C. Incubate reference mixture at 100 °C for 5 min immediately after mixing, then store in tightly closed microtubes at +4 °C until measurements (*see* **Note 6**).

4 Notes

1. As MCD binds hydrophobic compounds in reversible manner, crystallization of cholesterol when its colloid solutions with high (0.6–1 mM) concentrations are stored at lower temperatures may occur. Although samples with precipitated cholesterol may be heated, stirred, and sonicated to re-solubilize the lipid, it is still recommended to use fresh stocks for maximum reproducibility. No cholesterol precipitation is observed in MCD-based buffers with cholesterol concentration of 0.5 mM and below at 4 °C, which corresponds with earlier data on MCD-mediated solubilization of cholesterol in water/methanol systems, where rising MCD to cholesterol ratio increased the solubility of the latter [22].
2. One-time freezing of COX_{Ns} solution results in irreversible denaturation of the enzyme, while the sample stored on ice in a fridge for 30 days retains 47% of initial activity. Samples stored at room temperature lose 8–10% of activity daily.
3. Measurements in Britton–Robinson or sodium phosphate buffers yield almost similar results and saturation curves at similar pH values.
4. Parameters calculated for commercially available COX_{Sh} (specific activity 50.2 U mg^{-1} protein, pH_{opt} 6.5–7.0) are quite similar to the ones stated in the manufacturer's manual (specific activity ≥50 U/ mg protein, pH_{opt} 6.5), thus proving the correctness of the method. As the experimental data demonstrate good conformity to theoretical saturation curves (Fig. 2), we assume that the described protocol can be used for rapid and highly specific determination of cholesterol within the range of concentrations from 0.05 mM to 0.75 mM which is applicable in a pH range from 5.0 to 8.0 (*see* **Note 7**). Method is not applicable for measurements of minor (<0.05 mM) concentrations of cholesterol due to lack of sensitivity (Fig. 2b). Solutions with cholesterol content exceeding 0.75 mM may

provide incorrect results due to saturation of the enzyme. It is not recommended to increase COX amount in the reaction mixture to alleviate these limitations (*see* **Note 8**); samples with higher amounts of cholesterol (e.g., blood serum) must be diluted accordingly.

5. Instead of linear regression, one could consider using computational methods. Kinetic curves of COX could be expressed as fourth power polynomial functions with very good fit ($R^2 \geq 0.99$) (Fig. 2). This allows direct calculation of cholesterol concentration which corresponds to observed COX activity and thus possible errors associated with reciprocal plots could be avoided. Nevertheless, such approach is not recommended for routine cholesterol assays and is not discussed here extensively as it requires detailed saturation curves and implies time-consuming preliminary measurements.

6. No formation of green cation radical $ABTS^{\bullet+}$ is observed both in kinetic and end-point experiments. Subsequent chromatographic analysis of end-point samples shows conversion of cholesterol into cholest-3-en-4-one, which implies COX_{Ns} and COX_{Sh} activity and formation of hydrogen peroxide. Addition of H_2O_2 in final concentration of 0.5 mM into samples does not result in green coloration development and makes it evident that composition of cholate assay buffer prevents oxidation of ABTS by HRP. Although sodium cholate protocol proved to be inapplicable for ABTS/HRP determination of COX, we present the data here for consistency, to demonstrate the limitations of the proposed method, and emphasize the importance of the assay buffer composition for measurements with ABTS.

7. Although ABTS was shown to undergo enzymatic [24] or non-enzymatic [25] oxidation with the formation of $ABTS^{\bullet+}$ at pH as high as 10.5, alkaline environments should be avoided in measurements when the rate of this process and cation radical stability are important. $ABTS^{\bullet+}$ can participate in two-stage reverse reaction with H_2O_2 provided the latter present in excess, which results in reduced ABTS, O_2, and H_3O^+. The rate of reverse reaction between H_2O_2 and $ABTS^{\bullet+}$ depends on pH and was shown to increase several fold with the increase of pH from 3.5 to 7.0 [26]. If hydrogen peroxide concentration is relatively low, radical forms of ABTS are shown to be stable after 30 min of incubation at pH ranging from 1.0 to 7.0, but at pH 8.0 some discoloration occurred, which is acceptable for short kinetic measurements but could affect prolonged assays [27]. Concerning the rate of ABTS oxidation, we observe significant retardation of $ABTS^{\bullet+}$ formation in the presence of HRP and 0.44 mM hydrogen peroxide at pH higher than 6.0 (Fig. 4, solid line), and no coloration at

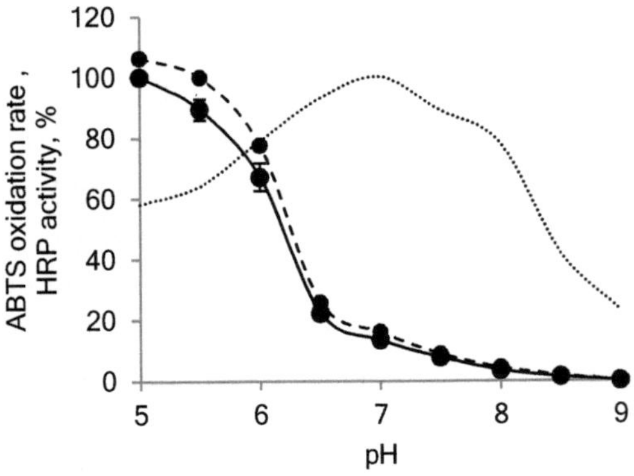

Fig. 4 Effect of pH on ABTS$^{•+}$ formation rate with HRP and 440 mM hydrogen peroxide. Solid line—5 U/mL HRP; dashed line—15 U/mL HRP; dotted line—optimum pH of soluble HRP [18]

all appears at pH $\geq$ 9.0. This could not be explained by HRP inactivation, as the enzyme has its optimum at pH 7.0–7.5 (Fig. 4, dotted line) [18] and addition of triple amount of HRP has little effect (Fig. 4, dashed line). In cholesterol measurements in alkalescent buffers up to pH 8.0 color development rate is limited by COX activity rather than by ABTS oxidation rate, the latter still exceeded that of cholesterol oxidation by an order of magnitude. This is not true for pH over 8.0; therefore, we do not recommend using the described method for COX and cholesterol measurements at pH range from 8.0 to 9.0 s and at pH below 5.0 where COX inactivation becomes significant. In pH range from 5 to 6.5 we recommend adding 10 U of HRP in each sample instead of 5 U to ensure that necessary excess of HRP activity over COX activity is preserved.

8. The process of ABTS oxidation by hydrogen peroxide is quite complex and, depending on pH and H_2O_2 concentration, may result in at least five different products, among which only cation radical ABTS$^{•+}$ is useful for spectrophotometric measurements. Preserving the correct stoichiometry, where H_2O_2 quantity is at least two times lower than that of ABTS, is pivotal for avoiding undesirable reactions [16] and substrate inhibition of HRP [15]. Here we report that maximum ABTS oxidation rate in phosphate assay buffer with 0.5 mM cholesterol and 1.4 mM ABTS occurred at H_2O_2 concentration of 0.44 mM (Fig. 5). At hydrogen peroxide concentrations exceeding 0.5 mM, drastic decrease in coloration development is observed, presumably due to process of conversion of cation radical back into ABTS along with undesired side product formation. Upon addition of millimolar quantities of H_2O_2, the absorption at 420 nm is rapidly bleached. In our

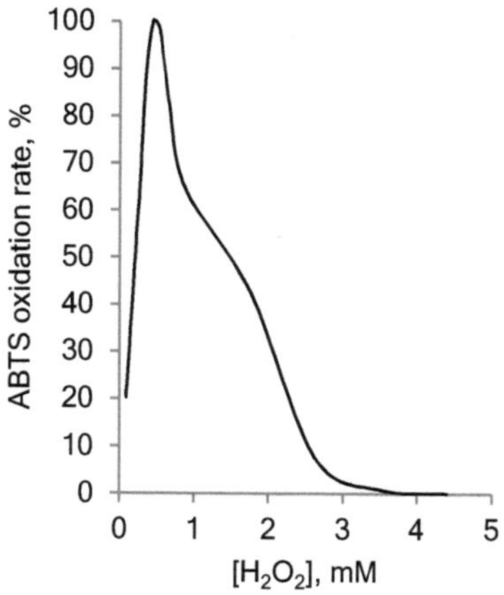

Fig. 5 Dependence of observed ABTS$^{\bullet+}$ formation rate on initial hydrogen peroxide concentration in phosphate assay buffer at pH 7.0 with 0.5 mM cholesterol

experiments, samples with more than 0.5 U of COX activity develop only pale green coloration or no coloration at all obviously as a result of H_2O_2 overproduction. Another point to consider is high molar absorption of ABTS and very high rate of its oxidation by HRP, which means almost instant increase of A_{420} value after addition of COX into sample. Depending on model of spectrophotometer, measurements of samples with absorption at 420 nm exceeding 1.5 units may become inaccurate. This may be an issue if too much COX activity is added into cuvette and excessive coloration develops during the mixing of the sample or at the beginning of measurement. Empirically, we found that using 0.02 U of COX for assay is most practical and grants most reproducible and reliable results. In any case, we do not recommend adding more than 0.1 U of COX for accuracy and convenience reasons. It should be noted, that even with elimination of H_2O_2 from reaction mixture, ABTS$^{\bullet+}$ still gradually decomposes as the back reaction still takes place due to possible involvement of solvent, but at much lower rate [16, 27]. This must be taken into account in prolonged experiments, but no corrections are normally necessary in kinetic studies where rates are measured within 2 min period [15].

Acknowledgments

This work is supported by the Russian Science Foundation, project No. 21-64-00024.

References

1. Owen DM (2015) Cholesterol depletion using methyl-β-cyclodextrin. Chapter 8. In: Methods in membrane lipids. Humana Press, New York, p 91
2. Warnick GR, Remaley AT (2001) Measurement of cholesterol in plasma and other body fluids. Curr Atheroscler Rep 3:404–411
3. Srisawasdi P, Jearanaikoon P, Kroll MH, Lolekha PH (2005) Performance characteristics of cholesterol oxidase for kinetic determination of total cholesterol. J Clin Lab Anal 19:247–252
4. Kumari L, Kanwar SS (2012) Cholesterol oxidase and its applications. Adv Microbiol 2:49–65
5. Richmond W (1973) Preparation and properties of a cholesterol oxidase from *Nocardia* sp. and its application to the enzymatic assay of total cholesterol in serum. Clin Chem 19: 1350–1356
6. Pesce MA, Bodourian SH (1976) Enzymatic rate method for measuring cholesterol in serum. Clin Chem 22:2042–2045
7. Shao M, Rao Z, Zhang X, Xu M, Yang T, Li H et al (2014) Bioconversion of cholesterol to 4-cholesten-3-one by recombinant *Bacillus subtilis* expressing *choM* gene encoding cholesterol oxidase from *Mycobacterium neoaurum* JC-12. J Chem Technol Biotechnol 90:1811–1920
8. Ye D, Lei J, Li W, Ge F, Wu K, Xu W et al (2008) Purification and characterization of extracellular cholesterol oxidase from *Enterobacter* sp. World J Microbiol Biotechnol 24: 2227–2233
9. Li B, Wang W, Wang F-Q, Wei D-Z (2010) Cholesterol oxidase ChoL is a critical enzyme that catalyzes the conversion of diosgenin to 4-ene-3-keto steroids in *Streptomyces virginiae* IBL-14. Appl Microbiol Biotechnol 85:1831–1838
10. Kasabe PJ, Mali GT, Dandgeet PB (2015) Assessment of alkaline cholesterol oxidase purified from *Rhodococcus* sp. PKPD-CL for its halo tolerance, detergent and organic solvent stability. Protein Expr Purif 116:30–41
11. Doukyu N, Nihei S (2015) Cholesterol oxidase with high catalytic activity from *Pseudomonas aeruginosa*: screening, molecular genetic analysis, expression and characterization. J Biosci Bioeng 120:24–30
12. Qin HM, Zhu Z, Ma Z, Xu P, Guo Q, Li A et al (2017) Rational design of cholesterol oxidase for efficient bioresolution of cholestane skeleton substrates. Sci Rep 7:16375
13. Ufot UF, Akpanabiatu MI, Cali K, Uffia ID, Udosen I (2015) pH-dependence of manganese (II) oxidation reaction by novel wild-type and mutants recombinant *Phlebia radiata* manganese peroxidase 3 (rPr-MnP3) enzymes. Am J Mol Biol 12:67–84
14. Scott SL, Chen W-J, Bakac A, Espenson JH (1993) Spectroscopic parameters, electrode potentials, acid ionization constants, and electron exchange rates of the 2,2′-azinobis (3-ethylbenzothiazoline-6-sulfonate) radicals and ions. J Phys Chem 97:6710–6714
15. Childs RE, Bardsley WG (1975) The steady-state kinetics of peroxidase with 2,2′-azino-di-(3-ethylbenzthiazoline-6-sulphonic acid) as chromogen. Biochem J 145:93–103
16. Kadnikova EN, Kostić NM (2002) Oxidation of ABTS by hydrogen peroxide catalyzed by horseradish peroxidase encapsulated into sol–gel glass. Effects of glass matrix on reactivity. J Mol Catal B Enzym 18:39–48
17. An H, Xiao T, Fan H, Wei D (2015) Molecular characterization of a novel thermostable laccase PPLCC2 from the brown rot fungus *Postia placenta* MAD-698-R. Electron J Biotechnol 18:451–458
18. Alshawafi WM, Aldhahri M, Almulaiky YQ, Salah N, Moselhy SS, Ibrahim IH et al (2018) Immobilization of horseradish peroxidase on PMMA nanofibers incorporated with nanodiamond. Artif Cells Nanomed Biotechnol 46: 973–981
19. Hou C, Ghéczy N, Messmer D, Szymańska K, Adamcik J, Mezzenga R et al (2019) Stable immobilization of enzymes in a macro- and mesoporous silica monolith. ACS Omega 4: 7795–7806
20. Herzog PL, Sützl L, Eisenhut B, Maresch D, Haltrich D, Obinger C et al (2019) Versatile oxidase and dehydrogenase activities of bacterial pyranose 2-oxidase facilitate redox cycling with manganese peroxidase *in vitro*. Appl Environ Microbiol 85:e00390–e00319
21. López CA, de Vries AH, Marrink SJ (2011) Molecular mechanism of cyclodextrin mediated cholesterol extraction. PLoS Comput Biol 7:e1002020
22. Charlton SA, Coym JW (2012) The use of methyl-β-cyclodextrin to solubilize cholesterol prior to coating onto a C18 stationary phase. J Chromatogr A 1266:69–75
23. Britton HTS, Robinson RA (1931) Universal buffer solutions and the dissociation constant of veronal. J Chem Soc 0:1456–1462

24. Huang Y, Lin J, Zou J, Xu J, Wang M, Cai H et al (2012) ABTS as an electron shuttle to accelerate the degradation of diclofenac with horseradish peroxidase-catalyzed hydrogen peroxide oxidation. Sci Total Environ 798: 149276
25. Ilyasov IR, Beloborodov VL, Selivanova IA, Terekhov RP (2020) ABTS/PP decolorization assay of antioxidant capacity reaction pathways. Int J Mol Sci 21:1131
26. Aliaga C, Lissi EA (1998) Reaction of 2,2-azinobis(3-ethylbenzothiazoline-6-sulfonic Acid) (ABTS) derived radicals with hydroperoxides. Kinetics and mechanism. Int J Chem Kinet 30: 565–570
27. Ferri M, Gianotti A, Tassoni A (2013) Optimisation of assay conditions for the determination of antioxidant capacity and polyphenols in cereal food components. J Food Compos Anal 30:94–101

Chapter 10

Measuring the Interaction of Sterols and Steroids with Proteins by Microscale Thermophoresis

Ola El Atab and Roger Schneiter

Abstract

Determining the affinity and specificity of protein–lipid interactions is crucial for understanding the physiological function and mode of action of signaling lipids, including steroids. Here we describe a method that relies on microscale thermophoresis (MST) to monitor the binding of sterols and steroids to proteins. The protein of interest is expressed as a polyhistidine-tagged fusion in *E. coli* and purified by affinity chromatography on a nickel-based resin. The purified protein is then labeled with a fluorescent dye and incubated with a serial dilution of the lipid ligand. The protein–ligand interaction is monitored by MST, which detects the fraction of the protein bound to the ligand based on its altered mobility in a thermal gradient. A binding curve fitted to the measured data points is then used to calculate the corresponding dissociation constant.

Key words Cholesterol, Steroids, Pregnenolone, Sterol-binding proteins, Pathogen-related yeast 1 (Pry1), CAP/SCP/TAPS superfamily, *Saccharomyces cerevisiae*

1 Introduction

Protein–lipid interactions govern many fundamental cellular and developmental processes, including the transport of lipids across membrane contact sites, steroidogenesis, and steroid signaling [1, 2]. Determining the affinity and specificity of these interactions can be experimentally challenging as the lipid-based ligand might have limited solubility in an aqueous buffer. Different methods have been established to monitor protein–lipid interactions, including the quenching of endogenous tryptophan fluorescence or binding of a radiolabeled lipid ligand [3, 4]. Here we describe an alternative, sensitive, and fast method to monitor protein–lipid interactions based on microscale thermophoresis (MST) [5, 6].

The protein of interest is expressed as a polyhistidine-tagged version in *E. coli* and then purified by affinity chromatography. The purified protein is labeled with a fluorescent dye and incubated with

Carlos Barreiro and José-Luis Barredo (eds.), *Microbial Steroids: Methods and Protocols*, Methods in Molecular Biology, vol. 2704, https://doi.org/10.1007/978-1-0716-3385-4_10,

a serial dilution of the ligand. Ligand binding is then measured on a microscale in glass capillaries. These capillaries are inserted into a dedicated device such as a Monolith (Nanotemper Technologies), which determines the fraction of the ligand-bound fluorescent protein upon diffusion in a temperature gradient.

Here we use a member of the Pathogenesis-Related Yeast (Pry) proteins to determine its interaction with cholesterol and pregnenolone. The yeast Pry proteins (Pry1-3) belong to a large protein superfamily known as CAP/SCP/TAPS/VALs proteins, and they have previously been shown to bind sterols in vitro and in vivo [7, 8]. CAP family members constitute an evolutionarily conserved and widespread protein superfamily, which is present in the animal, plant, fungal, and bacterial kingdoms (pfam PF00188). CAP proteins have been implicated in diverse biological processes, including immune defense, venom toxicity, reproduction, and cancer development. The characteristic feature of these proteins is a conserved 17–21 kDa CAP domain, which adopts a unique α-β-α sandwich fold, stabilized by disulfide bonds. Additional N- and C-terminal extensions with low sequence similarity are common among family members, and most CAP proteins contain a signal peptide directing secretion of these glycoproteins into the extracellular space [9–12].

Here, we use more water-soluble sulfated derivatives of cholesterol and pregnenolone to determine their interaction with the Pry1 protein. We show that a double mutant version of Pry1, $Pry1^{A292C\,P242C}$, is unable to bind sterols. In this mutant, the caveolin-binding domain (CBM, [13]), which is a flexible loop important for sterol binding, is clamped down by the introduction of two cysteine residues. When the protein is folded, these residues are close enough to form a disulfide bridge, thereby preventing the displacement of the loop required for the binding of sterols by Pry1 [14] (Fig. 1).

2 Materials

2.1 Equipment

1. Monolith NT.115 (Nanotemper Technologies, Munich, Germany).
2. MO. Control software (Nanotemper Technologies, Munich, Germany).
3. MO. Affinity Analysis software (Nanotemper Technologies, Munich, Germany).
4. Prism software (GraphPad, San Diego, CA).
5. Laptop.
6. Benchtop centrifuge.
7. 30 °C incubator.
8. Ultrasonic bath.
9. Vortex.

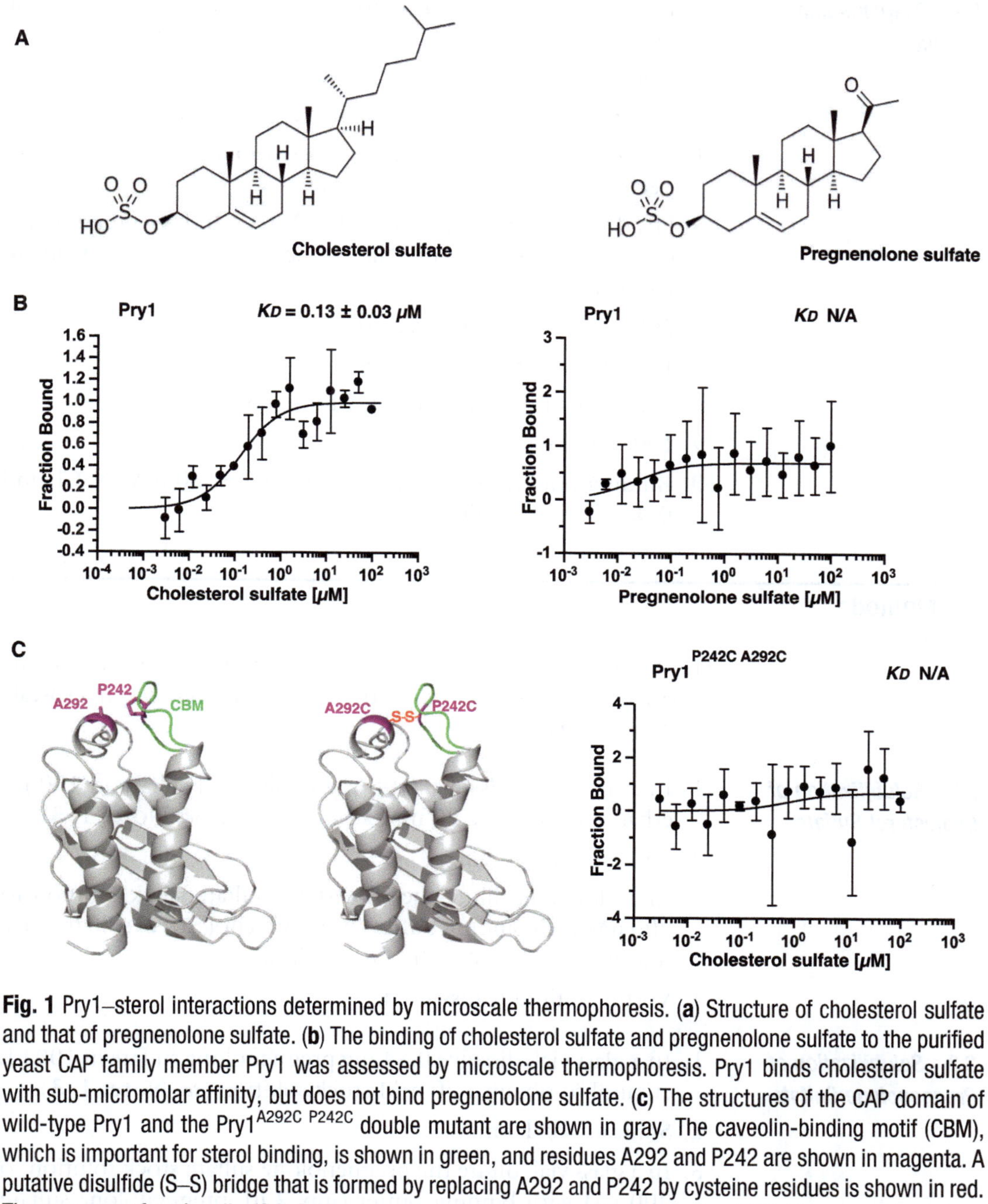

Fig. 1 Pry1–sterol interactions determined by microscale thermophoresis. (**a**) Structure of cholesterol sulfate and that of pregnenolone sulfate. (**b**) The binding of cholesterol sulfate and pregnenolone sulfate to the purified yeast CAP family member Pry1 was assessed by microscale thermophoresis. Pry1 binds cholesterol sulfate with sub-micromolar affinity, but does not bind pregnenolone sulfate. (**c**) The structures of the CAP domain of wild-type Pry1 and the Pry1$^{\text{A292C P242C}}$ double mutant are shown in gray. The caveolin-binding motif (CBM), which is important for sterol binding, is shown in green, and residues A292 and P242 are shown in magenta. A putative disulfide (S–S) bridge that is formed by replacing A292 and P242 by cysteine residues is shown in red. The presence of such a disulfide bridge is expected to block sterol binding by the CBM. The graph shows lack of binding of cholesterol sulfate by the Pry1$^{\text{P242C A292C}}$ double mutant version. All measurements were performed in triplicates, and the corresponding dissociation constants (K_D) are indicated. N/A not applicable

2.2 Reagents and Supplies

1. Purified His-tagged proteins: wild-type Pryl and a double mutant version, Pryl$^{A292C\ P242C}$ (produced in house).
2. Cholesterol sulfate (Sigma-Aldrich, St. Louis, MO).
3. Pregnenolone sulfate (Sigma-Aldrich, St. Louis, MO).
4. Monolith NT.115 standard glass capillaries (Nanotemper Technologies, Munich, Germany).
5. His-Tag Labeling Kit: RED-tris-NTA, 2nd Generation for protein labeling reagent (Nanotemper Technologies, Munich, Germany).
6. 200 μL PCR tube and cap strips.
7. 1.5 mL microtubes (Eppendorf).
8. 2 mL glass tubes (Interchim, Montluçon, France).
9. 100% ethanol.
10. Binding buffer: 20 mM Tris–HCl pH 7.5, 30 mM NaCl, and 0.05% Triton X-100.

3 Methods

The following is a description of a method to characterize the in vitro lipid-binding properties of a protein by microscale thermophoresis.

3.1 Solubilization of Cholesterol Sulfate

1. In a glass vial, dissolve 0.98 mg cholesterol sulfate in 100 μL ethanol to obtain a 20 mM stock solution (*see* **Notes 1–3**).
2. Sonicate for 10 min.
3. In a glass vial, dilute the cholesterol sulfate stock solution to 200 μM with binding buffer (mix 2 μL cholesterol sulfate stock solution in ethanol with 198 μL binding buffer).
4. Vortex until fully dissolved.

3.2 Solubilization of Pregnenolone Sulfate

1. In a glass vial, dissolve 0.84 mg pregnenolone sulfate in 100 μL ethanol to obtain a 20 mM stock solution (*see* **Notes 1–3**).
2. Sonicate for 10 min.
3. In a glass vial, dilute the pregnenolone sulfate stock solution to 200 μM with binding buffer (mix 2 μL pregnenolone sulfate stock solution in ethanol with 198 μL binding buffer).

3.3 Fluorescent Labeling of the His-Tagged Protein

1. Adjust the concentration of the purified His-tagged protein to 200 nM with binding buffer, in a final volume of 100 μL (*see* **Note 4**).
2. Adjust the concentration of the fluorescent dye solution provided with the His-Tag Labeling Kit to 100 nM, with binding buffer, in a final volume of 100 μL.

3. Mix 100 μL of the diluted His-tagged protein (200 nM) with 100 μL of the diluted fluorescent dye solution (100 nM). This gives a final concentration of the fluorescent-labeled protein of 100 nM (*see* **Note 5**).
4. Incubate for 30 min, in the dark, at room temperature.
5. After 30 min, centrifuge the fluorescent-labeled protein for 10 min at 4 °C, at 15000 × *g*.

3.4 Pretest of the Fluorescent-Labeled Protein

1. In a 1.5 mL Eppendorf tube, mix 10 μL of the fluorescent-labeled protein with 10 μL binding buffer.
2. Dip 2 glass capillaries in this diluted fluorescent-labeled protein solution (50 nM), insert the capillaries into the capillary holder, and run the pretest experiment on the Monolith NT.115.
3. This experiment is designed to test the quality of the fluorescent-labeled protein. There is no ligand present in this pretest (*see* **Note 6**).

3.5 Ligand-Binding Check

1. In a 1.5 mL Eppendorf tube, mix 25 μL of the fluorescent-labeled protein with 25 μL binding buffer.
2. In a second 1.5 mL Eppendorf tube, mix 25 μL of the fluorescent-labeled protein with 25 μL of the diluted cholesterol sulfate or pregnenolone sulfate solution.
3. Incubate the tubes for 20 min at 30 °C to allow for binding of the ligand to the fluorescent-labeled protein.
4. Centrifuge the tubes for 5 min at 4 °C, at 15000 × *g*.
5. Dip 4 glass capillaries in each one of the two Eppendorf tubes, insert the 8 capillaries into the capillary holder, and run the ligand-binding check experiment.
6. This experiment should reveal whether the ligand interacts with the fluorescent-labeled protein (*see* **Notes 7–10**).

3.6 Determining the Affinity of Ligand Binding

The ligand-binding reaction is carried out in a volume of 20 μL, containing a fixed concentration of the fluorescent-labeled protein (50 nM) and a varying concentration of the ligand, starting with 100 μM of either cholesterol sulfate or pregnenolone sulfate, diluted in binding buffer (*see* **Note 11**).

1. Label 2 PCR tube strips of 8 tubes each from 1 to 16.
2. Add 10 μL of binding buffer into tubes 2–16.
3. Add 20 μL of the 200 μM cholesterol sulfate or pregnenolone sulfate solution into tube 1.
4. Make a serial twofold dilution of the ligand by taking 10 μL from tube 1 and pipette it into tube 2. Then dilute from tube 2 into tube 3, etc. until you have reached tube 16.

5. Remove 10 μL from tube 16 and throw it away.
6. Add 10 μL of the fluorescent-labeled protein into all the 16 tubes (1–16); mix each one by pipetting up and down. The final volume in each tube is 20 μL. The concentration of the dye-labeled protein in each tube is 50 nM, and the concentration of the lipid ligand in tube 1 is 100 μM, in tube 2, 50 μM, etc.
7. Incubate for 20 min at 30 °C to allow for ligand binding.
8. Centrifuge for 1 min at 4 °C, at 15000 × *g*.
9. Dip 1 glass capillary into each one of the 16 PCR tubes.
10. Place the 16 glass capillaries into the capillary holder and run the affinity-binding experiment.
11. Calculate the dissociation constant (K_D) by plotting the fraction of the ligand-bound protein against the logarithm of the ligand concentrations, using the Prism software, option: 1, site-specific binding to fit a binding curve to the data points (*see* **Notes 12** and **13**).

4 Notes

1. It is recommended to use glass vials to dissolve hydrophobic ligands, such as cholesterol or pregnenolone, since such compounds have the tendency to stick to plastic.
2. As hydrophobic ligands such as lipids are very poorly soluble in water, it is advisable to screen for more water-soluble derivatives of the ligand of interest. This will allow to minimize the use of organic solvent in the binding reaction (i.e., ethanol, DMSO, etc.). If high volumes of organic solvents need to be used, it is recommended to add the organic solvent to binding buffer in which the protein is present to prevent its precipitation. In this case, it is important to verify by pretest experiment that the protein is not denatured.
3. Lipids are stored at −20 °C, in a tightly closed tube, to avoid oxidation. Proteins are stored at −80 °C after addition of 20% glycerol and flash frozen in liquid nitrogen.
4. It is not recommended to have imidazole in the reaction with the fluorescent dye as it will decrease the efficiency of the labeling reaction. Therefore, after the release of the His-tagged protein from the matrix with imidazole, the protein buffer should be exchanged to a buffer without imidazole, i.e., binding buffer, using a small desalting spin column for buffer exchange (i.e., Zeba™ Spin Desalting Column, Thermo Fisher Scientific, Waltham, MA).

5. For the His-Tag Labeling Kit, the Pico Red detector on the Monolith should be used. This detector is optimized for red fluorophores (excitation 600–640 nm, emission 660–725 nm). Affinity to GFP-tagged proteins can also be measured using the Nano Blue detector fluorophores (excitation 465–490 nm, emission 500–550 nm) of the Monolith. In this case there is no need for additional fluorescent labeling of the protein. Note, however, that the sensitivity of the red channel is much higher, as compared to that of the blue channel.
6. The pretest of the fluorescent-labeled protein is important to check the efficiency of the fluorescent labeling reaction, to detect protein aggregation or the binding of the fluorescent-labeled protein to the wall of the glass capillary. If the labeling strategy is not efficient or if the protein is not His-tagged, a different labeling strategy can be employed, i.e., RED-Maleimide (cysteine reactive) or RED-NHS (amine reactive) dye labeling reactions are available. If the protein is aggregating, centrifuge it at 4 °C, at high speed (15,000 × *g*) or optimize the buffer used for the binding reaction. The glass capillaries used in this experiment are the standard ones provided by Nanotemper Technologies. If the fluorescent compound adheres to the walls of the glass capillaries, optimize the composition of the binding buffer, and/or change the type of capillary used. To reduce adherence of the protein or the lipid ligand to the capillary wall, hydrophobic and premium capillaries are available from Nanotemper Technologies.
7. The ligand-binding check experiment is designed to detect the quality of binding between the ligand and the target protein. In case no binding is detected in this check experiment, the binding conditions must first be optimized before determining the ligand-binding affinity. Concentrations of the ligand and/or the target can be varied, or the composition of the ligand-binding buffer could be modified.
8. Since binding of cholesterol to Pry1, and possibly to other proteins as well, is strongly dependent on the pH, it is recommended to perform a set of preliminary assays using binding buffers covering a range of different pH values.
9. The Monolith uses microscale thermophoresis to quantify molecular interactions between a ligand and a target. Either the ligand or the target needs to contain a fluorescent label that is detectable by the instrument. The instruments can be obtained with different filter sets to allow for detection of red, blue, or green fluorophores. In addition, label-free protocols, which relay on the detection of the intrinsic fluorescence of proteins (UV channel), are available as well but need specialized instruments [15]. The measurement of the ligand–

target interaction is based on time-dependent changes in fluorescence intensity of the fluorophore-containing compound upon application of a temperature gradient (Nanotemper Technologies) [5, 6].

10. The laser power is set automatically to 100% and the starting temperature of measurement to 30 °C. These parameters can be changed and optimized by the user.
11. The ligand-binding affinity experiment measures quantitative binding between the ligand and the target. The K_D is immediately calculated by the program, MO. Affinity Analysis, and the data can be imported into Prism for further analysis.
12. If many of the data points are derived from the unbound state, try to increase the starting concentration of the ligand. On the other hand, if many of the data points are from the bound state, the starting concentration of the ligand may be lowered. In case the concentration of the ligand cannot be changed, the concentration of labeled protein can be varied.
13. This method for determining the affinity for a ligand is fast and safe and requires very low amounts of the purified protein. On the other hand, it is also very sensitive, and slight changes in the quality and/or the concentration of the protein will affect the resulting dissociation constant.

Acknowledgments

We thank Stéphanie Cottier for critically reading the manuscript. This work was supported by the Swiss National Science Foundation (31003A_173003) and (310030_207870).

References

1. Miller WL, Auchus RJ (2011) The molecular biology, biochemistry, and physiology of human steroidogenesis and its disorders. Endocr Rev 32:81–151
2. Prinz WA, Toulmay A, Balla T (2020) The functional universe of membrane contact sites. Nat Rev Mol Cell Biol 21:7–24
3. de Saint-Jean M, Delfosse V, Douguet D et al (2011) Osh4p exchanges sterols for phosphatidylinositol 4-phosphate between lipid bilayers. J Cell Biol 195:965–978
4. Darwiche R, Schneiter R (2017) A ligand-binding assay to measure the affinity and specificity of sterol-binding proteins in vitro. Methods Mol Biol 1645:361–368
5. Wienken CJ, Baaske P, Rothbauer U et al (2010) Protein-binding assays in biological liquids using microscale thermophoresis. Nat Commun 1:100
6. Seidel SA, Dijkman PM, Lea WA et al (2013) Microscale thermophoresis quantifies biomolecular interactions under previously challenging conditions. Methods 59:301–315
7. Choudhary V, Schneiter R (2012) Pathogen-Related Yeast (PRY) proteins and members of the CAP superfamily are secreted sterol-binding proteins. Proc Natl Acad Sci U S A 109:16882–16887
8. Darwiche R, El Atab O, Cottier S et al (2018) The function of yeast CAP family proteins in lipid export, mating, and pathogen defense. FEBS Lett 592:1304–1311
9. Gibbs GM, Roelants K, O'Bryan MK (2008) The CAP superfamily: cysteine-rich secretory

proteins, antigen 5, and pathogenesis-related 1 proteins--roles in reproduction, cancer, and immune defense. Endocr Rev 29:865–897
10. Cantacessi C, Gasser RB (2012) SCP/TAPS proteins in helminths--where to from now? Mol Cell Probes 26:54–59
11. Schneiter R, Di Pietro A (2013) The CAP protein superfamily: function in sterol export and fungal virulence. Biomol Concepts 4:519–525
12. Wilbers RHP, Schneiter R, Holterman MHM et al (2018) Secreted venom allergen-like proteins of helminths: conserved modulators of host responses in animals and plants. PLoS Pathog 14:e1007300
13. Choudhary V, Darwiche R, Gfeller D et al (2014) The caveolin-binding motif of the pathogen-related yeast protein Pry1, a member of the CAP protein superfamily, is required for in vivo export of cholesteryl acetate. J Lipid Res 55:883–894
14. Cottier S, Darwiche R, Meyenhofer F et al (2020) The yeast cell wall protein Pry3 inhibits mating through highly conserved residues within the CAP domain. Biol Open 9: bio053470
15. Seidel SAI, Wienken CJ, Geissler S et al (2012) Label-free microscale thermophoresis discriminates sites and affinity of protein–ligand binding. Angew Chem Int Ed 51:10656–10659

Part IV

Fermentation and Biocatalysis for Steroids Production

Chapter 11

Cultivation of *Mycolicibacterium* spp. Mutants in Miniaturized and High-Throughput Format to Characterize Their Growth, Phytosterol Conversion Ability, and Resistance to the Steroid Products

Simone Balzer Le, Anna Nordborg, Kjell Domaas Josefsen, Silje Malene Olsen, and Håvard Sletta

Abstract

This chapter describes methods for cultivation and characterization of the growth of *Mycolicibacterium* spp. mutants in a microbioreactor system in the presence of steroids and/or phytosterols followed by high-throughput mass spectrometry analysis to describe their ability to convert phytosterols into the target steroid androstenedione (AD). We focus on *Mycolicibacterium neoaurum* NRRL B-3805 Δ*kstD* which can convert phytosterol into androstenedione (AD) as one of its major steroid products, and mutants thereof with increased tolerance towards this end-product. By using BioLector 48-well plates with optodes at the bottom of each well, bacterial growth can be monitored online despite the turbidity of the growth medium resulting from non-dissolved phytosterol and steroid particles. To cope with the large number of samples that accumulate during growth experiments in microbioreactors and similar formats (e.g., microtiter plates), protocols for extraction and subsequent RapidFire–MS analysis are presented. This reduces the analysis time per sample to 10 s from 10 min required for regular LC–MS analysis.

Key words *Mycolicibacterium*, Phytosterol, Steroids, End-product inhibition, High-throughput, Microbioreactor cultivation, BioLector, RapidFire, Mass Spectrometry (MS)

1 Introduction

Development of an efficient bioprocess requires several systematic steps. These may include screening of strains and strain libraries, design of a suitable medium, and adaption of the process conditions to deal with particular challenges related to the strain, the substrate, and the product. Evaluation of multiple parameters and conditions can be time-consuming, and appropriate process development and analysis methods that can be used to perform and evaluate a large number of experiments to identify promising conditions and candidates are needed. For conversion of phytosterols to steroids using

Carlos Barreiro and José-Luis Barredo (eds.), *Microbial Steroids: Methods and Protocols*, Methods in Molecular Biology, vol. 2704, https://doi.org/10.1007/978-1-0716-3385-4_11,

mycobacteria, several challenges need to be addressed. Two of them are the limited bioavailability of the sterols for microbial conversion due to the complex cellular envelope of mycobacteria, and the end-product inhibition caused by the steroid products [1].

Efficient conversion requires high bioavailability of the hydrophobic sterols. The structure of the cellular envelope and the uptake systems play a key role to ensure access to the enzymes involved in the bioconversion process. To improve bioavailability, mutations to modify the cellular envelope of mycobacteria have been suggested [1, 2]. Another important obstacle to efficient microbial steroid bioconversion is the cellular toxicity of the steroid end-products [1]. Use of oil-based media [3] to physically separate the products and the cells and addition of methyl-β-cyclodextrin (MBC) or other cyclodextrins for steroid sequestration [4] are two methods applied for improved bioconversion. Another approach is to develop strains with increased resistance towards the toxic end-products [5]. An ideal strain should have both physiological functions to efficiently convert phytosterols to steroids and functions to eliminate the toxic effects of the products. Creating such a strain is a challenge. Targeting certain genes by genetic engineering is one approach, but it is also of interest to identify new possible targets for engineering. Non-targeted mutagenesis has the benefit that mutants can be generated with changes in several genes and regulatory functions simultaneously. Chemical mutagenesis procedures for mycobacteria have been described previously [5, 6]. Here, mutants with an ADD resistance phenotype were identified in agar-based screening using nutrient agar containing the target steroid ADD. Using such a preselection procedure, in agar-based or liquid medium, a mutant library can be enriched in mutants with increased steroid tolerance. In combination with today's sequencing capacities and available genome sequences, mutations can easily be identified to suggest new engineering targets.

To characterize these preselected mutants further, bioconversion studies can be performed in microtiter plates using protocols previously described [7]. This enables an in-depth characterization of mutants with the necessary biological replica in a high-throughput format. The problem is that the presence of non-dissolved phytosterols and steroids particles makes the medium turbid and this makes it difficult to monitor growth in liquid medium using standard optical density measurements. Indirect measures for biomass such as methylene blue conversion can be applied as described for filamentous organisms [8]. However, this method is not compatible with time-critical monitoring of growth when sampling needs to be timed or in high-throughput applications with small culture volumes. Similarly, determination of cell dry weight after washing with detergents to remove bulk phytosterols and steroids is not suitable for small volumes. Instead, cultivations in microbioreactors like the BioLector system [9] can be performed using special microreactor plates that enable online monitoring of biomass as scattered light intensity, pH, and dissolved oxygen (DO). Even subtle difference between mutants can be detected due to the possibility of closely following the experiment online. During miniaturized cultivations, samples can be taken to monitor phytosterol substrates and steroid products. To cope with the large number of samples, (partially) automated protocols for sample handling and rapid analysis help to increase the experimental throughput.

Phytosterols and steroids are often analyzed by gas chromatography (GC) or liquid chromatography (HPLC), with or without coupling to mass spectrometry (MS). Highly sensitive analysis by HPLC coupled to MS for the analysis of phytosterol to steroid conversion by *Mycolicibacterium* is reported in the chapter of this book entitled "Bioconversion of Phytosterols into Androstenedione by Mycolicibacterium." However, for efficient analysis of large number of samples, such as during screening to identify a high producer in a strain library, or during evaluation of large numbers of process parameters, shorter analysis times are needed. In HPLC, shorter analysis times can be achieved by using faster gradients, but the time frame is still minutes per sample. An alternative technique to HPLC–MS analysis, yielding shorter analysis times, relies not on separation of the steroids to be analyzed but instead on an ultrafast sample cleanup prior to mass spectrometry analysis. The RapidFire 365 instrument is an instrument for automated solid phase extraction that is compatible with sample plate formats up to 384 and even 1536-well plates. The coupling of RapidFire to mass spectrometric detection and use of multiple reaction monitoring (MRM) mode ensures specific detection of the selected analytes and thus enables rapid and highly specific analysis of large numbers of samples.

In this chapter, we describe protocols for cultivation of *M. neoaurum* NRRL B-3805 Δ*kstD* and mutants thereof in the BioLector system. The deletion of the *kstD* gene in this strain encoding 3-ketosteroid 1(2)-dehydrogenase leads to accumulation of the key pharmaceutical steroid intermediate AD since further conversion of AD to ADD is abolished (Fig. 1). This also means that the first step towards steroid ring degradation is prevented. Using the BioLector system, mutants can be characterized based on their ability to grow in the presence of AD and to convert phytosterol into AD and other steroid products. Samples from several experiments and time points can be collected for convenience. To process many samples simultaneously and reduce analysis time, methods developed for phytosterol and steroid extraction followed by RapidFire–MS analysis are presented. Based on these experiments, mutants can be selected for further characterizations and upscaling experiments described elsewhere [3, 10].

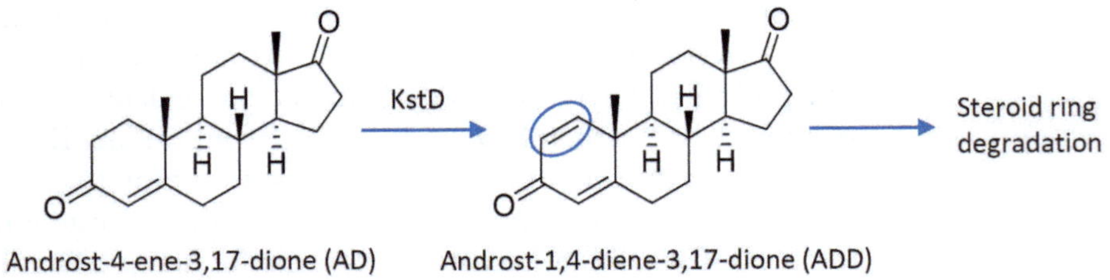

Fig. 1 Illustration of the enzymatic activity encoded by the *kstD* gene which is deleted in *M. neoaurum* NRRL B-3805 Δ*kstD*

2 Materials

The experiments described here require standard laboratory equipment such as shaking and static incubators, spectrophotometers to read DO_{660}, clean benches, multichannel and single channel pipettes, ventilation hoods, and freezers for sample storage. Protective gear is required. Consult the safety data sheets for handling and discarding harmful chemicals such as steroids and solvents. Special equipment and material are listed below.

2.1 Cultivation of Mycolicibacterium *and Sampling*

1. Microbioreactor system: BioLector I with the potential for 48 parallel cultivations (Beckman Coulter) (*see* **Note 1**).
2. 48-well microtiter plates (MTP) with round wells covered with gas-permeable sealing foil with evaporation reduction layer to minimize evaporation.
3. Polypropylene 2 mL 96 square deep-well plate for sample storage and solvent extraction. The MASTERBLOCK® (Greiner Bio-One) is well suited (*see* **Note 2**).
4. Electronic pipette with adjustable spacing to compensate for the different spacing between wells of BioLector 48-well and 96-well plates. The Voyager 8-channel pipette (50–1250 μL; INTEGRA BioSciences AG) and corresponding pipette tips (e.g., INTEGRA 1250 μL GRIPTIPS™) were well suited. Use 13.5 mm for BioLector plates and 9.0 mm for regular 96-well plates.
5. *M. neoaurum* NRRL B-3805 Δ*kstD* and mutants thereof (*see* **Note 3**).
6. M3 medium (*see* **Note 4**): Dissolve 5 g of glycerol, 5 g of yeast extract, 5 g of peptone, 1.5 g of $(NH_4)_2HPO_4$, 0.5 g of KH_2PO_4, and 0.5 g of K_2HPO_4 in 940 mL distilled water. Adjust pH to 7.0–7.2 and autoclave (121 °C, 20 min). Add 50 mL of M3 metal stock (see below) and 10 mL of 0.6 M NaOH to prepare the final medium. Addition of 0.05–0.1% Tween® 80 is optional (*see* **Note 5**).

7. M3 metal stock (*see* **Note 6**): Dissolve 4 g of $MgSO_4 \times 7\ H_2O$, 100 mg of $FeSO_4 \times 7\ H_2O$, 40 mg of $ZnSO_4 \times 7\ H_2O$, and 20 mL of HCl in 980 mL distilled water. Sterilize by passing through a 0.22 μm filter.
8. 0.6 M NaOH solution: Dissolve 2.4 g of NaOH pellets in distilled water and adjust to 100 mL. Be aware of the heat that develops and cool the solution to room temperature prior to final volume adjustment. Sterilize by passing through a 0.22 μm filter
9. AD stock: Prepare a 40 g/L stock solution of AD in methanol and filter sterilize (0.22 μm filter).
10. Phytosterol suspension: Prepare a 10 g/L suspension in M3 medium. For that, transfer phytosterol powder to an Erlenmeyer flask, e.g., 0.5 g. Add a drop of distilled water. Cover the flask and autoclave (121 °C, 20 min). Add sterile M3 medium (e.g., 50 mL) and sonicate in an ultrasonic bath, 3×5 min with manual shaking in between. This creates a fine suspension (*see* **Note** 7).
11. Tween® 80 stock: Prepare a 10% (v/v) solution in distilled water and sterilize by passing through a 0.22 μm filter.
12. Ultrasonic water bath and a laboratory support stand with clamps.

2.2 Equipment and Materials for Extraction and Mass Spectrometry Analysis

1. Liquid handling station for automated sample extraction and dilution such as the Biomek i7 Automated Workstation (Beckman Coulter) with corresponding software (Biomek 5 version). This should ideally be equipped with a shaker like BioShake 300 elm (QInstruments GmbH or equivalent). However, liquid handling steps can also be performed manually using multichannel pipettes.
2. Pipette tips compatible with liquid handling equipment such as Biomek AP96 P250 and P50 (Beckman Coulter).
3. Chemically resistant (silicone) covers for plates (e.g., Axygen® Impermamats).
4. Microplate heat sealer and suitable film, e.g., ALPS 3000™ Automated Microplate Sealer and Easy Pierce 20 μm heat sealing tape (Thermo Fisher). Make sure the tape has good solvent resistance.
5. Centrifuge with swing-out rotors and adapters for microtiter plates.
6. Extraction solvents: Ethyl acetate, acetonitrile, and isopropanol (all analytical or LC–MS grade).
7. RapidFire 365 instrument coupled to an Agilent 6490 QQQ (Agilent Technologies Inc.) equipped with atmospheric pressure chemical ionization (APCI) ionization source.

8. Cartridge packed with C8 material for reversed-phase mode solid phase extraction sample preparation using the RapidFire 365.
9. Reagents for eluent preparation: trifluoroacetic acid (TFA), methanol, isopropanol (all LC–MS grade), and distilled water.
10. Androst-4-ene-3,17-dione (AD; CAS 63-05-8) of high purity (analytical grade).
11. Androst-1,4-diene-3,17-dione (ADD; CAS 897-06-3) of high purity (analytical grade).
12. Stigmasterol (CAS 83-48-7) of high purity (analytical grade).
13. β-Sitosterol (CAS 83-46-5) of high purity (analytical grade).

3 Methods

3.1 Cultivation of Strains

1. Make precultures in shake flasks. Thaw a working ampule (*see* **Note 8**) at room temperature. When most of the glycerol stock has melted, transfer 1% of the stock to fresh M3 medium (e.g., 0.1–10 mL medium in a 125 mL shake flask; liquid volume ≤ one tenth of the flask volume). To make precultures with cell mass from agar plates, pick a single colony from a fresh agar plate with an inoculation loop and transfer it to 10 mL medium.
2. Incubate at 30 °C and 200 rpm in a shaking incubator (e.g., with a shaking amplitude of 25 mm) for one day (or more if needed) to obtain a culture that is still in the logarithmic growth phase.
3. The day before starting the microbioreactor experiment, transfer a suitable volume of AD stock solution (40 g/L) to a sterile Erlenmeyer flask. Cover the flask with a piece of sterile autoclave paper (previously sterilized by heat sterilization 121 °C for 20 min) in a covered glass beaker and place the Erlenmeyer flask at 37 °C overnight for the methanol to evaporate.
4. Add sterile M3 medium to reach a final AD concentration of 3 g/L. Briefly mix manually and use an ultrasonic bath to create a fine suspension. Mix in between.
5. While working in a clean bench, transfer 850 μL of half-strength M3 medium (*see* **Note 4**) from a single pipetting reservoir into each well using the Voyager pipette with 13.5 mm tip spacing. If the aim is to study phytosterol conversion, transfer 300 μL of phytosterol suspension and 550 μL of half-strength M3 medium to each well.
6. Add 500 μL of a 3 g/L AD stock solution to each well (*see* **Note 9**). For controls, add regular M3 medium instead.

Table 1
BioLector settings

Parameter	Value
Temperature	30 °C
Humidity	85%
Shaking frequency	900 rpm
Cycle time	15 min
Experiment time	48–96 h
Measurement data	Biomass G:25 Biomass G:35 Biomass G:40 pH pO_2
Lot number	*Can be found on each plate*

7. At last, add 150 μL preculture to each well.
8. Seal the plate with gas-permeable sealing foil with evaporation reduction layer.
9. Prepare the BioLector instrument according to the operation manual.
10. Turn on the BioLector instrument and connect to the instrument using the BioLection software (→ General setup → Synchronize). In the BioLection software follow these steps to create a new protocol for each experimental run (Choose → Experiment Setup → Protocol → New). Then enter the experimental settings according to Table 1 (*see* **Note 10**). Perform a calibration by choosing → Settings → Select the correct lot number and temperature for the experiment, i.e., thereby loading the corresponding calibration file (*see* **Note 11**).
11. Check that there is enough water in the water flask used for humidity control (should be 2/3 filled). If not, refill with sterile distilled water using a sterile pipette.
12. Insert and attach the BioLector plate.
13. Turn to the screen of the BioLector instrument. Under start assistant → Select your protocol. Make sure the settings are correct, and that humidity is set to 85% → press next and start the experiment.
14. During the experiment, parameters like biomass can be followed online to enable the experimenter to select suitable sampling times and determine an appropriate end of the experiment.

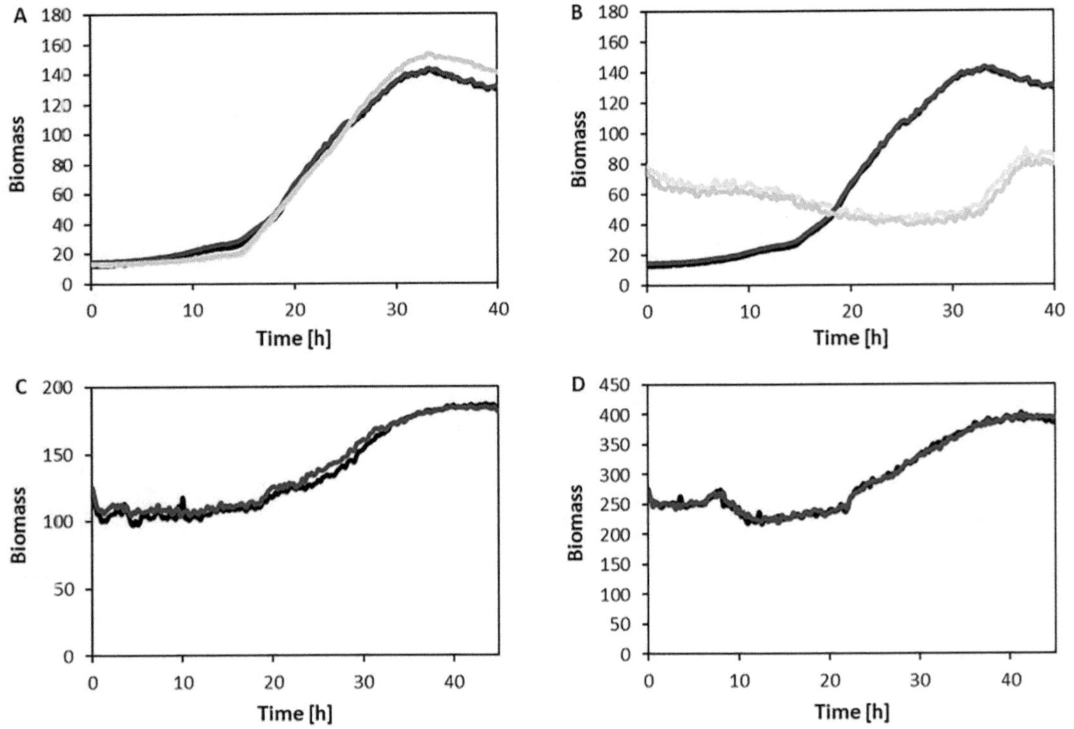

Fig. 2 Growth curves of *M. neoaurum* NRRL B-3805 Δ*kstD* cultivated in BioLector 48-well plates (round bottom) in M3 medium with different additives. (**a**) Regular M3 medium (half-strength) without (dark gray curves) or with 12 mM MBC (light gray curves) (**b**) Regular M3 medium (half strength) without (dark gray curves) or with 1 g/L AD (light gray curves). (**c**) M3 medium supplemented with 2 g/L phytosterol. (**d**) M3 medium supplemented with 5 g/L phytosterol. Curves from two parallel cultivations are presented

15. To stop the experiment, press "Stop" and take out the plate.
16. Data transfer: In the BioLection software → Data management → Transfer results from BioLector (a .csv file will be generated). Load the .csv file in the BioLection software on your personal computer (or any computer with Excel) → Calibration → Settings (select the correct Lot nr. and temperature) → calibrate → (*see* **Note 11**) Data Management → Export to an Excel template (an Excel file will be generated).
17. Extract the desired information and prepare plots according to your needs. Examples of growth curves generated during BioLector cultivations are shown in Fig. 2.

3.2 Sampling Procedures

1. During cultivation, the microbioreactor experiment can be paused to withdraw samples. Press the pause button and take out the plate for sample collection.
2. For collection of cell suspension for steroid analysis, take samples at several time points from the same well. Mix the content of the wells well by pipetting up and down several times.

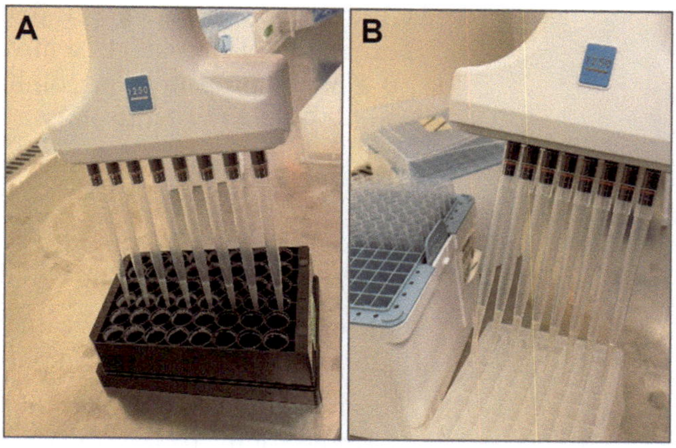

Fig. 3 Transfer of samples from BioLector plates (**a**) to 96-well storage microplates (**b**) using INTEGRA pipettes with adjustable tip spacing

Immediately withdraw 100 μL sample and transfer it to a polypropylene storage block (Fig. 3). For convenience, take samples from a row of 8 wells in a BioLector plate and transfer them to a column of 8 wells in a 96-well plate. In this way, the samples from two BioLector plates can be combined in one 96-well plate.

3. For collection of samples from phytosterol conversion experiments, we recommend using the sacrificial well approach. To do this, mix the content of the wells well by pipetting up and down several times and transfer the entire content of the well to a polypropylene storage block (Fig. 3). For convenience, collect samples from BioLector rows to 96-well plate columns. In this way, two sets of samples can be collected in one 96-well plate.
4. Cover the BioLector plate again. We recommend replacing the sealing foil at intervals, e.g., after 2 days.
5. Continue the cultivation.
6. Cover the storage block with aluminum sealing film and store at −20 °C until analysis.
7. Several storage blocks from multiple experiments with multiple samplings can be collected for combined sample extraction and analysis.

3.3 Extraction of Phytosterols and Steroids

1. Thaw the content of the storage blocks in a shaking incubator for microtiter plates (25 °C, 700_rpm, 3 mm amplitude) for 1 h.
2. Centrifuge for 5 min at 850 × *g* using a swing-out rotor and adapters for microtiter plates. This step collects the liquid at the bottom of each well.

3. Add an appropriate solvent: 0.9 mL ethyl acetate (primarily for extraction of steroids) or a mixture of 0.5 mL acetonitrile and 0.4 mL isopropanol (for combined extraction of phytosterols and steroids) (*see* **Note 12**).
4. Cover plates tightly with chemically resistant silicone covers (*see* **Note 13**).
5. Shake the plates thoroughly for 3 min (using a BioShake 3000 elm) with 2.0 mm mixing orbit at 1200 rpm).
6. Incubate the plates in a shaking incubator for microtiter plates (25 °C, 900 rpm, 3 mm amplitude) for 30 min.
7. Centrifuge for 20 min at 4800 × *g* using a swing-out rotor and adapters for microtiter plates. Two phases are formed here. The upper solvent phase contains phytosterols and steroids and is going to be used further.
8. Dilute the upper solvent phase 1:100 in methanol. For that, pipet 990 μL methanol into each well of a 96-polypropylene deep-well plate. Transfer 10 μL upper solvent phase (*see* **Note 14**).
9. Split the content of each well into four parallel wells on a 384-polypropylene deep-well plate. For that, transfer 90 μL to each of four quadrants as depicted in Fig. 4.
10. Seal the 384 deep-well plate with aluminum sealing foil using a microplate heat sealer at 170 °C for 4 s.
11. Store at −20 °C until MS analysis.

3.4 RapidFire–MS Analysis and Data Analysis

1. Prepare standards and eluents.
   (a) Stock standards—phytosterols: 1–2 mg/mL in chloroform. Weigh out accurate amounts of the individual phytosterols and add solvent (chloroform) to a phytosterol

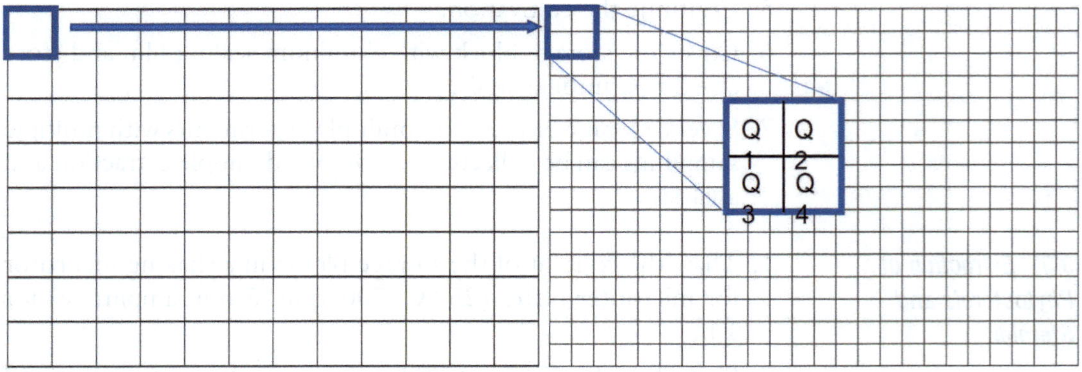

Fig. 4 Suggested transfer scheme of diluted solvent extract in preparation for RapidFire–MS analysis. The content from a single well in the 96 wp is transferred to four adjacent quadrants of a 384 wp (e.g., four times 90 μL)

concentration of 1–2 mg/mL (*see* **Note 15**). Freeze solutions at −20 °C. Stock standard—steroids: 1–2 mg/mL in ethyl acetate. Weigh out accurate amount of the individual steroids and add solvent (ethyl acetate) to a steroid concentration of 1–2 mg/mL (*see* **Note 15**). Freeze solutions at −20 °C.

(b) Working standards: 1000 μM in ethyl acetate. Dilute the individual stock standards to 1000 μM in ethyl acetate. Prepare a mixed working standard of all analytes at individual concentrations of 1000 μM in ethyl acetate. Freeze solutions at −20 °C.

(c) Calibration standards: 0, 0.01, 0.1, 1, 10, 100, and 1000 μM in methanol. Prepare calibration standards by diluting the mixed 1000 μM standard in methanol. Also prepare a blank of only methanol. Freeze solutions at −20 °C.

(d) Eluent A for RapidFire–MS analysis (per liter in distilled water): 0.1%v/v TFA. To 900 mL distilled water, add 1 mL of trifluoroacetic acid (analytical grade >99%) in a 1 L volumetric flask. Mix thoroughly and adjust the volume to 1 L with distilled water. Pour the resulting 0.1% TFA solution into an eluent bottle (*see* **Note 16**).

(e) Eluent B for RapidFire–MS analysis: 60:40 methanol/isopropanol (LC–MS grade). Add 600 mL of methanol and 400 mL of isopropanol to an eluent bottle and mix by inverting (*see* **Note 17**).

2. Establish the initial RapidFire operating conditions: Flow rate 1.25 mL/min., eluent A (90%), eluent B (10%). The cartridge used is a C8 pre-packed cartridge for reversed-phase mode solid phase extraction.

3. Establish the mass spectrometer operating conditions: APCI ionization. Positive mode. Multiple reaction monitoring. Gas temperature 250 °C, APCI heater 300 °C, gas flow rate 11 L/min, nebulizer 40 psi, and capillary 3500 V (+). MRM settings (also given in Table 2): AD (287→97), ADD (285→121), stigmasterol (395→297), and β-sitosterol (397→161). Use the following settings for all analytes: 4 V cell acceleration voltage, 380 V fragmentor, and 15 V collision energy (CE). Compounds included in the RapidFire–MS analysis are depicted in Fig. 5.

4. Elution program used for the RapidFire: Sample loading at 10% eluent B and elution at 99% eluent B. Elution time for all analytes is 0.13 min.

5. Inject a series of blank samples (methanol) and standards. Then inject the samples from the 384-well plate. Inject a set of

Table 2
Mass spectrometer instrument settings for RapidFire–MS analysis

Analyte	Precursor ion	Product ion	Elution time
Androst-4-ene-3,17-dione (AD)	287	97	0.13 min
Androst-1,4-diene-3,17-dione (ADD)	285	121	0.13 min
Stigmasterol	395	297	0.13 min
β-Sitosterol	397	161	0.13 min

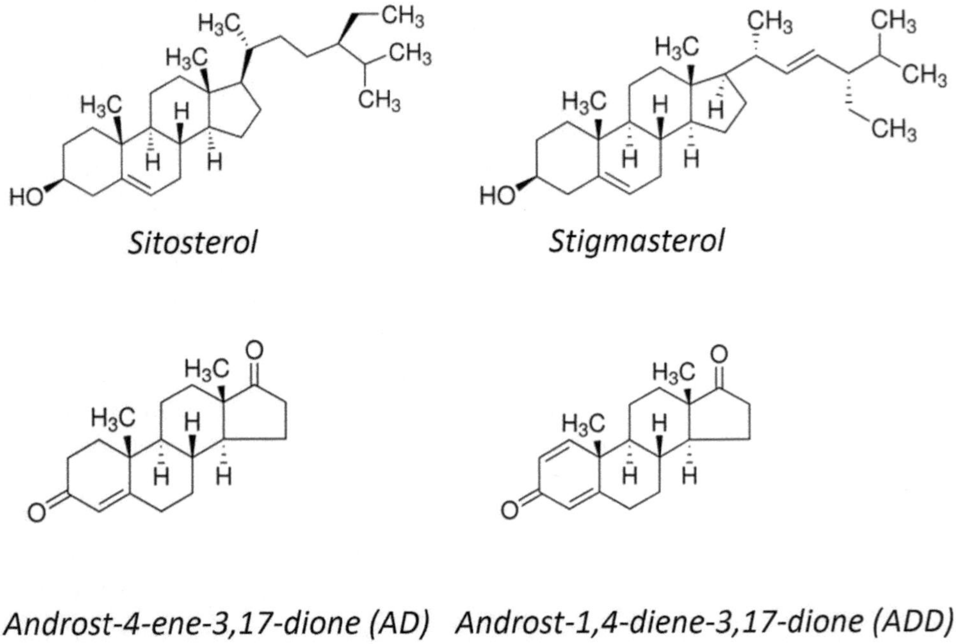

Fig. 5 Phytosterols and steroids included in RapidFire–MS analysis

standards and blanks throughout the sample analysis to monitor the instrument performance. End the analysis by injection of a series of standards and blanks.

6. Transfer the collected MS/MS data to the MS software.
7. Determine the peak area per concentration of the analyte using standards with concentration range 0–10 μM for steroids and 0–1000 μM for phytosterols (*see* **Note 18**).
8. Divide the peak area of the analyte in the sample with the peak area per concentration determined from the standards to obtain the concentration of the analyte in the analyzed sample. An example of output from RapidFire–MS analysis of samples from miniaturized bioconversion experiments in BioLector is shown in Fig. 6.

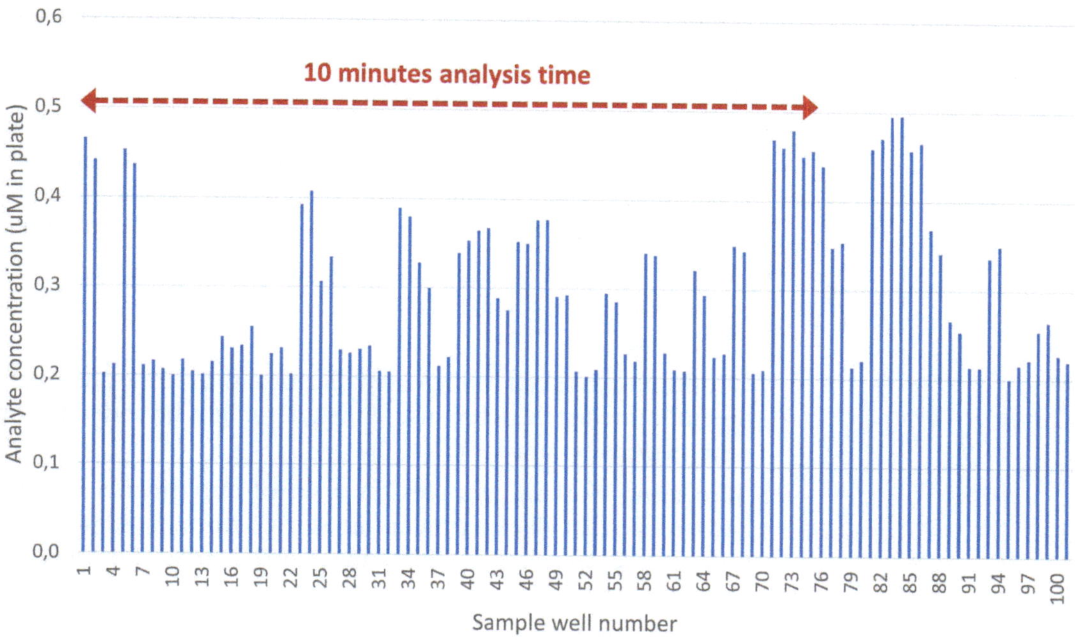

Fig. 6 Output from RapidFire–MS analysis of AD levels in samples from BioLector experiments. Analysis time per sample is <10 s

9. Determine the concentration of the analyte in the original culture by considering pre-concentrations and dilutions during sample workup.

4 Notes

1. Miniaturized cultivations may also be performed in microtiter plates as described elsewhere.
2. Polypropylene offers good resistance to the organic solvents used for phytosterol and steroid extraction. Avoid polystyrene for these purposes.
3. We have tested the protocol with preselected mutants from a N-methyl-N′-nitro-N-nitrosoguanidine (NTG) library of *M. neoaurum* NRRL B-3805 Δ*kstD*. The mutants were able to grow on M3 agar supplemented with 2.5 mg/mL AD. In liquid cultures, a lower concentration, 1 mg/mL AD, was sufficient to detect mutants with improved tolerance to the steroid relative to the parent strain.
4. The complex M3 described here may be replaced by semi-defined or even defined media described in [10]. The cultivation time must be adjusted and adaptation of the strains to the media is recommended. For precultures, full-strength M3 medium was used. During BioLector and other microtiter

cultivations, half-strength M3 medium made by mixing 1 volume M3 medium with 1 volume sterile distilled water can be used.

5. The presence of Tween® 80 and other detergents should be avoided due to interference with the subsequent MS analysis resulting in a background signal. Tween 80 might be added to the precultures, but not during cultivation in microbioreactors.
6. By adding $MgSO_4$ after heat sterilization of the medium, precipitation of $MgNH_4PO_4$ ($k_{sp} = 3 \cdot 10^{-13}$ at 25 °C) is minimized.
7. Concentrations can be increased. Perform the ultrasound treatment immediately before use. A magnet might be added to stir the suspension until use.
8. Preparing working ampules for a set of experiments has the advantage that all experiments are started from the same batch of glycerol stock and this can improve reproducibility of experiments [10].
9. The AD does not dissolve in M3. Make sure to mix well, e.g., by pipetting up and down several times before you add AD to the wells.
10. During cultivation, fermentation parameters are measured online. The frequency of data point acquisition can be varied. Here, 15 min intervals were chosen.
11. If the calibration data for the selected temperature are not listed yet, add it at the given temperature according to m2p instructions [11].
12. For these liquid handling steps, robotic scripts can be created based on the descriptions given here.
13. An extra lid attached with rubber band may be added on top. Silicon mats are re-usable.
14. When moving the extraction plates from the centrifuge to the workspace where the subsequent steps are performed, be careful not to disturb the two phases. If pipette tips with a volume <250 μL are used to transfer the methanol, split the intended 990 μL into smaller volumes (e.g., 4 times 230 μL and once 70 μL).
15. Solubility of especially the phytosterols is limited in the eluents used in the RapidFire–MS analysis. Therefore, always prepare the stock solutions at high concentrations in a good solvent (such as chloroform for phytosterols and ethyl acetate for steroids) and prepare only the diluted standards in ethyl acetate or methanol.
16. Prepare eluent A (0.1% TFA in water) fresh prior to analysis.

17. All reagents should be of analytical or HPLC grade unless otherwise stated. All solvents used for the HPLC eluents should be of mass spectrometric grade. Water used should be deionized, e.g., on a Milli-Q system.
18. The limit of detection and suitable calibration range can be different for different analytes. For steroids and phytosterols we used the concentration ranges 0–10 μM and 0–1000 μM, respectively. The limit of detection (LOD) for the two steroids (AD and ADD) was found to be 0.01 μM, while the LOD for sitosterol and stigmasterol was 0.1 μM and 1 μM, respectively. The higher LOD for the phytosterols is due to some carryover in the analysis. All concentrations are concentrations in the final sample plate.

Acknowledgments

This work was supported by a grant of the European Union program ERA CoBioTech Syntheroids. Funding was received from the European Union's Horizon 2020 Research and Innovation Programme under grant agreement no. [722361] and financed through the Research Council of Norway (project number 285849). The authors thank Andrea Draget Hoel, Marit Dyrendahl, and Susan Maleki for their technical support and Curia Global for supplying AD. They are also grateful to have had the Syntheroids consortium members as valuable discussion partners during the work.

References

1. Wang F-Q, Yao K, Wei D (2011) From soybean Phytosterols to steroid hormones. In: Soybean and health
2. Xiong L-B, Liu H-H, Song X-W, Meng X-G, Liu X-Z, Ji Y-Q, Wang F-Q, Wei D-Z (2020) Improving the biotransformation of phytosterols to 9α-hydroxy-4-androstene-3,17-dione by deleting embC associated with the assembly of cell envelope in Mycobacterium neoaurum. J Biotechnol 323:341–346
3. Martínez-Cámara S, Bahíllo E, Barredo J-L, Rodríguez-Sáiz M (2017) Scale-up of Phytosterols bioconversion into androstenedione. Methods Mol Biol Clifton NJ 1645:199–210
4. Donova MV, Nikolayeva VM, Dovbnya DV, Gulevskaya SA, Suzina NE (2007) Methyl-beta-cyclodextrin alters growth, activity and cell envelope features of sterol-transforming mycobacteria. Microbiol Read Engl 153: 1981–1992
5. Perez C, Falero A, Llanes N, Hung BR, Hervé ME, Palmero A, Martí E (2003) Resistance to androstanes as an approach for androstandienedione yield enhancement in industrial mycobacteria. J Ind Microbiol Biotechnol 30:623–626
6. Egorova OV, Gulevskaya SA, Puntus IF, Filonov AE, Donova MV (2002) Production of androstenedione using mutants of mycobacterium sp. J Chem Technol Biotechnol 77:141–147
7. Marques MPC, Fernandes P (2017) β-Sitosterol bioconversion to androstenedione in microtiter plates. Methods Mol Biol Clifton NJ 1645:167–176
8. Fischer M, Sawers RG (2013) A universally applicable and rapid method for measuring the growth of streptomyces and other filamentous microorganisms by methylene blue

adsorption-desorption. Appl Environ Microbiol 79:4499–4502

9. Kensy F, Zang E, Faulhammer C, Tan R-K, Büchs J (2009) Validation of a high-throughput fermentation system based on online monitoring of biomass and fluorescence in continuously shaken microtiter plates. Microb Cell Factories 8:31

10. Josefsen KD, Nordborg A, Sletta H (2017) Bioconversion of Phytosterols into androstenedione by mycobacterium. Methods Mol Biol Clifton NJ 1645:177–197

11. Calibration data BioLector https://www.m2p-labs.com/services/calibration-data/

Chapter 12

β-Sitosterol Bioconversion in Small-Scale Devices: From Microtiter Plates to Microfluidic Reactors

Marco P. C. Marques, Jorge Aranda Hernandez, and Pedro Fernandes

Abstract

Small-scale devices are routinely used as low-cost miniaturized bioreactors due to the large number of experiments that can be conducted simultaneously under similar conditions and replicate all functions of bench-scale reactors at dramatically smaller volumes. Microtiter plates, due to the standard footprint, can be integrated with liquid handling systems and associated equipment, expanding considerably their application and use. However, care has to be taken to operate the microtiter plates in optimized mixing and oxygen transfer conditions, preventing medium evaporation in prolonged experiment runs. Recently, to increase data quality, microbioreactors have emerged as an alternative to shaken systems. These systems offer higher degree of control over key process variables and when combined with sensing technology increase dramatically the reliability of translational process data. In this chapter, we describe the production of 4-androstene-3,17-dione (androstenedione (AD)), a key pharmaceutical steroid intermediate, by *Mycobacterium* sp. NRRL B-3805 via the selective cleavage of the side-chain of β-sitosterol using 24-well microtiter plates and microfluidic microbioreactors.

Key words Microtiter plate, Microbioreactor, *Mycobacterium* sp. NRRL B-3805, 4-Androstene-3,17-dione, AD, β-Sitosterol, Steroid, Evaporation, High-throughput screening

1 Introduction

Miniaturizing bioprocess unit operation devices is a well-established approach to find novel routes for process intensification and improved process economics. These devices, which allow a high level of parallelization while operating at reduced volumes, are especially attractive for fast and cost-effective process development.

When dealing with miniaturizing upstream unit operations, one can encounter shaken systems (Erlenmeyer flasks or microtiter plates), miniature, and microbioreactors [1–3]. Microtiter plates are widely used since they offer in a reduced footprint, typically

Carlos Barreiro and José-Luis Barredo (eds.), *Microbial Steroids: Methods and Protocols*, Methods in Molecular Biology, vol. 2704, https://doi.org/10.1007/978-1-0716-3385-4_12,

8.56 × 12.76 cm (*w* × *l*), the possibility to carry out 6–1536 experiments in working volumes ranging from 3 mL to 3 μL, respectively [4]. These devices are inexpensive when compared to other miniaturized devices and relatively simple to handle and due to the standard footprint enable automation and integration with liquid handling systems and associated equipment (e.g., HPLC and spectrometric plate readers). Furthermore, microtiter plates replicate all functions of bench-scale reactors, having monitoring capabilities of key process variables (e.g., pH and oxygen) [5], control of pH, and fed-batch operations [6], at similar operational conditions [7]. Microtiter plates have been extensively used for evaluating large biocatalyst libraries, for generation of design data (e.g., for medium screening, assessing pH and temperature influence in reaction yield and evaluating culturing and induction time) to be used in process and economic models, and for fermentation and biocatalysis process development [4, 8]. However, the data quality, i.e., the significance of the data for large-scale operations, obtained within these systems is still low.

Microfluidic devices (*microbioreactors*) have emerged in the last decades as alternatives to shaken systems, offering the same level of throughput but able to generate higher quality data at reduced costs [9]. The microbioreactors have characteristic dimensions in the submillimeter range and allow to manipulate fluids in the microliter scale. The fine control over microenvironment (spatial and temporal) is achieved by the laminar flow regime, imposed by the small dimensions. These devices can operate in several modes, i.e., batch fed-batch and continuous, under different process conditions [10–12]. Monitoring in microbioreactors of key process variables can be achieved by the incorporation of optical sensors, spectroscopy methods, and image analysis [9, 13].

In this chapter we show the use of microtiter plates and microbioreactors to perform multistep biocatalytic reactions using whole cells. The model system is the selective cleavage of the side-chain of β-sitosterol by *Mycobacterium* sp. NRRL B-3805 (currently *Mycobacterium neoaurum* NRRL B-3805) [14], leading to the synthesis of 4-androstene-3,17-dione (androstenedione; AD) as the major product and 1,4-androstadiene-3,17-dione (androstadienedione; ADD) as the minor product [15].

AD and ADD are key pharmaceutical steroid intermediates and their yearly reported market exceeds 1000 tons [16, 17]. This multistep reaction is especially challenging at the small scale due to the low aqueous solubility of β-sitosterol (need for in situ substrate supply strategies) and low reaction rates (prevention of significant medium evaporation is necessary).

2 Materials

Prepare all solutions with deionized water and analytical grade reagents. All reagents should be stored at room temperature (~22 °C) unless stated otherwise. No special equipment is required to perform these experiments in microtiter plates, besides an autoclave, a shaking incubator, and a vortex. Fabrication equipment (e.g., vacuum plasma chamber and micro-milling machine) is required for the fabrication of the microbioreactors. An HPLC is necessary for sample quantification. The use of personal protection equipment (PPE) is recommended throughout the entire work.

2.1 Culture Media and Equipment for Mycobacterium *sp. NRRL B-3805 Growth*

1. PDA culture plates: Add 1 L of water to a glass beaker. Weigh 42 g of potato dextrose agar (PDA) and transfer to the beaker. Mix with the aid of a magnetic bar. Transfer this solution to a 2 L screw capped bottle (*see* **Note 1**). Autoclave for 15 min at 121 °C (total cycle approx. 2 h). Allow the bottle to cool down until it is bearable to manipulate it without any heatproof gloves. Transfer approx. 20 mL of the autoclaved agar medium into a plastic Petri dish. Repeat this process until the entire autoclaved agar medium is used (*see* **Note 2**). Leave the plates without the cover in a safety cabinet until the agar starts to solidify (~1 h). After this period turn the Petri dishes upside down to prevent condensation and leave for a couple of hours to further solidify. After complete solidification place the lids on the Petri dishes and store them at 4 °C until use.
2. Sodium–potassium phosphate buffer (100 mM, pH 7): Add 1 L of water to a glass beaker. Weigh 14.20 g of Na_2HPO_4 (molecular weight 141.96 g/mol) and transfer to the beaker. Mix with the aid of a magnetic bar. Transfer this solution to a 2 L screw capped bottle. Repeat this process for KH_2PO_4 by adding 1 L of water to a glass beaker. Weigh 13.61 g of KH_2PO_4 (molecular weight 136.09 g/mol) and transfer to the beaker where it is mixed with a magnetic bar. Transfer this solution to a 2 L screw capped bottle. Transfer the Na_2HPO_4 solution to a large container. Mix with the aid of a magnetic bar and use a pH probe to measure continuously the pH. Add slowly the KH_2PO_4 solution until pH 7 is reached. Transfer the final buffer (100 mM, pH 7) to a 2 L screw capped bottle and store at room temperature until use.
3. Complex growth medium: Add 1 L of 100 mM sodium–potassium phosphate buffer pH 7 to a glass beaker. Weigh 10 g of yeast extract, 10 g of glycerol, 0.2 g of $MgSO_4{\cdot}7H_2O$, 1 g of micronized β-sitosterol, and 0.8 g of Tween® 20, and transfer to the beaker. Mix the solution with the aid of a magnetic bar. Transfer 20 mL of this solution to 100 mL shake flasks (denominated inoculum flasks) and 400 mL to 2 L shake flasks

(denominated growth flasks) (*see* **Notes 1** and **3**). Use cotton wool plugs to close the shake flasks and cover with aluminum foil. Autoclave for 15 min at 121 °C (total cycle approx. 2 h). Store at 4 °C until use.

4. Defined growth medium: Add 1 L of 0.1 M sodium–potassium phosphate buffer pH 7 to a glass beaker. Weigh 20 g of glycerol, 4 g of NH_4Cl, 0.2 g of $MgSO_4 \cdot 7H_2O$, and 0.8 g of Tween® 20, and transfer to the beaker. Mix with the aid of a magnetic bar. Transfer this solution to a 2 L screw capped bottle (*see* **Note 1**). Cap the bottles loosely and autoclave for 15 min at 121 °C (total cycle approx. 2 h). Tighten the bottle cap and store at 4 °C until use.
5. Ethanol β-sitosterol stock solution: Add 10 mL of absolute ethanol to a glass beaker. Weigh 99.53 mg of β-sitosterol and transfer to the beaker. Mix gently with the aid of a magnetic bar. Transfer this solution to a 10 mL screw capped bottle and store at 4 °C until use.
6. Poly(dimethylsiloxane) β-sitosterol stock solution: Add 10 mL of poly(dimethylsiloxane) (*see* **Note 4**) to a screw capped bottle. Weigh 145.15 mg of β-sitosterol and transfer to the bottle. Mix gently with the aid of a magnetic bar. Place the bottle in an oven at 100 °C. After 10 min, take the bottle out of the oven, using PPE. Mix with the aid of a magnetic bar. Repeat this process until β-sitosterol is completely dissolved.
7. Shaking incubator with orbital motion. The incubator needs to have controlled temperature and needs to accommodate 2 L shake flasks.

2.2 Microbioreactor Fabrication and Assembly

1. Computer-aided design (CAD) software.
2. Mastercam X4.
3. Computer numerical control (CNC) micro-milling machine (*see* **Note 5**).
4. Flute slot drills (e.g., 0.5, 1, and 2 mM).
5. Polycarbonate sheets.
6. Polydimethylsiloxane (PDMS).
7. Vacuum plasma chamber.
8. Oven.
9. Spin coater.
10. Silicon wafer.
11. Stainless steel plates (*see* **Note 6**).
12. Magnetic stirrer bar (*see* **Note 7**).
13. Motor with mounted magnet (*see* **Note 8**).
14. Bench top incubator.

15. Syringe pumps.
16. Plastic syringes.
17. 70% v/v ethanol: Add 70 mL of absolute ethanol to 30 mL reverse osmosis (RO) water in a screw capped bottle. Tighten the bottle cap and store at 4 °C until use.
18. M3 hex screws.
19. Flangeless fittings and ferrules, headless and flat bottom, for 1/16″ OD tubing.
20. Fluorinated ethylene propylene (FEP) tubing, 1/16″ OD.

2.3 AD Extraction and HPLC Analysis

1. Whatman® qualitative filter paper, Grade 1.
2. Eppendorf tube®.
3. Extraction solvent with internal standard: Add 1 L of n-heptane to a screw capped bottle. Weigh 0.2 g of progesterone and transfer to the bottle. Mix gently with the aid of a magnetic bar. When completely dissolved store at 4 °C (*see* **Note 9**).
4. HPLC equipment with UV detector.
5. LiChrospher® Si 60 column (particle size 5 μM; Merck).
6. Mobile phase, n-heptane and ethanol (9/1, v/v): Add 1.8 L of heptane and 0.2 L of absolute ethanol to a 2 L screw capped bottle. Mix gently with the aid of a magnetic bar. Flush the bottle for 20 min with helium to remove all air bubbles.
8. AD stock solutions (*see* **Note 10**): Add 8.6 mg of 4-androstene-3,17-dione and 15 mL of n-heptane to a 25 mL screw capped bottle. Mix gently with the aid of a magnetic bar. Prepare 10 Eppendorf tubes®, marked 1 to 10, containing each 500 μL of n-heptane. Transfer 500 μL of the initial solution present in the screw capped bottle to the first Eppendorf tube®. Transfer 500 μL from Eppendorf tube® 1 to Eppendorf tube® 2. Repeat this process until all the Eppendorf tubes® are filled. Transfer the all the solutions to HPLC vials.
7. ADD stock solution (*see* **Note 10**): Add 8.5 mg of 1,4-androstadiene-3,17-dione and 15 mL of n-heptane to a 25 mL screw capped bottle. Follow the same procedure as to prepare AD stock solutions.
8. Bench centrifuge.
9. Vortex.

2.4 Sterilized Materials

1. Box of blue (1 mL) and yellow (200 μL) pipet tips autoclaved (15 min at 121 °C, total cycle approx. 2 h).
2. Sterile microtiter plates.
3. Sterile Petri dishes.
4. RO water.

3 Methods

Carry out all procedures at room temperature (~22 °C) under aseptic conditions unless otherwise specified.

3.1 Microbioreactor Fabrication

The fabrication of the microbioreactors does not need to be performed under aseptic conditions.

1. Design the microbioreactor parts and membrane mold in a 3D CAD software (Fig. 1).
2. Generate the CNC code from the completed designs using either a CAM program (e.g., Mastercam X4) or directly on the 3D CAD software (*see* **Note 11**).
3. Set up the CNC micro-milling machine (*see* **Note 12**).
4. After milling, the microbioreactor parts and membrane mold should be cleaned with RO water and compressed air to remove polymer debris and dust.
5. Place a silicon wafer (*see* **Note 13**) onto the spin coater. Set the rotation speed to obtain the desired thickness of the PDMS layer. Drop 3 to 4 mL of PDMS (*see* **Note 14**) onto the center of the silicon wafer. Typically, the settings are: maximum acceleration to 500 rpm, spin for 20 s, maximum acceleration to 3000 rpm (30 μM thick membranes) or 700 rpm (100 μM thick membranes), spin for 30 s, and decelerate at maximum speed until full stop.
6. Cure the wafer for 1 h at 80 °C.
7. Cut membranes with a scalpel (*see* **Note 15**).

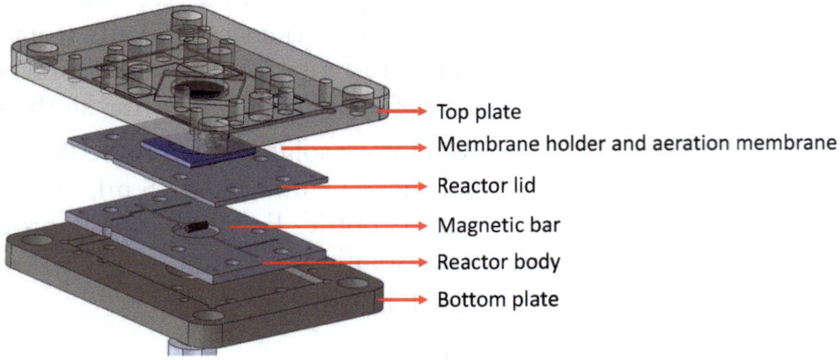

Fig. 1 Exploded view of a typical microbioreactor cassette comprising top and bottom plates and the reactor layers themselves. The reactor layers include the reactor body, the lid, and the aeration membrane

8. Drop PDMS onto a milled polycarbonate mold using cast-and-mold technique (*see* **Note 16**).
9. Remove the PDMS holder and together with the membrane using air plasma-assisted bonding for 2 min at 30 W and 500 mTorr.
10. Assemble the reactor body (containing the reactor chamber and various inlet and outlet channels) together with the lid and place between the two stainless steel plates. Screw to a maximum of 45 N m torque. Place the plates in the oven for 2 h at 135 °C. Allow the plates to cool before removing the bonded parts.
11. Place the magnetic stirrer bar in the reactor chamber.
12. Use the cassette components and assemble using the M3 hex screws.
21. Place the fittings and ferrules on the FEP tubing and connect to the assembled microbioreactor.
22. Connect the tubing to syringes.

3.2 Preparation of Resting Cells of *Mycobacterium* sp. NRRL B-3805

1. Defrost stock cultures (cell bank) of *Mycobacterium* sp. NRRL B-3805 and add 100 μL of the stock culture to a PDA culture plate.
2. Spread the culture using a loop according to the streaking method (*see* **Note 17**).
3. Place the culture plate upside down in a 30 °C incubator for at least 24 h.
4. Take an isolated colony of *Mycobacterium* sp. NRRL B-3805 (*see* **Note 18**) and add to the inoculum flasks containing complex growth medium.
5. Place the inoculum flasks (100 mL flask) in a shaking incubator at 200 rpm and 30 °C.
6. After 24 h (late exponential growth phase; *see* **Note 19**) transfer the entire content of the inoculum flask to growth flasks (2 L flasks) containing complex growth medium.
7. Place the growth flasks (2 L flask) in a shaking incubator at 200 rpm and 30 °C (*see* **Note 20**).
8. After about 72 h (late exponential growth phase; *see* **Note 19**), harvest the cells by vacuum filtration (*see* **Note 21**).
9. Thoroughly wash the cells on the filter with the prepared phosphate buffer (100 mM, pH 7).
10. Store the wet cell paste at −20 °C until use (*see* **Note 22**).

3.3 Bioconversion Trials Using Resting Cells of Mycobacterium sp. NRRL B-3805 in Microtiter Plates

1. Add 15 mL of phosphate buffer (100 mM, pH 7) to a glass beaker (*see* **Note 23**).
2. Weigh 0.7 g of the wet cell paste and transfer to the bottle.
3. Mix gently with the aid of a magnetic bar until yellowish solution is obtained.
4. Add 550 μL of the *Mycobacterium* sp. NRRL B-3805 solution to each well of a 24-well microtiter plate (*see* **Notes 24–26**).
5. Add 55 μL or 41 μL of the β-sitosterol stock solutions in ethanol or poly(dimethylsiloxane), respectively (*see* **Note 27**).
6. Cover the microtiter plate with a breathable sealing tape and place lid (*see* **Note 26**).
7. Place the microtiter plate in a shaking incubator at 250 rpm and 30 °C.
8. Use a sacrificial well approach by sampling the entire content of the microtiter plate well.
9. Transfer the sample to a 1.5 mL Eppendorf tube® and add 550 μL of the extraction solvent with internal standard.
10. Mix the content in a vortex for 30 s at maximum speed.
11. Place the Eppendorf tubes® in a centrifuge for 2 min, 4 °C and 10,000 × *g*.
12. Collect the organic phase (top phase) and place in a HPLC vial.
13. Analyze sampled in a HPLC (*see* **Note 28**).

3.4 Bioconversion Trials Using Resting Cells of Mycobacterium sp. NRRL B-3805 in Microbioreactors

1. Add 15 mL of phosphate buffer (100 mM, pH 7) to a glass beaker (*see* **Note 23**).
2. Weigh 0.7 g of the wet cell paste and transfer to the bottle.
3. Mix gently with the aid of a magnetic bar until yellowish solution is obtained.
4. Add 1.5 mL of the β-sitosterol stock solutions in ethanol (*see* **Note 27**) to the *Mycobacterium* sp. solution.
5. Fill a 10 mL syringe with the *Mycobacterium* sp. solution and one 10 mL syringe with 70% v/v ethanol solution.
6. Attach the syringes to the FEP tubing that are connected to the microbioreactor.
7. Start the 70% v/v ethanol solution to prime the microbioreactor at 20 μL/min (*see* **Note 29**).
8. Stop the 70% v/v ethanol solution and start the *Mycobacterium* sp. solution at 20 μL/min (*see* **Note 29**).
9. Stop the *Mycobacterium* sp. solution once the entire chamber is filled.
10. Place the microbioreactor in an incubator at 250 rpm and 30 °C (*see* **Note 30**).

11. Use a sacrificial well approach by sampling the entire content of the microbioreactor.
12. Transfer the sample to a 1.5 mL Eppendorf tube® and add 450 μL of the extraction solvent with internal standard.
13. Mix the content in a vortex for 30 s at maximum speed.
14. Place the Eppendorf tubes® in a centrifuge for 2 min, 4 °C, and 10,000 × *g*.
15. Collect the organic phase (top phase) and place in a HPLC vial.
16. Analyze the sample by HPLC (*see* **Note 28**).

3.5 Bioconversion Runs Using Growing Cells of Mycobacterium sp. NRRL B-3805 in Microtiter Plates

1. Prepare inoculum as in Subheading 3.2, **steps 1–5**.
2. After 24 h (late exponential growth phase; *see* **Note 19**), transfer the entire content of the inoculum flask to the defined medium bottle (500 mL bottle).
3. Gently mix the bottle.
4. Add 550 μL of the inoculated defined medium to each well of a 24-well microtiter plate (*see* **Notes 24–26**).
5. Add 55 μL or 41 μL of the β-sitosterol stock solutions in ethanol or poly(dimethylsiloxane), respectively (*see* **Note 27**).
6. Cover the microtiter plate with a breathable sealing tape and place lid (*see* **Note 26**).
7. Place the microtiter plate in a shaking incubator at 250 rpm and 30 °C (*see* **Note 30**).
8. Use a sacrificial well approach by sampling the entire content of the microtiter plate well.
9. Conduct similar sample processing and analysis as in Subheading 3.2, **steps 9–13**.

3.6 Bioconversion Trials Using Growing Cells of Mycobacterium sp. NRRL B-3805 in Microbioreactors

1. Prepare inoculum as in Subheading 3.2, **steps 1–5**.
2. After 24 h (late exponential growth phase; *see* **Note 19**), transfer the entire content of the inoculum flask to the defined medium bottle (250 mL bottle).
3. Gently mix the bottle.
4. Add 2 mL of the β-sitosterol stock solutions in ethanol (*see* **Note 17**).
5. Fill a 10 mL syringe with the *Mycobacterium* sp. solution, one 10 mL syringe with 70% v/v ethanol solution, and one 10 mL syringe with sterilized RO water.
6. Attach the syringes to the FEP tubing that are connected to the microbioreactor.
7. Start the 70% v/v ethanol solution to prime the microbioreactor at 20 μL/min (*see* **Note 29**).

8. Stop the 70% v/v ethanol solution and start the *Mycobacterium* sp. solution at 20 μL/min (*see* **Note 29**).
9. Stop the *Mycobacterium* sp. solution once the entire chamber is filled.
10. Place the microbioreactor in an incubator at 250 rpm and 30 °C (*see* **Note 30**).
11. After 12 h start the syringe of sterilized RO water at 0.20 μL/min.
12. Use a sacrificial well approach by sampling the entire content of the microbioreactor.
13. Conduct similar sample processing and analysis as in Subheading 3.3, **steps 9–13**.

4 Notes

1. PYREX® or other borosilicate glassware should be used for autoclaving since this type of glassware can withstand the stress of autoclave pressures and temperatures.
2. As a rule of thumb, a volume of 500 mL of potato dextrose agar can be used to fill 20 Petri dishes.
3. The medium can be stored at 4 °C for at least once a week. However, as a good practice, fresh medium should be prepared every time a new experiment is conducted.
4. Poly(dimethylsiloxane) (PDMS) suitable for high temperatures should be used to prepare the stock solutions [18]. The use of a "low temperature" PDMS is not recommended since the silicone can potentially solidify at room temperature.
5. The CNC micro-milling machine can be substituted by CO_2 laser machining or 3D printing. However, precision in the dimension of the fabricated feature and components will be compromised.
6. Stainless steel plates are used for thermal compression. The fabricated components are placed between the plates and subjected to compression. Therefore, one of these plates should be threaded.
7. The magnetic stirring bar must be plated with nickel and coated with parylene to avoid oxidation.
8. The motor with a mounted magnet can be substituted by a stirring plate. Care has to be taken to place the microbioreactor in the center of the plate to avoid any offset of the magnetic stirring bar.

9. The extraction solvent with internal standard should be allowed to come to room temperature (22 °C) and be mixed every time it is used; crystallization of progesterone can occur at low temperatures.
10. For the HPLC calibration curve, stock solution concentration of both AD and ADD can be used up to 1.5 mM. Higher concentrations can present deviation from linearity.
11. The CNC code (generally a GCODE file) is generated by a CAM software, which may either be stand-alone or an add-on for 3D CAD software. Commercial 3D CAD software have such add-ons available, as is the case of Autodesk™ or Solid-Works™. As a first step, the stock material needs to be aligned and anchored to the milling machine's stage. Note that the material stock must be larger than the footprint of the desired final part, to allow space for anchorage. The stock size can be defined in absolute terms as well as relative to the 3D build. In relative terms, a side offset of 10–25 mM is recommended; top and bottom offsets can be set to 0. This top face can be after defined in the software as the *XY* plane. Use the *Z*- and *X*-axes to define this plane and set the stock box point as the Origin. This will result in three stock points appearing in each corner of the build and as well as the middle. Always select the top middle point as the stock point. For each milling operation, e.g., circular holes, channels, and channel length and perimeter, there is a built-in routine in most CAM software; basic routines are Bore, Pocket, and Contour. Each operation must have an associated milling tool. Cutter parameters must be entered to the software: cutter diameter, shoulder length, flute length, overall length, and shaft diameter (Fig. 2). These can be found on the vendor sites (cutter specifications) or measured using a caliper. The feed and spindle speed must be selected according to the diameter of the tool (Table 1) and the material, though most modern CAM software can adjust most of the cutting parameters semi-automatically. Note that when machining materials with low melting temperatures, such as most polymers, lower spindle speeds are more adequate (5000–10,000 rpm). For cutting holes in the part, the bottom offset should be lower than the flute length of the selected tool for the cutting job. This ensures a clean cut. As a rule of thumb, the cutting depth, or pitch in this case, should be set to half the diameter of the selected cutter, at most. These steps/options should be repeated for all holes of different diameters or depths, which require different settings or tool sizes. For pockets, the first step is to select the bottom surface of the feature. The stock's top surface can be used to define the pocket's

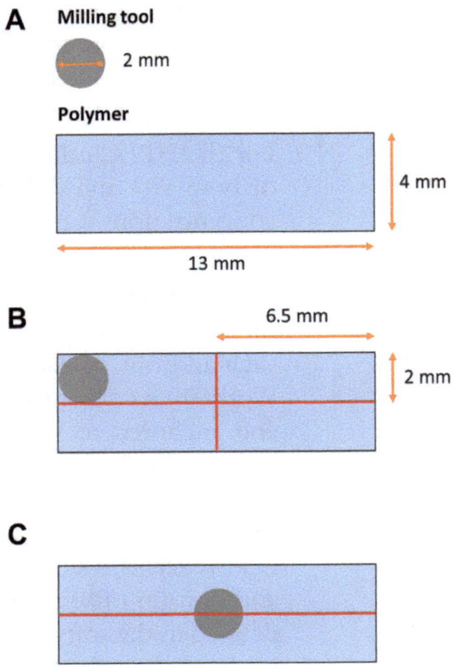

Fig. 2 Relocation of cutting tool on a CNC milling machine. (**a**) The CNC milling machine measures the cutting tool position at the center regardless of the tool diameter. By knowing the dimensions of the tool and the dimension of the device to be milled, it is possible to allocate the tool in a specific position accurate to 0.0001 mM. (**b**) Place manually the milling tool on one corner. (**c**) Move the tool manually to the center of the design. In this particular case, move the tool 5.5 mM right and 1 mM down

Table 1
Typical spindle speed and feeds for selected flute slot drills

	Flute slot drill diameter (mM)		
	0.5	**1.0**	**2.0**
Spindle speed (rpm)	14,000	10,000	5000
Lead-in feed rate (mM/min)	100	200	1000

height. When selecting passes, the maximum step-over should be adjusted according to the selected milling tool diameter; for a smoother finish a high degree of overlap is recommended. But some degree of overlap, at least a 10%, is recommended to prevent a rough finish. As with hole cutting operations, the maximum roughing step-down must not exceed half of the of the selected milling tool's diameter. The maximum roughing step-down is analogous to the pitch, but the term is use for linear motion. Finer finishing step-downs may be selected if a more polished finish is desired. These steps/options should be repeated for pockets with different dimension, requiring different tool sizes. For contour cutting, it is recommendable to use the largest tool possible. Contours rarely define functional device features and are time consuming. Larger tools allow for larger step-downs and significantly reduce time. The model top is selected as the top height and the top offset should be set to zero. The tolerance is usually adjusted to 0.01 mM to produce smooth contours, especially when these are curved. As with hole cutting, it is recommendable to set the contour bottom offset 0.1–0.2 mM below the stock's bottom surface to produce a clean cut. Before generating the final CAM file, each different milling operation should be renamed in the CAM software routine to include the required tool diameter as well as the name of the milling operation. This will reduce operator errors while operating the milling machine, particularly during tool changes (if tool changes are not automated). The order of the milling operations is important as they will be executed sequentially. Operations should be ordered to minimize tool changes, by grouping together all operations using a same tool size. Additionally, contour operations should be left last, particularly the final device cutoff, as this will release the part from its anchors and prevent posterior machining. Once the CAM file is exported, it is recommendable to confirm that the files have the required extension, generally either *.nc or *.tap format.

12. After turning on the milling machine controller, introduce the appropriate cutting tool in the spindle. The tool must have been previously chosen in the code. Do not introduce the cutting tool all the way up. Prepare the polymer to be milled as follows: (I) use a sacrificial layer of polymer of the same size as the polymer to be milled to cover the metal surface of the machine. (II) Mark in the polymer to be milled the reference point that has been selected in the CNC code. In this case, the top surface's center point. (III) Fix both polymers to the metal surface of the machine, screwing the anchors at 100 cNm. Make sure the polymer is completely flat in order to avoid breaking any cutting tools. This polymer sheets will tend to dome around the center. The tip of the cutting tool must match one of the chosen reference points. This can be done manually, in a stepwise manner by slowly lowering the tool to touch the drawn point and running a home operation. There are commercial milling machines that allow this movement with accuracies between 1 mM and 1×10^{-4} mM. When close to the mark, the cutting tool should be spinning. It is

recommended to use very small Z increments for the last millimeter, manually lowering the tool until the ink of the mark is removed. Now the machine's coordinates are in synchrony with the stock's physical coordinates. This technique can only be used if the reference point in the CNC code was set to be on the surface of the polymer. Note that a small white circle will be created surrounding the cutting tool. Only after this process can the CNC code ("G-code") be uploaded. If the polymer moves during milling, the cutting tool should be repositioned. *ZZ*-axis reposition can be performed using the sacrificial layer. For this, a precise knowledge of the thickness of the polymer to be milled is necessary. *XY*-axis repositioning can be done directly on the device by centering the cutting tool (Fig. 2).

13. To allow the easy release of PDMS layers, wafers are silanized. This is typically performed by exposing the wafers to vapors of trichloro(1H,1H,2H,2H-perfluorooctyl)silane in a vacuum desiccator for 1 h.
14. The PDMS is prepared by mixing the base to the curing agent in a 10:1 mixture. After the mixture is well mixed, it should be degassed in a vacuum desiccator up to 30 min, with several vacuum cycles.
15. Membrane thickness can be assessed using a stylus profiler.
16. The cast-and-mold technique is a fast and cost-effective way of prototyping PDMS (Sylgard 184, DOW) parts using basic lab equipment such as an oven, precision scale, and a desiccator. However, mold fabrication may require prototyping capabilities that can exceed those of a typical lab, such as CNC milling or laser ablation. Nevertheless, more cost-effective approaches such as 3D printing are also possible for mold fabrication. Most importantly, this technique makes non-destructive use of the mold, so outsourcing its fabrication may be a worthwhile one-time investment. Once a mold is obtained, which is an accurate negative geometrical representation of the desired part, it must be first thoroughly cleaned. Depending on the mold material, different solvents may be compatible. Polycarbonate is a good choice for mold fabrication due to its price, ready availability, and high melting point. Ethanol is recommended to clean polycarbonate molds before exposing them to the uncured PDMS. PDMS preparation, according to the manufacturer, is most commonly done at a 1:10 ratio of cross-linker/base solution. Thorough mixing for 1–2 min is required to produce a homogenous mix, key for homogenous curing. Mixing however, will introduce many microbubbles into the PDMS which need to be removed before curing.

This is done inside a vacuum desiccator, where 2–3 vacuum cycles of 5–10 min are recommended. A micropipette can be used to remove any lingering bubbles at the surface, by gently blowing air at them. Thoroughly mixed and degassed liquid PDMS can now be poured into the mold. In cases where de-molding agents or silanizing may be required, these are applied on the mold before pouring the PDMS. Again, pouring the liquid PDMS into the mold will introduce bubbles. Therefore, the PDMS-filled mold must be degassed in the dessicator as previously described. Often molds are open containers; however, PDMS molds often have a lid, in those cases where parts must have a defined height. It is good practice to design the latter type of mold so that the lid has no features in it and is just a sheet of material. This will make it easier to place the lid without introducing bubbles into the mold. For this purpose, first contact between lid and mold should be done by one of the lid vertices. The lid should then be lowered following a pivotal movement downwards, to ensure the bubbles and excess PDMS are pushed out of the mold before clamping it together. The final step is to bake the PDMS inside an oven. Curing times and temperatures for PDMS vary and can be consulted in the manufacturer's datasheets. Generally, curing times oscillate between 24 and 48 h at room temperature, to 1–2 h at 80 °C, where excess curing time is not a problem. It is recommendable to wait until the mold has cooled down before removing the PDMS part, as this will make it easier and prevent tearing of the parts.

17. The streaking method allows to obtain isolated colonies from an inoculum by streaking a plate several times in succession. This will create on a single plate several areas with increased dilution. For this method, a sterilized loop is necessary. The loop, containing the inoculum, is streaked across the surface of the sterile PDA culture plate. Following the initial streak, the subsequent streaks are made at 90° angles with the loop being sterilized in between. If a disposable loop is used, then the sterilization is unnecessary between streaks. A total of four streaks are advised.

18. *Mycobacterium* sp. NRRL B-3805 should present in PDA agar plates colonies yellowish in color, irregular form, raised elevation, rugose surface, and earthy smell. *Mycobacterium* sp. NRRL B-3805 can however change their phenotype according to specific growth conditions. It is therefore recommended that identification protocols should be carried out regularly [19].

Obtaining an isolated colony can be difficult with *Mycobacterium* sp. NRRL B-3805. For inoculation, collecting a loopful of several colonies is recommended.

19. Measuring the optical density (OD) in *Mycobacterium* sp. NRRL B-3805 cultures can be difficult since this bacterium forms aggregates. As an approximation, late exponential growth phase corresponds to an OD_{640} in the range of 0.6–0.9. A more reliable method of determining *Mycobacterium* sp. NRRL B-3805 is based on total protein or on dry cell weight. However, these methods are time consuming and data cannot be obtained in real time.
20. Flasks containing large volumes (e.g., growth flasks) should be placed in the incubator platform such that it is balanced and all weight is evenly distributed.
21. Whatman® qualitative filter paper, Grade 1 can be used to filter *Mycobacterium* sp. NRRL B-3805 cultures. The cake resistance will increase and filtration efficiency will decrease over time. To aid the filtration regularly and smoothly remove the cake from the filter with a spatula.
22. The wet cell paste contains about 190 mg dry cell weight per g of wet paste.
23. Different bioconversion conditions can be tested. For example, different salts can be added to the phosphate buffer to increase bioconversion yield [15].
24. Different filling volumes can be used [15]. However, k_La (volumetric oxygen mass transfer coefficient) will depend on filling volume [7]. As a rule of thumb, the lower the volume, the higher the oxygen transfer. Details are given in Fig. 3.
25. If lower volumes are used, special care should be given to the higher evaporation rate of the culture medium. The medium should be replenished every day with autoclaved deionized water. The amount to be replenished in each well can be determined by weighting the microtiter plate and dividing the value by the number of wells used in the experiment.

    A control well using standard bioconversion conditions should be used in each microtiter plate. We recommend 550 μL of the phosphate buffer (100 mM, pH 7) containing *Mycobacterium* sp. NRRL B-3805 cells supplemented with 55 μL of the ethanol β-sitosterol stock solution.
26. The breathable sealing tape will seal the microtiter plate, preventing medium evaporation and cross-contamination of the wells due to spilling. Detailed steps are given in Fig. 4.

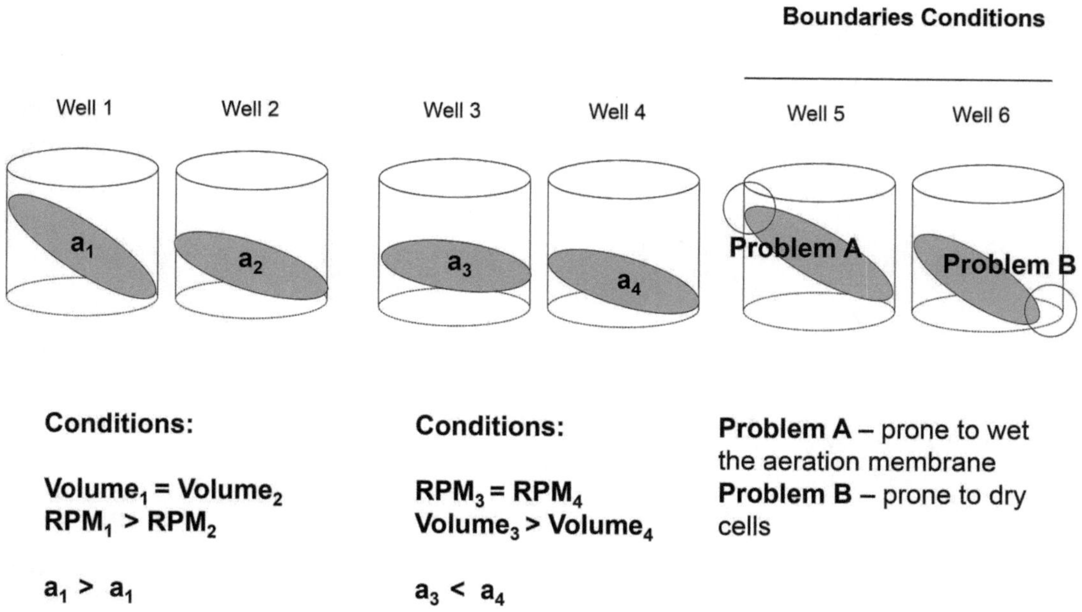

Fig. 3 Correlation of gas–liquid interface surface area (a_i) with the filling volume (volume) and the shaking frequency (RPM) of a microtiter plate well. Excessive or reduced filling volume results in either wetting of the aeration membrane or uncontrolled concentration of the media, respectively

27. Different organic solvent can be used or aqueous/organic ratio [20, 21].
28. Quantification of steroids should be performed by HPLC. A possible method consists of using a LiChrospher® Si 60 - column (particle size 5 μM) with mobile phase composed of n-heptane and ethanol (9:1, v/v). An isocratic elution of 1 mL/min should be used with detection at 215 nm (β-sitosterol) and 254 nm (4-androstene-3,17-dione, AD, and 1,4-androstadiene-3,17-dione, ADD).
29. Priming of the tubings can be done prior to the attachment of the tubing to the microbioreactors. This will avoid the introduction of air bubbles into the device.
30. Microtiter plates with integrated oxygen and pH sensors can be used to monitor bacterial growth [7, 22]. If the readout of the sensors is performed using fluorescence microplate readers, the medium evaporation rate can increase. Lowering the readout frequency can circumvent high evaporation rate of the culture medium. Nonetheless, it is advised to weigh the microtiter plate on an hourly basis and to replenish the medium with autoclaved deionized water, if necessary.

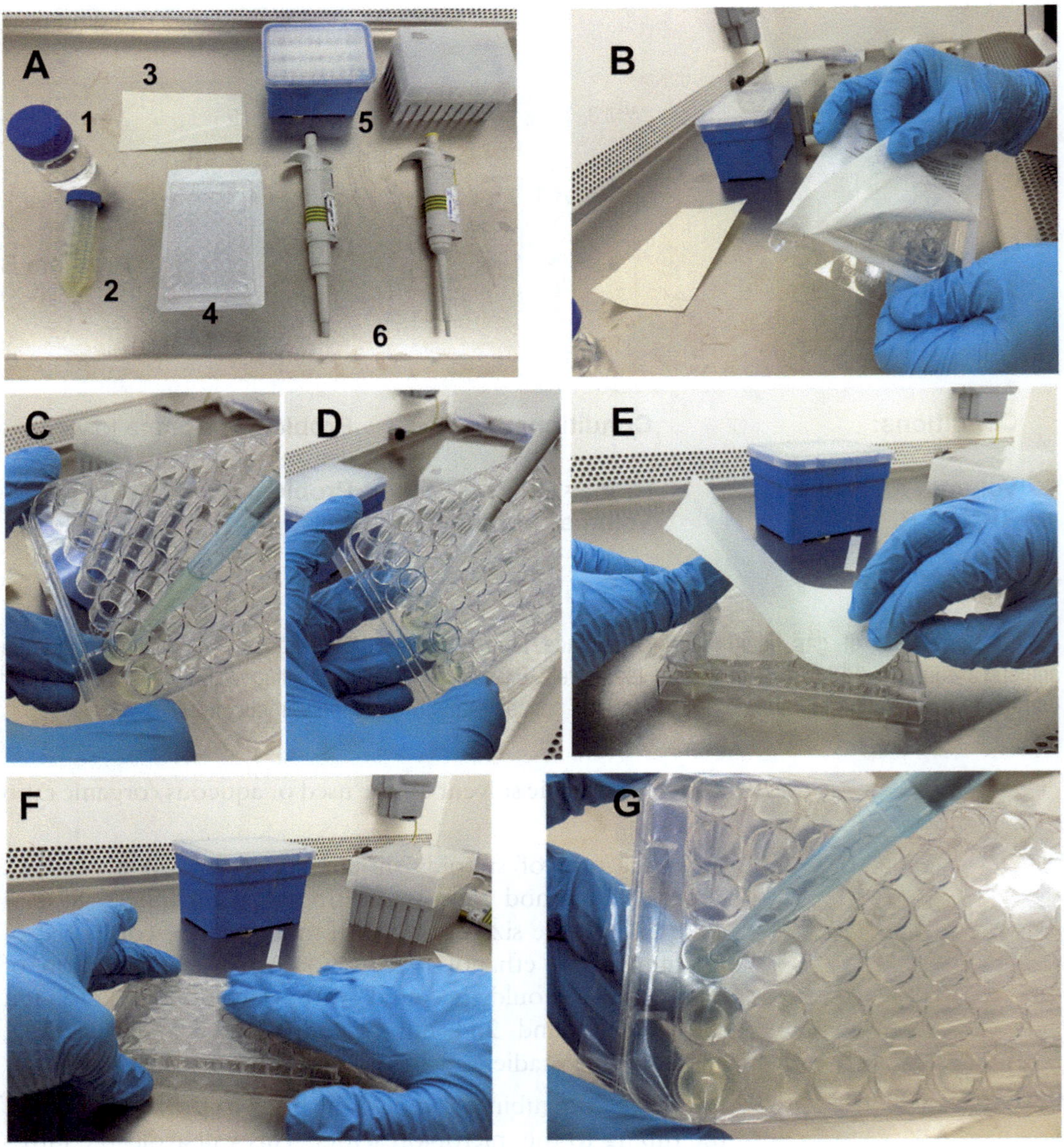

Fig. 4 Preparation of the microtiter plate. (**a**) Materials needed: (1) steroid solution in ethanol, (2) *Mycobacterium* sp. NRRL-3805 cell suspension, (3) aeration membrane, (4) microtiter plate, (5) pipette tips box, and (6) micropipettes. (**b**) Open the microtiter plate in the safety cabinet. Use gloves during handling. (**c**) Fill the wells with the mycobacterial cell suspension. Tilt the microtiter plate in such a manner that the tip of the pipette touches the bottom of the well. (**d**) Add the steroid solution to the wells. Immerge the tip in the mycobacterial cell suspension. (**e**) Place the sealing tape on top of the microtiter plate according to the manufacturer's instructions. (**f**) Certify that the sealing tape adequately covers all the wells. (**g**) When sampling, puncture the sealing tape with the tip and collect the entire volume of the well

References

1. Kirk TV, Szita N (2013) Oxygen transfer characteristics of miniaturized bioreactor systems. Biotechnol Bioeng 110:1005–1019
2. Marques MPC, Cabral JMS, Fernandes P (2009) High throughput in biotechnology: from shake-flasks to fully instrumented microfermentors. Recent Pat Biotechnol 3:124–140
3. Fernandes P, Carvalho F, Marques MPC (2011) Miniaturization in biotechnology: speeding up the development of bioprocesses. Recent Pat Biotechnol 5:160–173
4. Fernandes P, Cabral JMS (2006) Microlitre/millilitre shaken bioreactors in fermentative and biotransformation processes – a review. Biocatal Biotransformation 24:237–252
5. Kensy F, John GT, Hofmann B, Buchs J (2005) Characterisation of operation conditions and online monitoring of physiological culture parameters in shaken 24-well microtiter plates. Bioprocess Biosyst Eng 28:75–81
6. Funke M, Buchenauer A, Mokwa W, Kluge S, Hein L, Mueller C, Kensy F, Buechs J (2010) Bioprocess control in microscale: scalable fermentations in disposable and user-friendly microfluidic systems. Microb Cell Factories 9: 86
7. Marques MPC, Magalhães S, Cabral JMS, Fernandes P (2009) Characterization of 24-well microtiter plate reactors for a complex multistep bioconversion: from sitosterol to androstenedione. J Biotechnol 141:174–180
8. Lye GJ, Ayazi-Shamlou P, Baganz F, Dalby PA, Woodley JM (2003) Accelerated design of bioconversion processes using automated microscale processing techniques. Trends Biotechnol 21:29–37
9. Marques MPC, Szita N (2017) Bioprocess microfluidics: applying microfluidic devices for bioprocessing. Curr Opin Chem Eng 18:61–68
10. Davies MJ, Nesbeth DN, Szita N (2013) Development of a microbioreactor 'cassette' for the cultivation of microorganisms in batch and chemostat mode. Chim Oggi – Chem Today 31:46–49
11. Kirk TV, Marques MPC, Radhakrishnan ANP, Szita N (2016) Quantification of the oxygen uptake rate in a dissolved oxygen controlled oscillating jet-driven microbioreactor. J Chem Technol Biotechnol 91:823–831
12. Maldonado SL, Krull J, Rasch D, Panjan P, Sesay AM, Marques MPC, Szita N, Krull R (2019) Application of a multiphase microreactor chemostat for the determination of reaction kinetics of Staphylococcus carnosus. Bioprocess Biosyst Eng 42:953–961
13. Gruber P, Marques MPC, Szita N, Mayr T (2017) Integration and application of optical chemical sensors in microbioreactors. Lab Chip 17:2693–2712
14. Rodriguez-Garcia A, Fernandez-Alegre E, Morales A, Sola-Landa A, Lorraine J, Macdonald S, Dovbnya D, Smith MC, Donova M, Barreiro C (2016) Complete genome of *Mycobacterium neoaurum* NRRL B-3805, an androstenedione (AD) producer for industrial biotransformations of sterols. J Biotechnol 224:64–65
15. Marques M, de Carvalho C, Claudino M, Cabral J, Fernandes P (2007) On the feasibility of the microscale approach for a multistep biotransformation: sitosterol side chain cleavage. J Chem Technol Biotechnol 82:856–863
16. Fernandes P, Cabral JMS (2010) Steroid bioconversion. In: Flickinger MC (ed) Encyclopedia of industrial biotechnology – bioprocess, bioseparation, and cell technology, vol 7. John Wiley & Sons, New York
17. Nunes VO, Vanzelloti NC, Fraga JL, Pessoa F, Ferreira TF, Amaral P (2022) Biotransformation of phytosterols into androstenedione – a technological prospecting study. Molecules 27: 3164
18. Carvalho F, Marques MPC, de Carvalho CCCR, Cabral JMS, Fernandes P (2009) Sitosterol bioconversion with resting cells in liquid polymer based systems. Bioresour Technol 100:4050–4053
19. Ribn W (2012) Biochemical isolation and identification of mycobacteria. In: Jimenez-Lopez JC (ed) Biochemical testing. InTech. https://doi.org/10.5772/34309
20. Marques M, Carvalho F, Magalhães S, Cabral J, Fernandes P (2009) Screening for suitable solvents as substrate carriers for the microbial side-chain cleavage of sitosterol using microtitre plates. Process Biochem 44:556–561
21. Marques M, Carvalho F, de Carvalho C, Cabral J, Fernandes P (2010) Steroid bioconversion: towards green processes. Food Bioprod Process 88:12–20
22. Marques M, Cabral J, Fernandes P (2010) A microwell platform for the scale-up of a multistep bioconversion to bench-scale reactors: sitosterol side-chain cleavage. Biotechnol J 5: 402–412

Chapter 13

Biocatalysis of Steroids by *Mycobacterium* sp. in Aqueous and Organic Media

Carla C. C. R. de Carvalho and Pedro Fernandes

Abstract

Mycobacterium sp. can convert steroids such as β-sitosterol, campesterol, and cholesterol, by selective side-chain cleavage and oxidation of the C3 hydroxyl group to a ketone, into key intermediates that can be easily functionalized to yield commercially interesting pharmaceutical products. In aqueous systems, the biocatalysis is limited by the low solubility of the steroids in water. Several strategies have been introduced to tackle this limitation, e.g., formation of cyclodextrin–steroid complexes and generation of aqueous micro-dispersions with steroid particle size in the range of hundreds of nanometers. Still, the introduction of an organic phase acting as a substrate and/or product reservoir is a well-established and relatively easy to implement strategy to overcome the sparing water solubility of steroid molecules. However, the organic phase has to be carefully chosen to prevent tampering with the activity/viability of microbial cells.

In this chapter, we describe the methodology for the biocatalysis of β-sitosterol to 4-androstene-3,17-dione (AD) and 1,4-androstadiene-3,17-dione (ADD), both in aqueous and organic:aqueous systems. In the latter case, both traditional organic solvents and green solvents are proposed.

Key words Biotransformation, Bioconversion, Organic solvents, Two-phase systems, β-sitosterol, Cholesterol, Green solvents

1 Introduction

Steroids, like many other industrially relevant substrates, are poorly water-soluble and may often present an inhibitory or toxic effect on whole-cell biocatalysts [1–3]. Among these are substrates such as β-sitosterol, campesterol, and cholesterol (Fig. 1), which may be converted by *Mycobacterium* cells. The introduction of an organic solvent which will act as a reservoir for substrates and/or products is a well-established approach to circumvent this [4–7]. Nevertheless, organic solvents may influence the physiology and activity of the bacterial cells [2, 6–9]. Environmental, health, and safety concerns have led to the development of sustainable processes in a

Carlos Barreiro and José-Luis Barredo (eds.), *Microbial Steroids: Methods and Protocols*, Methods in Molecular Biology, vol. 2704,
https://doi.org/10.1007/978-1-0716-3385-4_13,

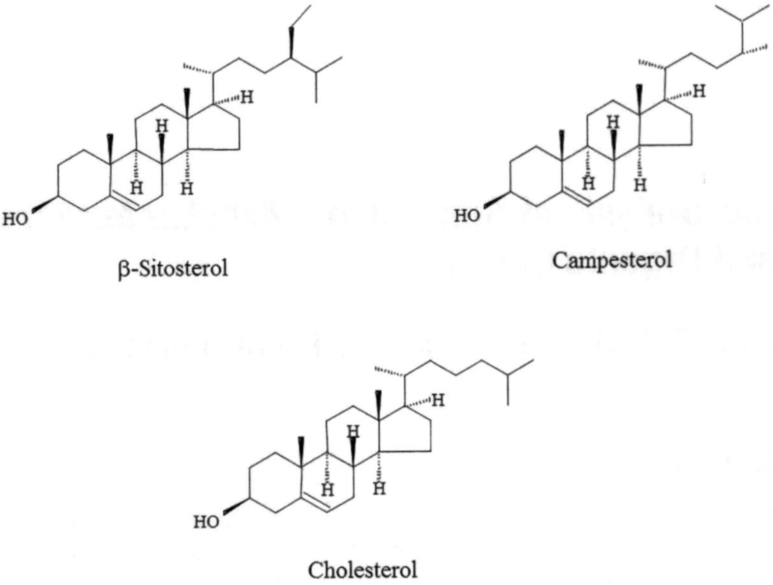

Fig. 1 Chemical structure of β-sitosterol, campesterol, and cholesterol

"Design for the Environment" approach [6, 10–12]. Steroid bioconversion by *Mycobacterium* cells has been successfully carried out in green solvents such as ionic liquids, polyethylene glycol, polypropylene glycol, silicone oil, and vegetable oil [7, 13, 14].

The bioconversion of steroids by *Mycobacterium* cells is often carried out with resting cells [15–19]. The cells are firstly produced by fermentation and then washed with phosphate buffer and recovered by filtration as a wet paste. During the bioconversions they remain in a vegetative state by lack of growth nutrients. These cells are able to remain viable for several days both in aqueous, biphasic and even single organic bioconversion systems [2, 20]. Nevertheless, other authors have proposed the production of steroids during the fermentative process [14, 21–28]. In these cases, solubilizing agents such as polypropylene glycol or silicone are added since they are biocompatible with *Mycobacterium* sp. growth.

Steroid productivity has been enhanced lately by improvements in metabolic pathway regulation [29], recombinant DNA and genetic engineering techniques [24, 30, 31], and changes in cell intracellular environment [28].

In this chapter, we present the side-chain cleavage of β-sitosterol (*see* **Note 1**) to 4-androstene-3,17-dione (AD) and 1,4-androstadiene-3,17-dione (ADD) as an example of a well-established steroid conversion system.

2 Materials

Prepare buffer solutions and defined growth media preferentially with ultrapure or deionized water. Store prepared buffers and culture media at 4 °C. Use ACS (American Chemical Society) and Reagent Grade solvents. Follow all safety handling and storage recommendations provided by the suppliers and diligently respect waste disposal regulations.

2.1 Media Preparation

1. Potato dextrose agar slants: Suspend 39 g of potato dextrose agar (Difco) in 1 L of water. Heat with agitation and boil for 1 min to completely dissolve the powder. Transfer 5–10 mL of molten agar to test tubes (the amount of medium will be dependent on the volume of the test tubes) and close the tubes with loose caps. Autoclave at 121 °C for 15 min. While the medium is hot, tilt the tubes over a surface so that a long-slanted surface is formed during solidification of the agar. After cooling, tighten the caps.
2. Growth medium (defined mineral medium): To prepare 1 L in water, mix 20 g of glycerol, 4 g of NH_4Cl, 0.8 g of Tween® 20 (*see* **Note 2**), 0.2 g of $MgSO_4.7H_2O$ and 1 g of β-sitosterol (*see* **Note 3**), and 100 mM of disodium/potassium phosphate buffer, pH 7.0 [13]. Autoclave at 121 °C for 20 min.
3. Disodium/potassium phosphate buffer, pH 7.0: To prepare 1 L of 100 mM buffer, dissolve 2.65 g of $NaH_2PO_4.H_2O$ and 14 g of K_2HPO_4 in two separate flasks, using 60–70% of the total volume of water. Add both solutions and measure the pH. If necessary, adjust the pH to 7.0 by adding the acid or base species. Adjust the volume to 1 L by adding water in a volumetric flask (*see* **Note 4**).
4. 50 mM phosphate buffer, pH 7.0: To prepare 1 L of buffer, dissolve 5.31 g of K_2HPO_4 and 2.65 g of KH_2PO_4 in two separate flasks using ca. 70% of the total volume of water. After adding both solutions, measure the pH and correct it by adding the necessary acid or base species. Adjust the volume of the mixture to 1 L by adding water in a volumetric flask (*see* **Note 4**).

2.2 Bioconversion Assay

1. Shaken microtiter plates with sealing tape (AeraSeal™, Breathe-Easy®, or equivalent) (*see* **Note 5**).
2. Erlenmeyer type flasks or equivalent plasticware/glassware.
3. Magnetically/mechanically stirred reactor (*see* **Note 6**).

2.3 Bacterial Strains

Mycobacterium strains able to convert steroids to AD and ADD (*see* **Note** 7).

1. *Mycobacterium neoaurum* NRRL B-3805 [32].
2. *Mycobacterium* sp. NRRL MB 3683 [22, 33].
3. *Mycobacterium* sp. NRRL B-3683 [34].
4. *M. neoaurum* VKPM strain As-1634 [23].
5. *M. neoaurum* NwIB-yV [15].

2.4 HPLC

1. HPLC equipment.
2. Lichrospher Si-60 column, 10 μm particle size, or equivalent.
3. Mobile phase for steroid detection at 215 nm: *n*-heptane and ethanol (98:2, v/v).
4. Mobile phase for steroid detection at 254 nm: *n*-heptane and ethanol (95:5, v/v).

3 Methods

3.1 Bacterial Growth

1. Grow *Mycobacterium* sp. NRRL B-3805 cells in potato dextrose agar slants, to use as inoculum in the fermentation step, for 3 days at 30 °C.
2. Grow *Mycobacterium* cells in orbital incubators at 30 °C and 200 rpm in 1 L Erlenmeyer flasks containing 200 mL of growth medium (*see* **Note 8**).
3. Monitor growth by measuring, e.g., DO_{600} of the culture medium with a spectrophotometer or by dry weight measurements. However, extensive cell aggregation may occur with *Mycobacterium* cells, thus preventing a reliable monitoring.
4. Harvest the cells by filtration with a 0.2 μm filter (*see* **Note 9**), wash the cells with 50 mM phosphate buffer pH 7.0 and store the cells as a wet paste (approximately 120–200 mg dry cell weight/g wet paste) at −20 °C until use.

3.2 Side-Chain Cleavage of Sitosterol in Aqueous Systems

1. Add the necessary volume of 50 mM potassium phosphate buffer, pH 7.0 (*see* **Note 10**) to reach a ratio of liquid to the total bioreactor volume from 1:10 to 1:5.
2. Weight the necessary wet biomass to achieve ca. 2 g/L of cell dry weight (approximately 10–18 g/L of wet biomass) (*see* **Note 11**) and add the cells to the phosphate buffer.
3. Start the bioconversion by adding β-sitosterol at a concentration of 1–2.5 mM (*see* **Note 12**).

4. Run the bioconversion assay for ca. 24 h at 30 °C and 200 rpm. Agitation may be achieved by an incubator or by magnetic or mechanical stirring.
5. Take samples along the assay and extract the dissolved steroids with an organic solvent such as *n*-heptane containing 0.2 g/L progesterone as internal standard (*see* **Note 13**).

3.3 Side-Chain Cleavage of Sitosterol in Organic–Aqueous Systems

1. Add the necessary volume of 50 mM potassium phosphate buffer, pH 7.0 (*see* **Note 10**) to reach a ratio of liquid to the total bioreactor volume from 1:10 to 1:5.
2. Weight the necessary wet biomass to achieve ca. 2 g/L of cell dry weight (approximately 10–18 g/L of wet biomass) (*see* **Note 11**).
3. Add from 1:3 to 1:1 (v/v) organic solvent:aqueous phase (phosphate buffer 50 mM, pH 7.0). As an organic phase, the following phthalates have been tested and are known to allow the bioconversion of β-sitosterol:bis(2-ethylhexyl) phthalate [2, 16], diethyl phthalate, and di-isodecyl phthalate [35] (*see* **Note 14**). Nevertheless, bis(2-ethylhexyl) phthalate allows the highest cellular activities [35]. Add β-sitosterol to achieve a concentration of 5 g/L, referred to the organic phase.
4. Run the conversion assay for ca. 120 h at 30 °C and 200 rpm. Agitation may be achieved by an incubator or by magnetic or mechanical stirring.
5. Take 300 μL samples along the assay, centrifuge them at 10,000 × *g* for 1 min and dilute 80 μL of the organic phase with 720 μL of *n*-heptane containing 0.2 g/L of an internal standard such as progesterone.

3.4 Side-Chain Cleavage of Sitosterol Using Green Solvents

1. Use 50 mM potassium phosphate buffer, pH 7.0 (*see* **Note 10**) as aqueous phase.
2. Weight the necessary wet biomass to achieve ca. 2 g/L of cell dry weight (approximately 10–18 g/L of wet biomass) (*see* **Note 11**).
3. Add 5 g/L of β-sitosterol in a green solvent such as a liquid polymer (e.g., poly(methylphenylsiloxane), polyethylene glycol, or polypropylene glycol) [13, 36]. Use a 1:1 green solvent:aqueous phase ratio.
4. Carry out the experiment at 35 °C and 200 rpm, and periodically take 300 μL samples from the polymer phase and extract the β-sitosterol with an organic solvent such as *n*-heptane containing 0.2 g/L progesterone as internal standard.

3.5 Determination of Side-Chain Cleavage Activity

1. Perform steroid and sterol quantification by HPLC with 1 mL/min isocratic elution.
2. As a mobile phase, prepare either (i) *n*-heptane and ethanol (98:2, v/v) for steroid detection at 215 nm or (ii) *n*-heptane and ethanol (95:5, v/v) for steroid detection at 254 nm.

4 Notes

1. Besides β-sitosterol, other sterols may be used as a substrate to yield the same steroid products.
2. Tween is intended at decreasing the formation of clusters by the *Mycobacterium* cells during growth. Tween 80 may also be used for the same purpose (*see*, e.g., [37]).
3. The substrate of the bioconversion, β-sitosterol, should be added during growth to stimulate the corresponding metabolic pathway.
4. To prevent the growth of microorganisms, a 100× concentrated solution may be prepared, which may be stored at 4 °C until being diluted for use.
5. In order to minimize evaporation, bioreactors should be closed with sealing tape while allowing for gas transfer, as defective aeration hampers sterol side-chain cleavage [32].
6. The working volume should be maintained at approximately 15% of the total volume of the reactor vessel. When working with organic solvents such as phthalates, glass reactors should be used.
7. Several studies have proposed the production of mutants to separate AD production from that of ADD. In 1972, a mutant was selected by UV irradiation, *Mycobacterium* sp. NRRL B-3683, capable of converting sterols into ADD [38]. Another mutant, unable to 1-dehydrogenate steroids, *Mycobacterium* sp. NRRL B-3805, allowed in the production of AD from sterols [38]. *M. neoaurum* mutants able to produce large amounts of AD using an atmospheric and room temperature plasma treatment were also selected [39]. *Mycobacterium fortuitum* subsp. *fortuitum* NCIM 5239 able to convert soya sterols to AD was selected by combining mitomycin C and UV treatment [21]. Nitrosoguanidine mutagenesis was used to select the *Mycobacterium* sp. NRRL B-3805S and B-3683 mutants, one able to produce AD and the other ADD, from phytosterol [40].
8. The "volume of liquid/volume of Erlenmeyer" ratio should not be higher than 0.2 to prevent oxygen limitation during growth.

9. Since *Mycobacterium* sp. cells may form large cell aggregates, even when Tween 80 is used, a filter with a lower particle retention capacity may be used to facilitate the filtration process.
10. To increase *Mycobacterium* sp. activity and stability, the addition of the following salts to the reaction medium (in g/L) was proposed: 0.75 of $CaCl_2$, 0.01 of $CuSO_4$, 0.01 of $CoCl_2{\cdot}6H_2O$, 0.03 of $Fe_2(SO_4)_3$, 0.01 of $MnSO_4{\cdot}H_2O$, 0.01 of $ZnSO_4{\cdot}7H_2O$, and 10.0 of $MgSO_4{\cdot}7H_2O$ [41].
11. The wet paste may be weighted frozen for easier handling.
12. β-Sitosterol concentration may be increased with the addition of a water-miscible cosolvent such as dimethylformamide, methanol, acetone, or 1,2-propanediol. The amount of cosolvent should be maintained below 1.5–5% (v/v) to prevent biocatalyst inhibition.
13. If a concentration of β-sitosterol above its solubility in aqueous phase is used, suicidal assays must be performed. In this case, parallel assays are made, and the entire content of one bioreactor is extracted with an organic solvent such as *n*-heptane containing an internal standard (e.g., progesterone) for analysis in each data point.
14. Wear protection gloves, clothes, and glasses when handling phthalates. Use a fume hood and always follow manufacturer's safety instructions when handling organic solvents.

Acknowledgments

The work was partially funded by national funds from FCT in the scope of the project UIDB/04565/2020 and UIDP/04565/2020 of the Research Unit iBB-Institute for Bioengineering and Biosciences and of the project LA/P/0140/2020 of the i4HN-Associate Laboratory Institute for Health and Bioeconomy.

References

1. Schmid A, Dordick JS, Hauer B, Kiener A, Wubbolts M, Witholt B (2001) Industrial biocatalysis today and tomorrow. Nature 409: 258–268
2. de Carvalho CCCR, Cruz A, Angelova B, Fernandes P, Pons MN, Pinheiro HM, Cabral JMS, da Fonseca MMR (2004) Behaviour of *Mycobacterium* sp. NRRL B-3805 whole cells in aqueous, organic-aqueous and organic media studied by fluorescence microscopy. Appl Microbiol Biotechnol 64:696–701
3. Batth R, Nicolle C, Cuciurean IS, Simonsen HT (2020) Biosynthesis and industrial production of androsteroids. Plants 9:1144
4. van Schie MMCH, Spöring J-D, Bocola M, Domínguez de María P, Rother D (2021) Applied biocatalysis beyond just buffers – from aqueous to unconventional media. Options and guidelines. Green Chem 23: 3191–3206

5. de Carvalho CCCR (2017) Whole cell biocatalysts: essential workers from Nature to the industry. Microb Biotechnol 10:250–263
6. Grundtvig IPR, Heintz S, Krühne U, Gernaey KV, Adlercreutz P, Hayler JD, Wells AS, Woodley JM (2018) Screening of organic solvents for bioprocesses using aqueous-organic two-phase systems. Biotechnol Adv 36:1801–1814
7. Rokade R, Ravindran S, Singh P, Suthar J (2018) Microbial biotransformation for the production of steroid medicament. In: Vijayakumar R, Raja SS (eds) Secondary metabolites – sources and applications. IntechOpen, London, p 115
8. Heipieper HJ, Neumann G, Cornelissen S, Meinhardt F (2007) Solvent-tolerant bacteria for biotransformations in two-phase fermentation systems. Appl Microbiol Biotechnol 74: 961–973
9. de Carvalho CCCR (2010) Adaptation of *Rhodococcus* to organic solvents. In: Alvarez MH (ed) Biology of *Rhodococcus*. Springer, Berlin, Heidelberg, p 368
10. Little AD (2001) Making EHS an integral part of process design. American Institute of Chemical Engineers, CWRT, CCPS, New York
11. Tao J, Kazlauskas RJ (2011) Biocatalysis for green chemistry and chemical process development. John Wiley & Sons, Inc., Hoboken
12. Sheldon RA, Woodley JM (2018) Role of biocatalysis in sustainable chemistry. Chem Rev 118:801–838
13. Marques MPC, Carvalho F, de Carvalho CCCR, Cabral JMS, Fernandes P (2010) Steroid bioconversion: towards green processes. Food Bioprod Process 88:12–20
14. Nunes VO, Vanzellotti ND, Fraga JL, Pessoa FL, Ferreira TF, Amaral PF (2022) Biotransformation of phytosterols into androstenedione – a technological prospecting study. Molecules 27:3164
15. Gao XQ, Feng JX, Wang XD, Hua Q, Wei DZ (2015) Enhanced steroid metabolites production by resting cell phytosterol bioconversion. Chem Biochem Eng Q 29:567–573
16. Cruz A, Fernandes P, Cabral JMS, Pinheiro HM (2001) Whole-cell bioconversion of β-sitosterol in aqueous–organic two-phase systems. J Mol Catal B Enzym 11:579–585
17. Donova MV, Egorova OV (2012) Microbial steroid transformations: current state and prospects. Appl Microbiol Biotechnol 94:1423–1447
18. Hu Y, Wang D, Wang X, Wei D (2020) A recycled batch biotransformation strategy for 22-hydroxy-23,24-bisnorchol-4-ene-3-one production from high concentration of phytosterols by mycobacterial resting cells. Biotechnol Lett 42:2589–2594
19. Wang D, Zhang J, Cao D-D, Wang X, Wei D (2021) Identification and in situ removal of an inhibitory intermediate to develop an efficient phytosterol bioconversion process using a cyclodextrin-resting cell system. RSC Adv 11: 24787–24793
20. Angelova B, Fernandes P, Spasova D, Mutafov S, Pinheiro HM, Cabral JMS (2006) Scanning electron microscopy investigations on bis(2-ethylhexyl)phthalate treated *Mycobacterium* cells. Microsc Res Tech 69:613–617
21. Gulla V, Banerjee T, Patil S (2010) Bioconversion of soysterols to androstenedione by *Mycobacterium fortuitum* subsp. *fortuitum* NCIM 5239, a mutant derived from total sterol degrader strain. J Chem Technol Biotechnol 85:1135–1141
22. Kutney JP, Milanova RK, Vassilev CD, Stefanov SS, Nedelcheva NV, Kutney P, Milanova K (1999) High yield microbial conversion of phytosterol to androstadienedione and androstenedione, in presence of solubilizer, e.g. silicone. WO9949075-A; EP1066399-A; WO9949075-A1. (Forbes Medi-Tech Inc; Akzo Nobel Nv; Organon Nv)
23. Rodina NV, Molchanova MA, Voishvillo NE, Andryushina VA, Stytsenko TS (2008) Conversion of phytosterols into androstenedione by *Mycobacterium neoaurum*. Appl Biochem Microbiol 44:48–54
24. Li X, Chen T, Peng F, Song S, Yu J, Sidoine DN, Cheng X, Huang Y, He Y, Su Z (2021) Efficient conversion of phytosterols into 4-androstene-3,17-dione and its C1,2-dehydrogenized and 9α-hydroxylated derivatives by engineered Mycobacteria. Microb Cell Factories 20:158
25. Zhao W, Xie H, Zhang X, Wang Z (2022) Crystal substrate inhibition during microbial transformation of phytosterols in Pickering emulsions. Appl Microbiol Biotechnol 106: 2403–2414
26. Andriushina VA, Rodina NV, Stytsenko TC, Luu DH, Druzhinina AV, Iaderets VV, Voishvillo NE (2011) Conversion of soybean sterols into 3,17-diketosteroids using actinobacteria *Mycobacterium neoaurum*, *Pimelobacter simplex*, and *Rhodococcus erythropolis*. Prikl Biokhim Mikrobiol 47:297–301
27. Liu Y-J, W-t J, Song L, Tao X-Y, Zhao M, Gao B, Meng H, Wang F-Q, Wei D-Z (2022) Transformation of phytosterols into pregnatetraenedione by a combined microbial and chemical process. Green Chem 24:3759–3771

28. Zhang Y, Zhou X, Wang X, Wang L, Xia M, Luo J, Shen Y, Wang M (2020) Improving phytosterol biotransformation at low nitrogen levels by enhancing the methylcitrate cycle with transcriptional regulators PrpR and GlnR of *Mycobacterium neoaurum*. Microb Cell Factories 19:13
29. Chang H, Zhang H, Zhu L, Zhang W, You S, Qi W, Qian J, Su R, He Z (2020) A combined strategy of metabolic pathway regulation and two-step bioprocess for improved 4-androstene-3,17-dione production with an engineered *Mycobacterium neoaurum*. Biochem Eng J 164:107789
30. Fernández-Cabezón L, Galán B, García JL (2018) New insights on steroid biotechnology. Front Microbiol 9:958
31. Zhao Y-Q, Liu Y-J, Ji W-T, Liu K, Gao B, Tao X-Y, Zhao M, Wang F-Q, Wei D-Z (2022) One-pot biosynthesis of 7β-hydroxyandrost-4-ene-3,17-dione from phytosterols by cofactor regeneration system in engineered *Mycolicibacterium neoaurum*. Microb Cell Factories 21:59
32. Cruz A, Angelova B, Fernandes P, Cabral JMS, Pinheiro HM (2004) Study of key operational parameters for the side-chain cleavage of sitosterol by free mycobacterial cells in bis-(2-ethylhexyl) phthalate. Biocatal Biotransformation 22:189–194
33. Sedlaczek L, Gorminski BM, Lisowska K (1994) Effect of inhibitors of cell-envelope synthesis on beta-sitosterol side-chain degradation by *Mycobacterium* sp. NRRL-MB-3683. J Basic Microbiol 34:387–399
34. Llanes N, Hung B, Falero A, Perez C, Aguila B (1995) Glucose and lactose effect on AD and ADD bioconversion by *Mycobacterium* sp. Biotechnol Lett 17:1237–1240
35. Cruz A, Fernandes P, Cabral JMS, Pinheiro HM (2004) Solvent partitioning and whole-cell sitosterol bioconversion activity in aqueous-organic two-phase systems. Enzym Microb Technol 34:342–353
36. Carvalho F, Marques MPC, de Carvalho CCCR, Cabral JMS, Fernandes P (2009) Sitosterol bioconversion with resting cells in liquid polymer based systems. Bioresour Technol 100:4050–4053
37. Sih CJ (1988) Process for preparing steroids. US 4755463 A
38. Marsheck WJ, Kraychy S, Muir RD (1972) Microbial degradation of sterols. Appl Microbiol 23:72–77
39. Liu C, Zhang X, Rao Z-M, Shao M-L, Zhang L-L, Wu D, Xu Z-H, Li H (2015) Mutation breeding of high 4-androstene-3,17-dione-producing *Mycobacterium neoaurum* ZADF-4 by atmospheric and room temperature plasma treatment. J Zhejiang Univ Sci B 16:286–295
40. Huang C-L, Chen Y-R, Liu W-H (2006) Production of androstenones from phytosterol by mutants of *Mycobacterium* sp. Enzym Microb Technol 39:296–300
41. Marques MPC, de Carvalho CCCR, Claudino MJC, Cabral JMS, Fernandes P (2007) On the feasibility of the microscale approach for a multistep biotransformation: sitosterol side chain cleavage. J Chem Technol Biotechnol 82: 856–863

Chapter 14

Scale-Up of Phytosterols Bioconversion into Androstenedione

Sonia Martínez-Cámara, Manuel de la Torre, José-Luis Barredo, and Marta Rodríguez-Sáiz

Abstract

Phytosterols, coming as a by-product of vegetable oils or wood pulp, contain the cyclopentanoperhydrophenanthrene nucleus and can be bioconverted into steroid intermediates by removing the C17 side chain. This chapter shows the scale-up, from flask to bioreactor, of phytosterols bioconversion into 4-androstene-3,17-dione (androstenedione; AD) using *Mycolicibacterium neoaurum* B-3805. Due to the fact that phytosterols and AD are nearly insoluble in water, two-phase systems and the use of chemically modified cyclodextrins have been described as methods to solve it. Here, we use a water–oil two-phase system that allows the bioconversion of up to 20 g/L of phytosterols into AD in 5 L and 20 L bioreactors.

Key words Bioconversion, Phytosterols, Fermentation, 4-Androstene-3,17-dione, Androstenedione, AD, *Mycolicibacterium neoaurum*, *Mycobacterium neoaurum*, Two-phase system

1 Introduction

Androst-4-ene-3,17-dione (androstenedione; AD) is a key intermediate in the synthesis of steroids such as testosterone, estradiol, progesterone, cortisone, prednisone, or prednisolone, which have great interest for the pharmaceutical companies [1–3]. In fact, the global market for steroids was significantly large in 2020, and it is expected to reach the value of US $17 billion by the end of 2025 with an annual growth rate of around 3%. Key factors responsible for driving revenue growth are the increasing number of applications in medical issues and changes in the market dynamics due to the outbreak of the COVID-19 pandemic. In fact, dexamethasone treatment was probed to reduce mortality in patients with COVID-19 who were receiving respiratory support.

Carlos Barreiro and José-Luis Barredo (eds.), *Microbial Steroids: Methods and Protocols*, Methods in Molecular Biology, vol. 2704, https://doi.org/10.1007/978-1-0716-3385-4_14,

Anabolic steroids, including AD, androsta-1,4-diene-3,17-dione (androstadienedione; ADD), and derivatives, are used to treat medical conditions such as late puberty, hormonal imbalance, breast cancer in woman, or muscle loss of patients suffering from AIDS and cancer. This type of steroids marks a faster growth rate in developed countries, especially in North America due to an increase in their questioned use as enhancer for athletic performance. According to MarketWatch, the global AD market was valued at USD 190 million in 2018 and is estimated to reach USD 210 million by 2025 [4]. The growing demand of the world market for these and other steroidal products has focused pharmaceutical research on producing high-value steroid precursors through the use of efficient eco-friendly technologies with low production costs, as an alternative to the more environmentally unfriendly chemical processes [5].

The success of a fermentation-based process depends on the process cost (raw materials, utilities, manpower, indirect costs), the fermentation yield, and the market price. In this way, the use of phytosterols, a by-product of oil or wood pulp industry with similar characteristics to cholesterol such as structural behavior and functionality, as a starting material to produce steroid intermediates with *Mycolicibacterium* sp. in a single fermentation step, allows for the development of highly competitive processes in the global market [6–8].

Mycolicibacterium sp., traditionally known as *Mycobacterium* sp., possesses the ability to catabolize naturally occurring 3-β-hydroxy-Δ5-sterols (such as cholesterol, β-sitosterol, etc.) or a mixture of sterols (phytosterols) to carbon dioxide and water, being able to accumulate a variety of steroidal intermediates during the degradation process [9–11]. New mutants have been obtained by classical mutagenesis or synthetic biology to block the catabolic pathway of phytosterol, favoring the accumulation of steroid intermediates of pharmaceutical value (Fig. 1) [12, 13]. Among them, *Mycolicibacterium neoaurum* NRRL-3805 (previously known as *Mycobacterium neoaurum* NRRL-3805), an AD producing strain obtained by mutagenesis from a soil isolate producing ADD, is found [14, 15]. This mutant transforms phytosterols into AD with a low content of ADD because of the partial failure of the dehydrogenation at C1. The genome has been recently sequenced showing a circular chromosome of 5,421,338 bp containing 4844 protein-coding gene [16].

However, the industrial biotechnological processes for AD and other steroids are severely limited by the poor water solubility of phytosterols and the inhibition or toxicity that both, substrate and products, can cause to the microorganism [17]. Steroid-producing companies are focused on dealing with these obstacles, including

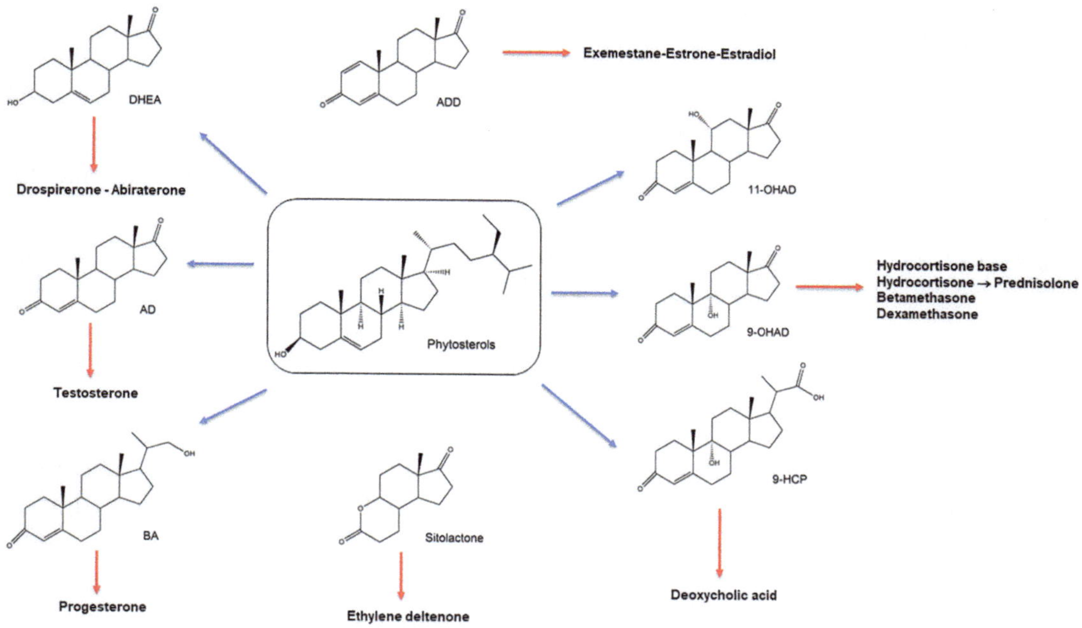

Fig. 1 Steroid bioconversions (blue arrows) using phytosterols as a starting material, and chemical processes (red arrows) for production of steroid Active Pharmaceutical Ingredients (APIs) from steroid intermediates

the use of biphasic systems, surfactants, liquid polymers, cyclodextrins, ionic liquids, organic solvents, resting cells, in situ product recovery, etc. to increase the phytosterol loading and achieve a more competitive process [4, 18–22].

Carrying out a large-scale experimental fermentation is therefore costly, so pilot scale (from 5 to 40 L) is a powerful tool before industrializing the laboratory results. A pilot plant mimics the industrial-scale fermentation process on a smaller scale, being cost-effective, environmentally aware, and the best option to explore potential changes in a production operation. Potential critical parameters of the AD bioconversion process, such as bioreactor design, raw materials, development of the vegetative phases, phytosterol loading, type of fermentation (batch, semi-batch, continuous, etc.), pH, oxygen, temperature, or viscosity, can be deeply studied at pilot scale before scaling up to industrial.

Finally, the pilot-scale research allows the evaluation of the costs of different AD upstream and downstream technologies and to take a decision on process scale-up to comply with the quality commercial requirements.

In this chapter, we show a protocol for the bioconversion of phytosterols into AD at 5 L and 20 L scale, emphasizing those steps considered as critical points of the process.

2 Materials

2.1 Vegetative Stages

2.1.1 First Vegetative and Second Vegetative Stages (Flask Scale)

1. *Mycolicibacterium neoaurum* B-3805 (CECT 3018; DSM 2967; NCIMB 11678).
2. Erlenmeyer flasks: 500 mL for the first vegetative stage and 2 L for the second vegetative stage.
3. Growth medium: 54 g/L of molasses, 5.4 g/L of $NaNO_3$, 0.6 g/L of $NH_4H_2PO_4$, and 20 g/L of soya oil. pH adjusted to 7.0.
4. Flow laminar cabinet.
5. Orbital shaker.

2.1.2 Third Vegetative Stage (5 L Bioreactor)

1. Growth medium: 54 g/L of molasses, 5.4 g/L of $NaNO_3$, 0.6 g/L of $NH_4H_2PO_4$, and 20 g/L of soya oil. pH adjusted to 7.0.
2. Fermentor Bioflo 3000, or equivalent, with a 5 L glass vessel (New Brunswick Scientific).
3. Inoculation flask: sterile Erlenmeyer 2 L flask with silicone tube and stainless-steel needle.

2.2 AD Bioconversion at 5 L Scale

1. Bioconversion medium: 2 g/L of NH_4NO_3, 1 g/L of KH_2PO_4, 2 g/L of K_2HPO_4, 0.2 g/L of KCl, 0.3 g/L of $CaCl_2$, 150 g/L of soya oil, and 20 g/L of phytosterol from pine tree (Arboris). Add 1 mL/L from a micronutrient solution (11 g/L of $ZnSO_4$, 1 g/L of $FeSO_4$, 6 g/L of $MnSO_4$, 0.04 g/L of $CuSO_4$, 0.3 g/L of $CoSO_4$, 0.03 g/L of H_3PO_4, and 0.001 g/L of KI). pH adjusted to 7.0.
2. Fermentor Bioflo 3000, or equivalent, with a 5 L glass vessel (New Brunswick Scientific).
3. Blender.

2.3 AD Bioconversion at 20 L Scale

1. Bioconversion medium: 2 g/L of NH_4NO_3, 1 g/L of KH_2PO_4, 2 g/L of K_2HPO_4, 0.2 g/L of KCl, 0.3 g/L of $CaCl_2$, 150 g/L of soya oil, and 20 g/L of phytosterol from pine tree (Arboris). Add 1 mL/L from a micronutrient solution (11 g/L of $ZnSO_4$, 1 g/L of $FeSO_4$, 6 g/L of $MnSO_4$, 0.04 g/L of $CuSO_4$, 0.3 g/L of $CoSO_4$, 0.03 g/L of H_3PO_4, and 0.001 g/L of KI). pH adjusted to 7.0.
2. Fermentor Biostat C, or equivalent, with a stainless-steel vessel of 20 L (about 15 L working volume) (Sartorius).
3. Sterile tubing system for vegetative culture transfer.

2.4 AD Analysis

2.4.1 AD Analysis by HPLC

1. Androstenedione (4-androstene-3,17-dione, AD) 99.5% purity (Sigma-Aldrich).
2. Butyl acetate.
3. Acetonitrile HPLC grade.
4. Analytical balance.
5. HPLC equipment. Waters Alliance e2695 HPLC system with Waters 2998 Photodiode Array (PDA) Detector, or equivalent (Waters).
6. Column RP C18 Nucleosil 250 × 4.6 mM, 5 μM (with guard column), or equivalent.
7. Mobile phase: Acetonitrile/water (52/48) with acetic acid 0.01%.

2.4.2 AD Analysis by TLC

1. Ethyl acetate.
2. TLC chamber for 20 × 20 cm aluminum sheets.
3. TLC Silica gel 60 F354 aluminum sheets 20 × 20 cm (Merck).
4. Mobile phase: Heptane/ethyl acetate (1/1).
5. Development solution: 20% H_2SO_4.
6. Heater plate.

2.5 General Equipment

1. Desktop centrifuge.
2. Swinging head centrifuge.
3. Microcentrifuge.
4. pH meter.
5. Autoclave.
6. Microscope.
7. Graduate tubes.
8. Magnetic stirrer.

3 Methods

3.1 Vegetative Stages

The vegetative stages are necessary in any fermentation process to obtain enough biomass for the successful productive stage. The number of seed stages and incubation time depends on the growth rate of the microorganism and on the size of the bioreactor. In the case of AD bioconversion process, about 10% (v/v) seeding is required to inoculate the productive stage. Therefore, a bioconversion process in a 5 L bioreactor with a working volume of 3 L will need 300 mL of vegetative stage.

From a vial with 1 mL of frozen (−80 °C) suspension as initial seeding, two growth stages (first vegetative and second vegetative) are necessary to obtain enough biomass to inoculate the fermentative stage. The process in a 20 L bioreactor with a working volume of 15 L includes an additional seed stage (third vegetative) to get 1.5 L.

3.1.1 *Mycolicibacterium neoaurum* B-3805 First Vegetative Stage (100 mL)

1. Take a colony of *M. neoaurum* B-3805 from a Petri plate or 1 mL of frozen suspension kept at −80 °C, and seed a 500 mL flask with 100 mL of growth medium (*see* **Notes 1** and **2**).
2. Incubate in an orbital shaker with 5 cm stroke at 31 °C and 250 rpm for 68–74 h (*see* **Note 3**).
3. Take a sample of the broth and measure the pH. The expected value after 72 h incubation should be about 6.8 ± 0.2.
4. Transfer 10 mL of the broth to a graduated tube and centrifuge the sample in a swinging head centrifuge at 4000 × *g* for 10 min.
5. Calculate the pellet volume (Packed Cell Volume; PCV) by expressing the biomass content as a percentage of the centrifuged volume. The optimal PCV to transfer to the second vegetative stage is around 5–6%.

3.1.2 *M. neoaurum* B-3805 Second Vegetative Stage (300 mL)

1. Take 30 mL of the first vegetative culture and seed a 2 L flask with 300 mL of growth medium (*see* **Note 2**).
2. Incubate in an orbital shaker of 5 cm stroke at 31 °C and 200 rpm for 24–30 h (*see* **Note 4**).
3. Take a sample of the broth and measure pH and PCV according to the Subheading 3.1.1. The expected pH after 24 h incubation is about 6.8 ± 0.2 and PCV >5%.
4. Transfer the total culture broth to the third vegetative stage or to the bioconversion stage (5 L bioreactor).

3.1.3 *M. neoaurum* B-3805 Third Vegetative Stage (5 L)

The bioconversion process in a 20 L bioreactor needs a third vegetative stage to obtain enough biomass to inoculate the productive stage. This culture can be performed in a bioreactor instead of using several 2 L flasks, thus allowing to increase the parameter controls and to reduce the contamination risk during the transfer of the cultures from several flasks to the bioreactor.

Media and Bioreactor Preparation

1. Weigh the raw materials of the growth medium and dissolve them sequentially in warm water (final volume 3 L).
2. Adjust pH to 7.0 using 20% NaOH.
3. Transfer the medium to the bioreactor vessel, assemble all bioreactor parts, and autoclave at 121 °C for 30 min with submerged probe.
4. After sterilization, place the bioreactor on its control console, connect all accessories, and set the next fermentation parameters: temperature 31 °C, air flow rate 2.4 L/min, and agitation 450 rpm.

5. Calibrate the DO_2 probe assuming a theoretical rate of 100% before inoculating (maximal oxygen supply at the initial aeration, pressure, and agitation). A rate of 0% is obtained by briefly disconnecting the lead cable from the DO electrode and set to zero.

Incubation

1. Transfer 300 mL of a well-grown *M. neoaurum* B-3805 inoculum obtained according to Subheading 3.1.2 to a sterile inoculation flask in sterile conditions (*see* **Note 2**).
2. Inoculate the 5 L bioreactor by inserting the needle of the inoculation flask through the septum previously sterilized with a sporicidal solution.
3. Keep the incubation parameters defined in this Subheading and start the dissolved oxygen control over 30% (*see* **Note 5**).
4. Take samples every 24 h in order to monitor *M. neoaurum* growth according to the procedure described in Subheading 3.1.1.
5. The culture is ready to inoculate the bioconversion stage when pH is about 6.8 ± 0.2 and PCV > 5%.

3.2 AD Bioconversion at 5 L Scale

3.2.1 Media and Bioreactor Preparation

1. Weigh the raw materials of the bioconversion medium, except phytosterol and oil, and dissolve them sequentially in warm water (*see* **Note 6**).
2. Adjust pH to 7.0 with 20% NaOH.
3. Weigh oil and phytosterol separately, add them to the vessel of the bioreactor, and mix (*see* **Note 7**).
4. Add the bioconversion medium previously prepared to the bioreactor vessel. The final culture medium should be 3 L.
5. Assemble all bioreactor parts and autoclave at 121 °C for 30 min with submerged probe.
6. After sterilization, place the bioreactor in its control console (*see* **Note 8**). Connect all accessories and set the following fermentation parameters: temperature 31 °C, air flow rate 2.4 L/min, and agitation 800 rpm.
7. Calibrate the DO_2 probe according to Subheading 3.1.3.

3.2.2 Bioconversion of Phytosterol to AD

1. Inoculate the 5 L bioreactor with 300 mL of a well-grown *M. neoaurum* B-3805 inoculum (optimal pH and PCV parameters are 6.8 ± 0.2 and >5%, respectively) according to the procedure described in the Subheading 3.1.2.

2. Incubate according to the following fermentation parameters: temperature 31 °C, air flow rate 2.4 L/min, agitation 800 rpm, and dissolved oxygen over 30% (*see* **Note 9**).
3. Take samples every 8 h to check the absence of microbial contamination (*see* **Note 2**), pH (*see* **Note 10**), growth rate measured as PCV (Subheading 3.1.1) (*see* **Note 11**), phytosterol by TLC, and AD by TLC and HPLC (Subheading 3.4).
4. Keep the bioconversion process until the residual concentration of phytosterol in the broth is under 1 g/L (analyzed by TLC). The bioconversion should be completed in about 120 h (*see* **Note 12**).

3.3 AD Bioconversion at 20 L Scale

3.3.1 Media and Bioreactor Preparation

1. Weigh the raw materials of the bioconversion medium, except phytosterol and oil, and dissolve them sequentially by stirring in 4–5 L of warm water. Adjust pH to 7.0 with 20% NaOH.
2. Place the pH and pO2 probes in the corresponding inputs of the fermentation vessel and calibrate the pH probe with the appropriate buffer standard solutions (7.01 and 4.01 buffers).
3. Assembly the air filter and check that the position of the air entry is set to "Sterilization" position.
4. Place the dissolved culture medium inside the bioreactor vessel, and check the pH with the bioreactor probe.
5. Adjust the volume to 12.5 L with distilled water and start stirring at 200 rpm.
6. Add the oil and phytosterol. The final volume should be approximately 15 L (*see* **Note 13**).
7. Close all the ports of the bioreactor vessel and perform a pressure test to check that the vessel is properly sealed (*see* **Note 14**). After that, open the air output to release pressure.
8. Start the sterilization cycle according to the following sterilization parameters: temperature 121 °C, agitation 200 rpm, time 30 min, and final temperature 31 °C (*see* **Note 15**).
9. Once the sterilization process is finished, set the following fermentation parameters: temperature 31 °C, agitation 450 rpm, air flow rate 15 L/min, and back pressure 0.8 bar.
10. Calibrate the DO_2 probe according to Subheading 3.1.3.

3.3.2 Bioconversion of Phytosterol to AD

1. Inoculate the bioconversion medium with 1.5 L of *M. neoaurum* B-3805 third vegetative (Subheading 3.1.3).
2. Incubate the following fermentation parameters defined in Subheading 3.3.1 and start the dissolved oxygen control over 30% (*see* **Note 16**).

3. Take samples every 8 h to check the absence of microbial contamination (*see* **Note 2**), pH (*see* **Note 10**), growth rate measured as PCV (Subheading 3.1.1) (*see* **Note 11**), phytosterol by TLC, and AD by TLC and HPLC (Subheading 3.4) (*see* **Note 17**).
4. Keep the bioconversion process until the residual concentration of phytosterol in the broth is under 1 g/L (analyzed by TLC). The bioconversion should be completed in about 96–100 h.
5. Harvest the bioconversion broth and begin the downstream process to purify AD.

3.4 Quantification of Sterols and AD

3.4.1 Analysis by HPLC

Sample Extraction

1. Take 20 mL of the bioconversion broth and mix with a magnetic stirrer for 10 min.
2. Take 100 μL of the stirred sample and mix with 900 μL of butyl acetate in a 2.2 mL microtube.
3. Stir at 50 °C for 20 min.
4. Centrifuge at 10,000 × *g* for 10 min.
5. Take a sample of 200 μL of the supernatant and prepare a second dilution in 800 μL of acetonitrile HPLC grade. The final dilution of the broth sample is 1/50.

Standard Preparation

1. Weigh 10 mg of AD standard in a 10 mL volumetric flask and bring up to 70% of the final volume with acetonitrile HPLC grade.
2. Sonicate until complete dissolution and bring up to 10 mL to achieve a concentration of 1 mg/mL.
3. Dilute 1 mL in a 50 mL volumetric flask with acetonitrile HPLC grade to achieve a final concentration of 20 mg/L.

Parameter Analysis

1. Analyze 10 μL of the sample according to the following HPLC conditions: temperature 50 °C, flow 1 mL/min, and wavelength 240 nm.
2. Check the products according to their retention time (RT): ADD 5.4 min, testosterone 6.5 min, and AD 6.7 min (*see* **Note 18**).

3.4.2 Analysis by TLC

Sample Extraction

1. Take 20 mL of the bioconversion broth and mix with a magnetic stirrer for 10 min.
2. Take 100 μL of the stirred sample and mix with 900 μL of ethyl acetate in a 2.2 mL microtube (*see* **Note 19**).
3. Stir at 50 °C for 20 min.
4. Centrifuge at 10,000 × *g* for 10 min.
5. Transfer the supernatant to another microtube and discard the pellet.

TLC Development

1. Cut a TLC sheet of 8–10 cm in length, and draw a line at about 1 cm from the bottom with a pencil.
2. Put 10 μL of the supernatant sample. If several samples are running, keep a 1 cm distance between samples and on each side. Let the solvent evaporate for 5 min (*see* **Note 20**).
3. Place the TLC sheet into the tilt chamber (at about a 45° angle in relation to the wall chamber).
4. Fill the TLC chamber up to 0.5 cm with mobile phase and close the chamber.
5. Run the TLC until the mobile phase reaches to the upper side of the TLC sheet (0.5 cm before the edge).
6. Dry the sheet for 5–10 min at room temperature.
7. Develop the TLC by spreading 20% H_2SO_4 onto the surface with a cotton swab or spraying it.
8. Let it dry and then heat it on a heating plate or in an oven at 100 °C until color development (*see* **Note 21**).

4 Notes

1. The preparation of a Working Cell Bank (WCB) stored at −80 °C is desirable to adequately preserve the strain and standardize the seeding procedure.
2. Seed a TSA Petri plate and incubate at 31 °C for 24–48 h to detect potential microbial contaminations.
3. If necessary, the incubation time can be extended up to 80 h without any negative effect.
4. It is important to avoid wetting the plugs, so you can incubate at 180 rpm if necessary.
5. It is important to keep the DO_2 over 30% to guarantee an optimal oxygen supply for growth. An agitation cascade associated to DO_2 level is useful; when DO_2 is going down, the agitation is going up to keep DO_2 over 30%.
6. To calculate the volume of this solution, subtract the volume corresponding to oil from the final volume (3 L).
7. The use of a blender will facilitate the breakdown of the phytosterol crystals, improving the suspension of oil–phytosterol. Place the oil and phytosterol inside the glass vessel and mix with the blender for 2–5 min.
8. Once the autoclave sterilization occurs, let the media cool down while applying strong agitation; otherwise, the oil–phytosterol mixture can solidify at the bottom, and it will not be available for *M. neoaurum* during the bioconversion process.

9. Up until 60–70 h, the broth has a heterogeneous aspect because of the presence of phytosterol globules. From 60–70 h onward, the broth takes a more homogeneous appearance, and the DO_2 rises to reach values of about 100% at the end of the bioconversion.
10. If there is a significant difference between the pH values measured offline and online, it is recommended to recalibrate the online pH.
11. In this case, the packed cells are located in the oily phase at the top of the tube due to the hydrophobic nature of the mycobacterial cellular surface.
12. Phytosterols are not soluble in water, and it is usual to find some phytosterol stuck to the vessel wall or baffles, which is not available for the bioconversion, decreasing the AD bioconversion yield. To minimize this effect, it is advisable to disconnect the engine daily and manually shake the glass vessel, trying to break the unavailable phytosterol free.
13. In this case it is not necessary to mix previously the oil and phytosterol with a blender, because the medium sterilization process occurs with agitation.
14. Apply the pressure test, close the air output, and start to put air inside (5 L/min) until it reaches a pressure of 1 bar; then stop the air flow and wait for at least 5 min. The pressure should remain constant during that time.
15. During the cooling phase of the sterilization process, beware when changing the filter position from "Sterilization" to "Fermentation," as some foaming may occur. To minimize it, change the filter position when the temperature reaches 60 °C. If temperature is lower, the phytosterol may crystallize inside the air sparger and block the air entry.
16. The bioconversion process at 20 L scale is similar to the one described at 5 L scale, but the third vegetative used to inoculate the 15 L broth has normally more biomass content, and the physical parameters can be slightly different (impeller size, back pressure, oxygen transfer rate, etc.) [22]. At 20 L scale, the heterogeneous aspect of the bioconversion broth changes at 50–60 h (about 10–20 h earlier than at 5 L scale), and the bioconversion is completed after 96–100 h.
17. To take a sample, sterilize with steam the sampling valve for 15 min. Then, discard the first 50 mL of sample and place 50–60 mL in a sterile container. Finally, sterilize the sampling valve again for 15 min with steam.
18. Butyl acetate retention time is 4.5 min. It could cause confusion and hide some other HPLC chromatogram peaks.

19. The use of organic solvents with low boiling point, such as ethyl acetate, is highly recommendable for the sample extraction to be analyzed by TLC. These solvents evaporate quickly once the sample on the TLC sheet is uploaded, avoiding the sample from spreading.
20. For a proper TLC development, put 3 μL of the sample as a small drop on the line and let the solvent evaporate. Repeat the operation over the same drop until completing the sample loading.
21. It is advisable to include standards of phytosterol and AD of known concentration in the TLC sheet to help identify and quantify the product bands [23].

Acknowledgments

This work was fully supported by grants of the European Union program ERA-IB [MySterI (EIB.12.010)] and ERA CoBioTech for the project Syntheroids. MySterI funding was received through the APCIN call of the Spanish Ministry of Economy and Competitiveness (MINECO, Spain) (PCIN-2013-024-C02-01). Syntheroids funding was received from the European Union's Horizon 2020 research and innovation program under grant agreement No [722361] and financed through the Center for the Industrial Technology Development (CDTI, Spain) (EXP-00108753/SERA-20181033). The authors want to thank the European Union programs ERA-IB and ERA CoBioTech, the Spanish Ministry of Economy and Competitiveness (MINECO, Spain), and the Center for the Industrial Technology Development (CDTI, Spain), as well as the MySterI Consortium (INBIOTEC, Pharmins Ltd., University of York, SINTEF, Technische Universität Dortmund, and Gadea Biopharma S.L.) and Syntheroids Consortium (INBIOTEC, Pharmins Ltd., SINTEF, Technische Universität Dortmund and Bionice S.L).

References

1. Fernandes P, Cruz A, Angelova B, Pinheiro HM, Cabral JMS (2003) Microbial conversion of steroid compounds: recent developments. Enzym Microb Technol 32:688–705
2. Batthi H, Khera RA (2012) Biological transformations of steroidal compounds: a review. Steroids 77:1267–1290
3. Fernández-Cabezón L, Galán B, García JL (2018) New insights on steroid biotechnology. Front Microbiol 9:958. https://doi.org/10.3389/fmicb.2018.00958. PMID: 29867863; PMCID: PMC5962712
4. Nunes VO, Vanzellotti NC, Fraga JL, Pessoa FLP, Ferreira TF, Amaral PFF (2022) Biotransformation of phytosterols into androstenedione-a technological prospecting study. Molecules 27(10):3164. https://doi.org/10.3390/molecules27103164
5. Feng J, Wu Q, Zhu D, Ma Y (2022) Biotransformation enables innovations toward green synthesis of steroidal pharmaceuticals. ChemSusChem 15(9):e202102399. https://doi.org/10.1002/cssc.202102399. Epub 2022 Feb 15. PMID: 35089653

6. Moreau RA, Whitaker BD, Hicks KB (2002) Phytosterols, phytostanols, and their conjugates in foods: structural diversity, quantitative analysis, and health-promoting uses. Prog Lipid Res 41:457–500
7. Donova M (2022) Microbial steroid production technologies: current trends and prospects. Microorganisms 10(1):53. https://doi.org/10.3390/microorganisms10010053
8. Kutney JP, Milanova RK, Vassilev CD, Stefanov S, Nedelcheva NV (2000) Process for the microbial conversion of phytosterol to androstenedione and androstadienedione. US 6071714. US Patent
9. Donova MV (2007) Transformation of steroids by actinobacteria: a review. Appl Biochem Microbiol 43:1–14
10. Rodina NV, Molchanova MA, Voishvillo NE, Andryushina VA, Stytsenko TS (2008) Conversion of phytosterols into androstenedione by *Mycobacterium neoaurum*. Appl Biochem Microbiol 44:48–54
11. Pérez C, Falero A, Duc HL, Balcinde Y, Hung BR (2006) A very efficient bioconversion of soybean phytosterols mixtures to androstanes by mycobacteria. J Ind Microbiol Biotechnol 33:719–723
12. Egorova OV, Gulevskaya SA, Puntus IF, Filonov AE, Donova MV (2002) Production of androstenedione using mutants of *Mycobacterium* sp. J Chem Technol Biotechnol 77:141–147
13. Zhang XY, Peng Y, Su ZR, Chen QH, Ruan H, He GQ (2013) Optimization of biotransformation from phytosterol to androstenedione by a mutant *Mycobacterium neoaurum* ZJUVN-08. J Zhejiang Univ Sci B 14:132–143
14. Marsheck WJ, Kraychy S, Muir RD (1972) Microbial degradation of sterols. Appl Microbiol 23(1):72–77
15. Marsheck WJ, Heights A, Kraychy S (1973) Selective microbiological preparation of androst-4-ene-3,17-dione. Patent US 3759791
16. Rodríguez-García A, Fernández-Alegre E, Morales A, Sola-Landa A, Lorraine J, Macdonald S, Dovbnya D, Smith MC, Donova M, Barreiro C (2016) Complete genome sequence of '*Mycobacterium neoaurum*' NRRL B-3805, an androstenedione (AD) producer for industrial biotransformation of sterols. J Biotechnol 224:64–65
17. Malaviya A, Gomes J (2008) Androstenedione production by biotransformation of phytosterols. Bioresour Technol 99:6725–6737
18. Hesselink PGM, Vliet SV, Vries HD, Witholt B (1989) Optimization of steroid side chain cleavage by *Mycobacterium* sp. in the presence of cyclodextrins. Enzym Microb Technol 11: 398–404
19. Carvalho F, Marques MPC, de Carvalho CCCR, Cabral JMS, Fernandes P (2009) Sitosterol bioconversion with resting cells in liquid polymer based systems. Bioresour Technol 100:4050–4053
20. Cruz A, Fernandes P, Cabral JMS, Pinheiro HM (2001) Whole-cell bioconversion of β-sitosterol in aqueous-organic two-phase systems. J Mol Catal B Enzym 11:579–585
21. Stefanov S, Yankov D, Beschkov V (2006) Biotransformation of phytosterols to androstenedione in two phase water-oil systems. Chem Biochem Eng Q 20:421–427
22. García-Ochoa F, Gómez E (2009) Fermentor scale-up and oxygen transfer rate in microbial processes: an overview. Biotechnol Adv 27(2): 153–176
23. Gulla V, Banerjee T, Patil S (2008) Quantitative TLC analysis of steroid drug inter-mediates formed during bioconversion of soysterols. Chromatographia 68:663–667

Chapter 15

Bioconversion of Phytosterols into Androstenedione by *Mycolicibacterium*

Kjell Domaas Josefsen, Anna Nordborg, Simone Balzer Le, Silje Malene Olsen, and Håvard Sletta

Abstract

The chapter describes the bioconversion of phytosterols into androstenedione (AD) by *Mycolicibacterium* spp. in shake flasks and fermenters, as well as LC–MS-based methods for analysis of phytosterols and steroid products. Phytosterols are derived as by-products of vegetable oil refining and manufacture of wood pulp. They contain the same four-ring nucleus as steroids and may be converted to high-value steroids by removing the sidechain at C17 and minor changes at other sites in the ring structure. Many bacteria, including *Mycolicibacterium* spp., can degrade phytosterols. Mutants of *Mycolicibacterium* spp. unable of ring cleavage can, when growing on phytosterols, accumulate the steroid intermediates androstenedione (AD) and androstadienedione (ADD). The practical challenge with microbial conversion of phytosterols to steroids is that both the substrate and the product are virtually insoluble in water. In addition, some steroids, notably ADD, may be toxic for the cells. Two main strategies have been employed to overcome this challenge: the use of two-phase systems and the addition of chemically modified cyclodextrins. The latter method is used here. Defined cultivation and bioconversion media for both shake flask and fermenter are given, as well as hints how to minimize the practical problems due to the water-insoluble phytosterol. Sampling, sample extraction, and quantification of substrates and products using LC–MS analysis are described.

Key words *Mycolicibacterium*, Phytosterols, Steroids, Transformation, Analysis, Fermentation, LC, MS

1 Introduction

Phytosterol is a collective term for plant-derived sterols and stanols. Stanols are saturated sterols without double bonds in the sterol ring structure. They are triterpenes and important structural components of plant membranes. Phytosterols are structurally related to cholesterol but have different sidechain configurations (Fig. 1). More than 100 types of phytosterols have been reported in plant species, but the most abundant are β-sitosterol, stigmasterol, and campesterol [1]. Commercially phytosterols are isolated from

Carlos Barreiro and José-Luis Barredo (eds.), *Microbial Steroids: Methods and Protocols*, Methods in Molecular Biology, vol. 2704, https://doi.org/10.1007/978-1-0716-3385-4_15,

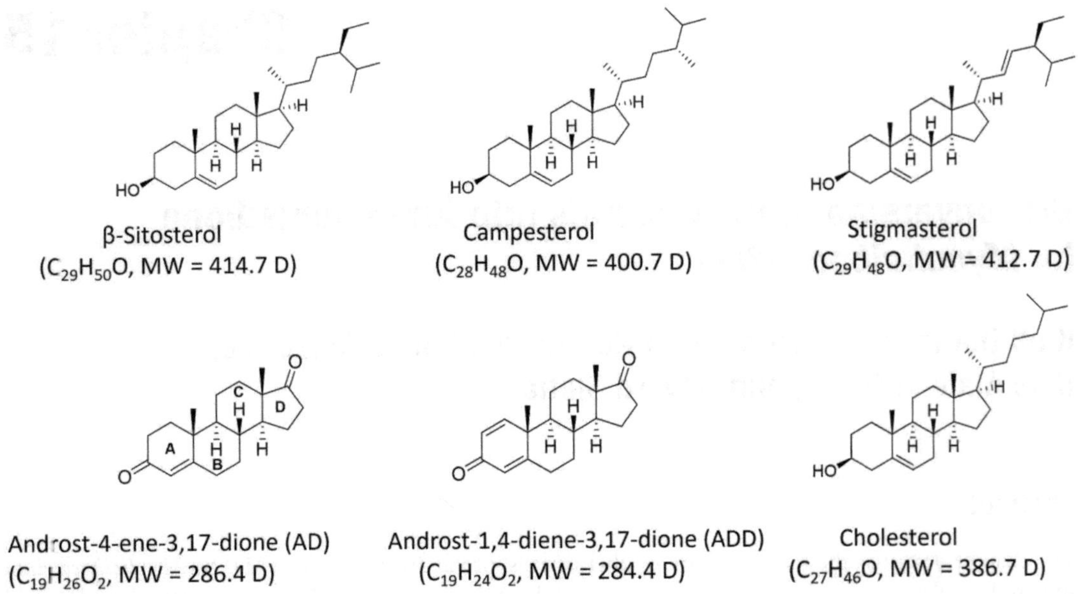

Fig. 1 Chemical structure of phytosterols and steroids. β-sitosterol, campesterol, stigmasterol, cholesterol, androstenedione (AD), and androstadienedione (ADD). The conventional designation of the rings (A–D) is indicated in the AD structure

vegetable oils (soybean oil, rapeseed oil, sunflower oil, etc.) as a by-product of traditional vegetable oil refining and from the so-called tall oil, a by-product of the manufacture of wood pulp [2]. These commercial products contain a mixture of different sterols and stanols and the composition reflects the source and the production process (Table 1). Phytosterols contain the same four-ring nucleus as steroids and may be converted to androst-4-ene-3,17-dione (androstenedione, AD) or androst-1,4-diene-3,17-dione (androstadienedione, ADD) by removing the sidechain at C17 and minor modifications at other sites in the structure.

AD and ADD are steroid intermediates used in the pharmaceutical industry for production of anabolic drugs. The conversion of phytosterols into AD or ADD may be achieved chemically, enzymatically, and with microorganisms, including *Mycolicibacterium* spp. Here we focus on microbial conversion. For a recent review of chemical and enzymatic processes, see [3]. Theoretically, and according to their molecular mass, 1.0 kg β-sitosterol can yield 0.69 kg AD.

1.1 Microbial Transformation of Phytosterols to Steroids

The initial steps in the microbial degradation of phytosterols are outlined in Fig. 2. The key to microbial conversion of phytosterols to steroids is to prevent ring opening. The two key reactions in the ring opening process are the addition of an OH-group in the 9α-position by a 9α-hydroxylase and the introduction of a double bond between C1 and C2 in ring A by an 1,2-dehydrogenase.

Table 1
Composition (%) of different commercial phytosterols [2]

Phytosterol/stanol	Tall oil	Vegetable oil
β-Sitosterol	36–80	43–52
β-Sitostanol	6–34	0–3
Campesterol	4–25	24–28
Campestanol	0–14	0–2
Stigmasterol	<1	14–24
Brassicasterol	<1	0–9
Minor sterols	0–2	<3

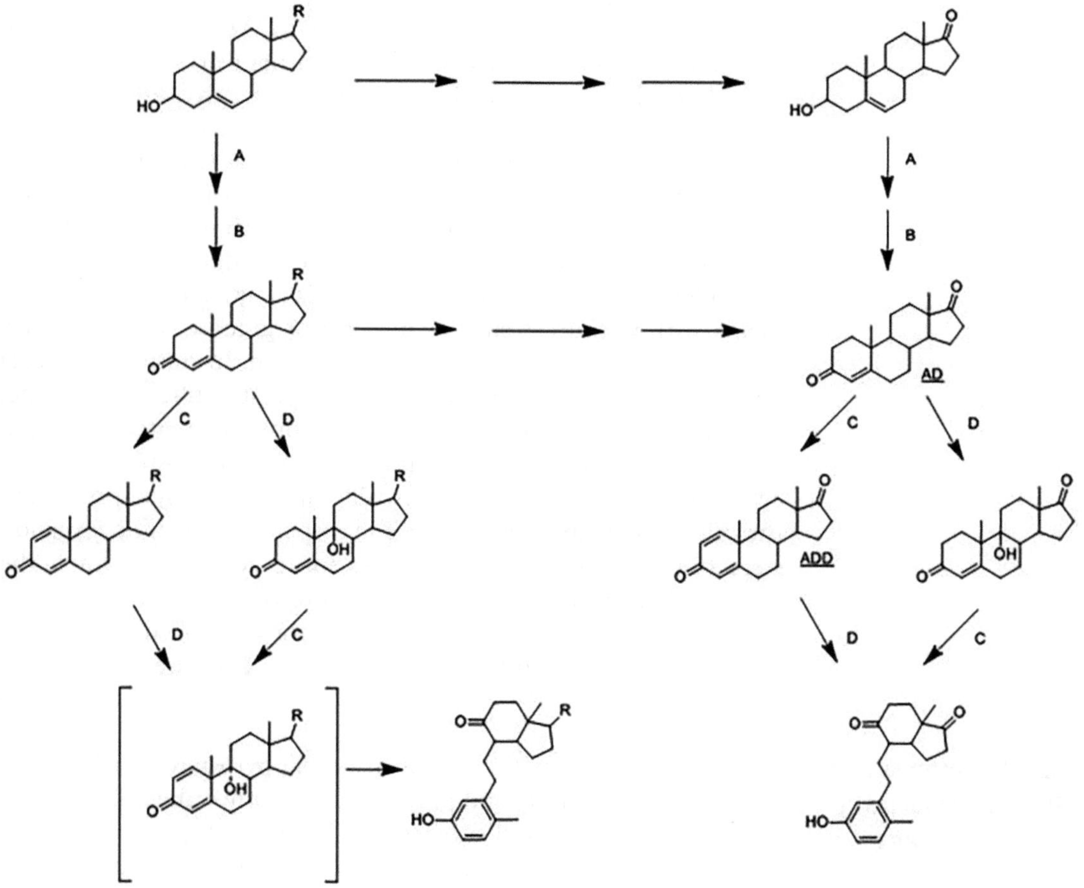

Fig. 2 Outline of the initial steps in bacterial degradation of phytosterols including the ring opening. Based on [4] and references therein. A = β-hydroxysteroid dehydrogenase, B = 3-ketosteroid isomerase, C = 9α--hydroxylase, D = 1,2-dehydrogenase

When these reactions are combined, they cause aromatization and cleavage of ring B [4]. The activity of these two enzymes is relatively independent of the sidechain at C17. Similarly, the multi-step enzymatic degradation of the sidechain at C17 is relatively independent of the chemical structure at ring A [4]. Thus, if the activity of the 9α-hydroxylase and/or the 1,2-dehydrogenase can be inhibited, the degradation of the four-ring sterol nucleus can be prevented, and AD and/or ADD will accumulate. Initially, in the 1960s, research focused on the use of inhibitory agents such as 8-hydroxyquinoline, Ni^{2+} and α,α-dipyridyl or on chemical modification of the sterol structure itself to prevent attack; for a review see [5]. However, focus soon shifted to mutants devoid of 9-α-hydroxylase and/or 1,2-dehydrogenase, and several mutants unable or only poorly able of ring cleavage were isolated and patented during the 1970s. Mutants of *Mycolicibacterium* spp. were prominent among these.

The bioconversion of phytosterols to AD with mutants of *Mycolicibacterium* spp. may be achieved either by (i) first producing the cell mass on another carbon substrate, e.g., glycerol or glucose, followed by a bioconversion phase with phytosterol as C-source, or by (ii) cultivating the mutants on phytosterol as C-source from the start. Both strategies have their advantages and disadvantages. Cultivation of the mutants can be performed both in shake flasks and fermenters, but pH control can be a challenge in shake flasks, particularly when *Mycolicibacterium* is cultivated on defined media with C-sources such as glycerol or glucose. Here pH may become very low (<4) due to acid production. In cultivation media containing complex ingredients such as yeast extract or peptone, pH changes are more moderate. The bioconversion rate is not rapid and will, depending on the concentration of phytosterol and the cell concentration from start, usually take days. Thus, producing the cell mass on another C-source may speed up the bioconversion step, but the process also becomes more complicated than the one-pot cultivation process with phytosterol as C-source from the start. It should also be considered that the cells need a transition period before they can attack phytosterols if cultivated on another C-source.

1.2 Phytosterols and Steroids Are Almost Insoluble in Water

The main challenge with bioconversion of phytosterols to AD(D) is that both the substrate and products are poorly water soluble. The estimated solubility of β-sitosterol at 25 °C is 20 μg/L [6], while the solubility of the related molecule cholesterol is about 26 μg/L at 30 °C [7]. However, the measured "solubility" of mixed phytosterols is considerably higher, possibly due to micelle formation. Yuan et al. [8] found a "solubility" of approx. 11 mg/L for mixed phytosterols at 30 °C. For a bio-converting bacterium, it is the access to the substrate that is critical, not if the compound is truly dissolved or not. However, 11 mg/L is also a very low concentration and in systems where phytosterol is dispersed in water-based media, direct contact between the cells and the phytosterol is probably required for efficient substrate uptake. Small, finely dispersed phytosterol particles and thus a large surface to volume ratio are an advantage. Fine phytosterol particles can easily be made with a mortar or otherwise. Unfortunately, during heat sterilization, the finely dispersed phytosterol particles often aggregate into larger particles that are difficult to disperse again. An alternative is to dissolve the phytosterols in a water-miscible solvent, e.g., ethanol [9], sterilize the solution, and slowly drip it into the sterile medium. The phytosterol will then precipitate as small particles. The disadvantage is that if the solvent concentration becomes too high, this will affect the cells negatively.

The solubility of steroids in water is typically 20–50 mg/L [10]. The solubility of testosterone and AD at 25 °C is 23 and 58 mg/L, respectively [11]. Again, less rigorous solubility studies indicate a much higher "solubility," possibly due to micelle formation or other phenomena. Khomutov et al. [12] found a solubility of 9α-hydroxy-4-androstene-3,17-dione and ADD at 30 °C of 632 and 464 mg/L, respectively.

ADD has been shown to have toxic effects on mycobacteria and other bacteria [13–15] and AD is suspected of having toxic effects [16]. ADD is often produced as a by-product also when AD is the main product. When cells convert phytosterols to steroids in a water-based system, it may be difficult for the cells to "get rid of" the formed water-insoluble steroids. The cells may export them from the interior, but the hydrophobic outer layers on mycobacteria and the cell membrane will be a more hydrophobic environment than the water phase, and steroids may aggregate around the cells rather than dissolve or disperse in the water phase. This is exactly the place you do not want a cell toxic compound.

To overcome the challenge of a water-insoluble substrate and an almost water-insoluble and possibly toxic product, several strategies have been employed. The two main approaches are two-phase systems and the use of cyclodextrins.

1.3 Two-Phase Systems

Several different aqueous-organic two-phase systems have been studied. The principle is that the phytosterols are dissolved in a water-immiscible organic substance such as vegetable oils [17, 18], ionic liquids [19], polypropylene glycol [18, 19], silicone oils [18, 19], or phthalates [20–22]. The phytosterol–oil mixture is added to the medium and stirring or shaking disperses the oil as small droplets that the cells can adhere to and thus get in contact with the phytosterol. The produced steroids may dissolve in the same oil phase and thus be removed from the cell surface. Important aspects to consider are the solubility of both phytosterols and steroids in the oil, the compatibility of the oils with the cells, the viscosity of the medium with oil, and the effects on downstream processing and purification of the product. Notably, emulsion formation due to the presence of surface-active biological material and intense mixing may complicate downstream separation. In large scale, the cost of the oil and effects on the working as well as the surrounding environment must also be considered.

1.4 Cyclodextrins

Cyclodextrins (CDs) are cyclic oligosaccharides consisting of six α-CDs, seven β-CDs, eight γ-CDs, or more glucopyranose units linked by α-(1,4) bonds [23]. They form a torus-like three-dimensional structure with a hydrophobic inner cavity and a hydrophilic outer surface. The hydrophobic cavity provides a microenvironment into which appropriately sized non-polar moieties can enter to form inclusion complexes. One or two guest molecules can be entrapped by one, two, or three CDs [23] (and references therein). Complex formation is a dimensional fit between host cavity and guest molecule. No covalent bonds are broken or formed during the formation, and the binding of the guest molecule within the host CD is not fixed or permanent but a dynamic equilibrium. Binding strength depends on how well the "host–guest" complex fits together and on local interactions between surface atoms.

Native CDs have limited water solubility. The solubility of β-CD is 18.5 g/L (16 mM) [24]. By chemical modification, the CDs can become much more water-soluble. The water solubility of randomly methylated β-cyclodextrin (Me-β-CD) and 2-hydroxypropyl-β-cyclodextrin (HP-β-CD) is >500 and >600 g/L, respectively [24]. Both these modified CDs have been used to solubilize steroids and possibly also to some extent phytosterols in microbial phytosterol bioconversion processes. Unfortunately, the cost of these modified β-CDs is high, in the order of 300–350 USD/kg (bulk prices) versus 5 USD/kg for native β-CD [24]. This makes them too costly for industrial use although attempts have been made to recirculate HP-β-CD [25]. However, in laboratory scale where the aim is to study and understand processes and phenomena, they are very useful.

The modified β-CDs may in high concentrations have a negative effect on the growth of *Mycolicibacterium* [26, 27]. Thus, the concentration of modified β-CDs in the medium from the start should not be too high (<50 g/L Me-β-CD, <70 g/L HP-β-CD). More can be added during the bioconversion process if needed. A general guideline is that at least one Me-β-CD/HP-β-CD molecule should be available for each steroid molecule produced. As one phytosterol molecule theoretically can generate one steroid molecule, this means that on weight basis at least 3.3 g Me-β-CD or 3.9 g HP-β-CD should be added for each g phytosterol.

1.5 Cultivation of Mycolicibacterium

The cultivation of *Mycolicibacterium* spp. and bioconversion of phytosterols have been performed in several different media. Generally, we may separate cultivation media into three groups: rich media where all or most components are complex (e.g., yeast extract, peptone, tryptone, etc.), semi-defined media where one or a few complex medium components are added to an otherwise defined medium, and defined media where all constituents are pure chemicals. Defined media are an advantage for basic studies as all carbon sources and their quantities are exactly known, but cell growth is usually more rapid on rich or semi-defined media. Preparation of and cultivation on defined media is more challenging than preparation and use of more complex media. When using defined media, precipitates may be formed during preparation and the pH changes during cultivation are usually larger.

1.6 Sampling

The poor solubility of both the substrate and the product(s) makes it difficult to obtain representative samples from the cultures, in particular from fermenters as clumps of phytosterols and cells aggregate at various locations in the fermenter and are almost impossible to disperse in the water phase before sampling. In shake flask studies, an alternative is to harvest a whole shake flask at every sampling point, especially if the aim is to obtain a good mass balance. However, the latter strategy implies relatively large samples that may complicate the extraction process.

1.7 Analysis

Phytosterols and steroids can be extracted from the samples with ethyl acetate, diethyl ether, hexane, or chloroform and analyzed by gas chromatography (GC), TLC [28], or HPLC [29], with or without coupling to mass spectrometry [30, 31]. When analyzing by HPLC using UV, identification of the analytes is based on comparison with retention times for standards [32] and combination with mass spectrometry has therefore been used to aid analyte identification [33]. Mass spectrometric detection gives more reliable identification than UV-based detection, and use of multiple reaction monitoring (MRM) mode ensures specific detection of the selected analytes and greatly improved the limit of detection. In HPLC coupled to mass spectrometry, electrospray ionization has been used for both steroids and phytosterols, but ionization efficiency may be a problem for some steroids and derivatization protocols have been established to help overcome these issues [34]. For phytosterols, atmospheric pressure chemical ionization (APCI) has also been used [35, 36]. Due to the desire to analyze steroids and phytosterols simultaneously, APCI was chosen as the ionization method in this work.

2 Materials

2.1 Cell Culture

Note that for this work, *Mycolicibacterium* media are not supplemented with Tween® 20 since it will interfere with the subsequent LC–MS analysis.

1. A *Mycolicibacterium* strain obtained from a culture collection (e.g., *Mycolicibacterium* sp. NRRL B-8054 or *Mycolicibacterium neoaurum* NRRL B-3805).
2. Semi-defined growth medium with nitrate as nitrogen source (per liter in distilled water or similar quality): 10 g of glycerol (99.5%), 10 g of yeast extract, 1.94 g of $NaNO_3$, 1.49 g of K_2HPO_4, and 1.27 g of KH_2PO_4. Mix these components in 994 mL distilled water, pH adjusted to 7.0–7.2 (*see* **Note 1**), and sterilize (121 °C, 20 min.) to create the "bulk medium." After sterilization, add the following filter sterilized (0.2 μm) solutions (per liter final medium): 1.0 mL of SINTEF's trace mineral solution (see below), 1.0 mL of 0.6 M NaOH (*see* **Note 2**), and 4.0 mL of $MgSO_4$ solution (*see* below and **Note 3**)

   (based on recipe from Prof. Marina V. Donova, GK Skryabin Inst. of Biochemistry and Physiology of Microorganisms, Russian Acad. Sci., Pushchino, Moscow Region, Russia).
3. Solid semi-defined growth medium: Add 15 g/L of agar to the semi-defined growth medium.
4. SINTEF's defined cultivation medium for shake flasks: 7.5 g of $NaNO_3$, 3.2 g of K_2HPO_4, 2.0 g of NaH_2PO_4, 1.0 g of K_2SO_4, 0.6 g of citric acid · H_2O, 0.50 g of sodium glutamate · H_2O, 0.50 g of glutamine, and 25.1 g of TES [N-[Tris(hydroxymethyl)-methyl]-2-aminoethanesulfonic acid] sodium salt. Dissolve these components in about 800 mL distilled water, adjust the pH to 7.0–7.2 (*see* **Note 1**), and adjust the volume to 900 mL with distilled water. After heat (121 °C, 20 min.) or filter (0.2 μm) sterilization, add 15 mL of 65% (w/v) glycerol solution (see below), 40 mL of $MgSO_4$ solution (*see* below and **Note 3**), 15 mL of SINTEF's trace mineral solution (see below), 15 mL of 0.6 M NaOH (*see* **Note 3**), and 15 mL Wolfe's vitamin solution (see below). A slight precipitate may be observed in the final medium.
5. SINTEF's defined bioconversion medium for shake flasks (*see* **Note 4**): 7.5 g of $NaNO_3$, 3.2 g of K_2HPO_4, 2.0 g of NaH_2PO_4, 1.0 g of K_2SO_4, 0.6 g of citric acid · H_2O, 0.50 g of sodium glutamate · H_2O, 0.5 g of glutamine, and 25.1 g of TES [N-[Tris(hydroxymethyl)-methyl]-2-aminoethanesulfonic acid] sodium salt. Dissolve these components in about 600 mL distilled water, adjust pH to 7.0–7.2 (*see* **Note 1**),

and adjust the volume to 685 mL with distilled water. After heat (121 °C, 20 min.) or filter (0.2 μm) sterilization, add 200 mL of phytosterol suspension (see below), 15 mL of 65% (w/v) glycerol solution (see below), 40 mL of $MgSO_4$ solution (*see* below and **Note 3**), 15 mL of SINTEF's trace mineral solution (see below), 15 mL of 0.6 M NaOH (*see* **Note 2**), and 15 mL of Wolfe's vitamin solution (see below).

6. SINTEF's defined cultivation medium for fermenter (*see* **Note 5**): 10 g of $NaNO_3$, 3.2 g of K_2HPO_4, 2.0 g of NaH_2PO_4, 1.0 g of K_2SO_4, 0.6 g of citric acid · H_2O, 0.50 g of sodium glutamate · H_2O, and 0.5 g of glutamine. Dissolve these components in 894 mL distilled water and adjust pH to 7.0–7.2 (*see* **Note 1**). After heat (121 °C, 20 min.) sterilization, add 15 mL of 65% (w/v) glycerol solution (see below), 40 mL of $MgSO_4$ solution (*see* below and **Note 3**), 15 mL of SINTEF's trace mineral solution (see below), 15 mL of 0.6 M NaOH (*see* **Note 2**), 6 mL of $CaCl_2$ solution (see below), and 15 mL of Wolfe's vitamin solution (see below). A slight precipitate may be observed in the final medium. Mix 10 mL of antifoam (e.g., Antifoam 204 (Sigma)) with 90 mL of distilled water and heat-sterilize (121 °C, 20 min). From the start 5 mL of this mixture is added per liter medium. More is added as required during the fermentation (*see* **Note 6**). pH is adjusted and controlled at pH 7.0–7.2 with 3 M HCl and 3 M NaOH. pH will mostly decrease during the fermentation.
7. SINTEF's defined bioconversion medium for fermenter (*see* **Note 7**): 10 g of $NaNO_3$, 3.2 g of K_2HPO_4, 2.0 g of NaH_2PO_4, 1.0 g of K_2SO_4, 0.6 g of citric acid · H_2O, 0.50 g of sodium glutamate · H_2O, and 0.5 g of glutamine. Dissolve these components in 560.3 mL of distilled water and adjust pH to 7.0–7.2. After heat (121 °C, 20 min.) sterilization, add 15 mL of 65% (w/v) glycerol solution (see below), 40 mL of $MgSO_4$ solution (*see* below and **Note 3**), 15 mL of SINTEF's trace mineral solution (see below), 15 mL of 0.6 M NaOH (*see* **Note 2**), 6 mL of $CaCl_2$ solution (see below), 15 mL of Wolfe's vitamin solution (see below), and 250 mL of phytosterol suspension. Use 13.3 mL of washing water (see below) to transfer the remaining particles of phytosterol from the flask. Mix 10 mL of antifoam (e.g., Antifoam 204 (Sigma)) with 90 mL distilled water and heat-sterilize (121 °C, 20 min). From the start 5 mL of this mixture is added per liter of medium. More is added as required during the fermentation (*see* **Note 6**). pH is adjusted and controlled at pH 7.0–7.2 with 3 M HCl and 3 M NaOH. pH will mostly decrease during the fermentation.
8. Heat-sterilized phytosterol suspension. Phytosterol (20 g) and 125.8 g β-methyl cyclodextrin are mixed per liter of distilled water. This leads to a molar ratio of 0.5 mole of sterol

(calculated for the molecular weight of campesterol) per mole of β-methyl cyclodextrin. For practical reasons, combine smaller quantities, e.g., 2 g phytosterol and 12.58 g β-methyl cyclodextrin, in 100 mL distilled water in a small bottle or an Erlenmeyer flask and heat-sterilize (121 °C, 20 min.). Submerge the suspension in an ultrasonic water bath and sonicate for at least 3 × 5 min with manual shaking in between. This results in a fine suspension that can be added to the final medium.

9. Heat-sterilized washing water. Distilled water (50 mL) is heat-sterilized in a small bottle or an Erlenmeyer flask and used to wash out remaining particles of phytosterol in the flask above before it is transferred to the fermenter. It is an advantage if the water is still hot after heat sterilization.

10. SINTEF's trace mineral solution (per liter): 2.27 g of $FeCl_2$ (beads), 0.37 g of $CaCl_2 \cdot 2H_2O$, 0.27 g of $CuCl_2 \cdot 2H_2O$, 0.21 g of $ZnCl_2$, 0.15 g of $MnCl_2 \cdot 2H_2O$, 17 mg of $CoCl_2 \cdot 6H_2O$, 10 mg of $Na_2MoO_4 \cdot 2H_2O$, 50 mL of conc. HCl (37%), and distilled water (or similar quality) up to 1000 mL (*see* **Note 8**).

11. $MgSO_4$ solution (per liter): 50 g of $MgSO_4 \cdot 7H_2O$ dissolved in distilled water. Sterilize the solution separately to minimize precipitation of Mg phosphates and Ca sulfate. The solution may be sterilized by filtration (0.2 μm) or heat (121 °C, 20 min.) and added sterile to the final medium.

12. Glycerol solution (65% (w/v)): 650.0 g of glycerol (99.5%) and up to 1000 mL with distilled water. Glycerol may be added directly to the base medium before it is sterilized, but a separate solution makes it easier to, if desired, decrease or increase the amount of C-source in the medium. Sterilize the solution by heat (121 °C, 20 min). The presence of glycerol in bioconversion media will, both in our and others experience, often enhance the conversion of phytosterol to steroids.

13. Wolfe's vitamin solution: Wolfe's vitamin solution is commonly used in defined media for cultivation of microorganisms [37]. It may also be acquired commercially as a ready-made solution. The composition is given in Table 2.

    Dissolve all vitamins in approximately 900 mL of distilled water and adjust the volume to 1000 mL. Split the vitamin solution into suitable volumes, e.g., 10–50 mL, and freeze (−20 °C) these smaller volumes. In this way you avoid thawing the whole solution when only a small volume, e.g., 10 mL, is required. After thawing, the solution is filter sterilized (0.2 μm) immediately before use.

14. Autoclave for heat sterilization.

15. Ultrasonic water bath.

Table 2
Composition of Wolfe's vitamin solution

Compound	Amount
Pyridoxine · HCl	10.0 mg
Thiamine · HCl	5.0 mg
Riboflavin	5.0 mg
Nicotinic acid	5.0 mg
Calcium D-pantothenate	5.0 mg
Thioctic acid	5.0 mg
Biotin	2.0 mg
Folic acid	2.0 mg
Vitamin B_{12}	0.1 mg
Distilled water to	1000 mL

16. Petri dishes (9 cm) for preparing agar plates.
17. Static incubator (at 30 °C).
18. 250 mL and 500 mL baffled shake flasks.
19. Shaking incubator (at 30 °C).
20. Spectrophotometer and 1 cm cuvettes.
21. Light microscope.
22. Stirred tank bioreactor system with a vessel volume of 1–10 L.
23. Centrifuge with rotors fitting 1.5 mL Eppendorf tubes, 15 mL and 50 mL tubes.
24. Whirl mixer.

2.2 Sterol and Steroid Analysis

1. Eluent A for LC–MS analysis (per liter in distilled water): 25 mM formic acid. Add 0.94 mL of formic acid (analytical grade >99%) to distilled water in a 1 L volumetric flask. Mix thoroughly the mixture and dilute with distilled water. Transfer the resulting 25 mM formic acid solution to an eluent bottle (*see* **Note 9**).
2. Eluent B for LC–MS analysis: Methanol (LC–MS grade) (*see* **Note 10**).
3. Extraction solvent: Ethyl acetate (analytical or LC–MS grade).
4. Androst-4-ene-3,17-dione (AD).
5. Androst-1,4-diene-3,17-dione (ADD).
6. 3β-Hydroxyandrost-5-ene-17-one (DHEA).
7. 11α-Hydroxyandrost-4-ene-3,17-dione (11-α-OH-AD).

8. Testosterone.
9. Campesterol.
10. Stigmasterol.
11. β-Sitosterol.
12. Brassicasterol.
13. Stock standards—phytosterols: 1–2 mg/mL in chloroform. Weigh out accurate amounts of the individual phytosterols and add solvent (chloroform) to a phytosterol concentration of 1–2 mg/mL. Store solutions at −20 °C (*see* **Note 11**).
14. Stock standards—steroids: 1–2 mg/mL in ethyl acetate. Weigh out accurate amounts of the individual steroids and add solvent (ethyl acetate) to a steroid concentration of 1–2 mg/mL. Store solutions at −20 °C (*see* **Note 11**).
15. Working standards: 1000 μM in ethyl acetate. Dilute the individual stock standards to 1000 μM in ethyl acetate. Prepare a mixed working standard of all analytes at individual concentrations of 1000 μM in ethyl acetate. Store solutions at −20 °C.
16. Calibration standards: 0, 0.1, 0.5, 1, 5, 10, 25, 50, and 100 μM in methanol. Prepare calibration standards by diluting the mixed 1000 μM standard in methanol. Also prepare a blank of only methanol. Store solutions at −20 °C.
17. Fume hood.
18. Whirl mixer.
19. Heating block for reagent tubes.
20. Rotary shaker.
21. LC–MS equipment: High-pressure gradient pumps equipped with degasser, autosampler, and column thermostat, connected to a triple quadrupole MS. For example, Agilent 1200 system connected to an Agilent triple quadrupole equipped with an atmospheric pressure chemical ionization (APCI) ion source.
22. LC-column Zorbax SB-C8 (4.6 ×100, 3.5 μm) or equivalent.

3 Methods

3.1 Preparation of Standard Inoculum

1. Open an ampoule of freeze-dried culture of *Mycolicibacterium* (*see* **Note 12**).
2. After opening the ampoule, add a few drops of sterile semi-defined growth medium to disperse the freeze-dried cells.
3. Seed a shake flask of 250 mL with 50 mL of semi-defined growth medium.

4. As a backup, transfer some of the dispersed cells to semi-defined agar medium and incubate the agar plates at 30 °C.
5. Incubate the shake flask at 30 °C, 150–200 rpm, for 48–96 h, until late logarithmic growth phase (*see* **Note 13**).
6. When the culture is in late logarithmic growth phase, transfer a small volume (1–3 mL) to a new 500 mL shake flask with 100 mL defined cultivation medium (100 mL) (*see* **Note 14**).
7. Incubate the shake flask at 30 °C, 150–200 rpm for 48–96 h until a suitable culture density is obtained (*see* **Note 15**).
8. Transfer a suitable volume of culture, e.g., 60 mL, to a sterile flask with screw cap and add 40 mL sterile 50% (v/v) glycerol solution. Mix well.
9. Distribute the mixture of culture and glycerol in sterile cryotubes (1–3 mL per tube, at least 20–25 cryotubes) and freeze at −80 °C (*see* **Notes 16** and **17**).

3.2 Purity Check of Standard Inoculum

1. Select at random 2–3 cryotubes with frozen standard inoculum. Thaw them rapidly (*see* **Note 18**) and transfer 1 mL thawed culture from each cryotube to a 500 mL shake flask with 100 mL semi-defined medium.
2. Incubate the shake flasks at 30 °C and 150–200 rpm for 1–3 days until the culture is in late logarithmic growth phase.
3. Transfer a small volume (0.1–0.5 mL culture) to a new shake flask (500 mL) with 100 mL semi-defined medium and again incubate until late logarithmic growth phase.
4. Repeat the transfer procedure one or two times to give any contaminant present in the original inoculum a chance to increase to a significant fraction of the culture.
5. From the late logarithmic growth phase culture, make a tenfold dilution series in defined cultivation medium without C-source, and plate aliquots (0.1 mL) from suitable dilutions on semi-defined agar medium.
6. Incubate the agar plates at 30 °C until colonies appear (2–4 days) and inspect the plates for suspicious or atypical colonies that may belong to a contaminating organism in the culture (*see* **Note 19**).

3.3 Shake Flask Cultivation

1. Seed a shake flask of 500 mL with 100 mL of defined cultivation medium for shake flasks with 1–3 mL standard inoculum (*see* **Note 18**).
2. Incubate the shake flask at 30 °C, 150–200 rpm, for 48–96 h, until late logarithmic growth phase (*see* **Note 20**).
3. Use the shake flask culture to inoculate shake flasks for bioconversion studies or fermenters.

3.4 Shake Flask Bioconversion of Phytosterol to AD

1. Seed a shake flask of 500 mL with 100 mL of defined bioconversion medium for shake flasks with 2–30 mL of the shake flask culture above.
2. If a large inoculum is used (>5 mL), centrifuge down the cells (3200 × *g*, 10 min, 15 °C) and disperse them in a small volume (2–5 mL) of defined bioconversion medium without phytosterol before transferring them to the shake flask (*see* **Note 21**).
3. Incubate the shake flask at 30 °C, 100 rpm, for 48–120 h (*see* **Note 20**).

3.5 Fermenter Cultivation

1. Prepare the defined cultivation medium in a 1–10 L fermentation vessel and sterilize at 121 °C for 20 min.
2. Inoculate from a shake flask with defined cultivation medium. An inoculum volume of 10–30 mL/L fermenter medium is often suitable.
3. Incubate at 30 °C and pH 7.0–7.2 (*see* **Note 22**).
4. Control the dissolved oxygen level at 20–30% of saturation by stirring rate.
5. If foaming becomes a serious problem, add some antifoam from a sterile anti-foam solution.

3.6 Fermenter Bioconversion of Phytosterol to AD

1. Prepare the defined bioconversion medium in a 1–10 L fermentation vessel and sterilize at 121 °C for 20 min.
2. Inoculate from a shake flask or an inoculum fermenter. If the inoculum volume is small (≤30 mL/L fermenter medium), transfer the culture directly to the fermenter. If the inoculum volume is large, centrifuge down the cells (3200 × *g*, 10 min., 15 °C) and resuspend them in a small volume (10–30 mL/L fermenter medium) of defined bioconversion medium before inoculation (*see* **Note 21**).
3. Incubate at 30 °C and pH 7.0–7.2 (*see* **Note 22**).
4. Control dissolved oxygen level at 20–30% of saturation by stirring rate.
5. If foaming becomes a serious problem, add some antifoam from a sterile antifoam solution.

3.7 Sample Collection

1. Collect samples of 5 mL from the cultures at suitable intervals (12–24 h) to monitor the conversion of phytosterols to steroids (*see* **Note 23**).
2. If sampling a shake flask, try to disperse as much as possible of the cells and phytosterols on the glass wall in the culture by manual shaking of the flask before sampling. Sterile inoculation loops may be used to manually detach lumps of cells and/or phytosterol if necessary.

3. If sampling a laboratory fermenter, first shake the fermenter manually if possible (*see* **Note 22**). Thereafter, increase the stirring to high speed (1000–1500 rpm) for a short period (15–30 s, watch out for foaming) to disperse as much as possible of the phytosterol in the water phase before sampling.
4. Transfer the sample to a centrifuge tube.
5. Centrifuge the sample immediately (18,000 × *g*, 10 min, 10 °C) or place on ice until centrifugation to minimize further bioconversion.
6. Transfer the supernatant to a new tube.
7. Extract the cell-phytosterol pellet and the supernatant as described below, either immediately or freeze (−20 °C) both fractions for later extraction.

3.8 Sample Extraction and Preparation

1. Resuspend the pellet in the centrifugation tube in 0.5 mL distilled water and transfer the suspension to a glass extraction tube (*see* **Note 24**). Wash the centrifugation tube with 0.5 mL distilled water to ensure complete transfer of the pellet and transfer the washing water to the extraction tube. If the pellet has been frozen, thaw it first, but avoid heating it to room temperature to minimize bacterial activity in the pellet.
2. Add 10 mL ethyl acetate to the extraction tube. Close the tube with the cap and mix vigorously (whirl mixer).
3. From the supernatant, transfer a 1.0 mL aliquot to the same type of glass extraction tube as above. If the supernatant has been frozen, it should be thawed, heated to room temperature, and mixed well before the aliquot is transferred.
4. Add 10 mL ethyl acetate to the extraction tube. Close the tube with the cap and mix vigorously (whirl mixer).
5. Place the extraction tubes in a rotary shaker for 20 min. (room temperature).
6. After extraction, centrifuge the tubes (3300 × *g*, 10 min., room temperature) to aid layer separation.
7. The ethyl acetate layer (top) is collected and reduced in volume under a stream of air (at elevated temperature, 40 °C).
8. The dried samples are reconstituted in ethyl acetate and diluted in methanol. Typically, 1+9, 1+99, and 1+999 dilutions are prepared for each sample.

3.9 LC–MS Analysis

1. Establish the initial LC operating conditions: Flow rate 0.5 mL/min., eluent A (20%), eluent B (80%), column temperature, 45 °C.
2. Establish the mass spectrometer operating conditions: APCI ionization. Positive mode. Multiple reaction monitoring. Gas

Table 3
MRM settings for steroids and phytosterols

Analyte		M	Precursor	Product	Fragmentor (V)	CE (V)	RT (min)
DHEA	$C_{19}H_{28}O_2$	288.4244	289.4	271.4	100	4	3.95
			289.4	253.3	100	10	
AD	$C_{19}H_{26}O_2$	286.4085	287.4	97.3	100	15	3.75
11-α-OH-AD	$C_{19}H_{26}O_3$	302.41	303.4	267.2	100	15	2.71
Testosterone	$C_{19}H_{28}O_2$	288.4244	289.4	97.3	100	15	3.95
ADD	$C_{19}H_{24}O_2$	284.39	285.4	121,3	100	15	3.43
Campesterol	$C_{28}H_{48}O$	400.68	383,4	161,3	140	15	7.66
Stigmasterol	$C_{29}H_{48}O$	412.69	395.4	297,3	140	15	7.61
β-Sitosterol	$C_{29}H_{50}O$	414.71	397.4	161,3	140	15	7.95
Brassicasterol	$C_{28}H_{46}O$	398.66	381.4	297.3	140	15	7.29

DHEA 3β-hydroxyandrost-5-ene-17-one, *AD* Androst-4-ene-3,17-dione, *ADD* androst-1,4-diene-3,17-dione, *11-α-OH-AD* 11α-hydroxyandrost-4-ene-3,17-dione

temperature 250 °C, APCI heater 300 °C, gas flow rate 4 L/min, nebulizer 40 psi, and capillary 3500 V (+). All parameters are optimized for the selected analytes. MRM settings are given in Table 3.

3. Allow the LC system and the mass spectrometer to equilibrate. This typically takes up to 2 h.
4. Inject a blank sample (methanol). The injection volume (2 μL) is used also for samples and standards. Gradient for elution of standards and samples are:

Time (min)	A (%)	B (%)
0	20	80
1	20	80
4	1	99
7	1	99
8	20	80
10	20	80

Post-run before next run 1 min.

5. Inject a series of standards (0–100 μM).
6. Inject samples.

7. For every 20 samples, inject a 10 μM standard and a blank sample.
8. Finally inject a series of standards.

3.10 Calculation of Results

1. Use the standards to calculate the peak area per concentration of the analyte. This is usually performed by the LC–MS software. Plot the calibration graph.
2. Divide the peak area of the analyte in the sample with the peak area per concentration determined from the standards to obtain the concentration of the analyte in the analyzed sample.
3. The concentration of the analyte in the original culture is calculated by considering pre-concentrations and dilutions during sample workup.

4 Notes

1. The shake flask medium is designed to yield approx. 4.5 g cells (d.w.) per liter. pH may decrease to a final pH <4 due to acid production.
2. The trace mineral solution contains HCl (0.6 M) to ensure that the trace minerals remain dissolved during storage. NaOH is added to neutralize the acid in the trace mineral solution so that pH in the medium does not change significantly.
3. Precipitation of $MgNH_4PO_4$ ($k_{sp} = 3 \cdot 10^{-13}$ at 25 °C) is always a challenge when preparing defined cultivation media. By adding $MgSO_4$ after heat sterilization of the medium, precipitation is minimized. The effect of this salt on pH is limited. If $MgSO_4$ is added to the medium before filter sterilization, precipitated components may be removed during filtration. The $MgSO_4$ solution may be stored in a closed bottle for several weeks but should be filter sterilized immediately before use. Even with these precautions, a slight precipitation may be observed in the final medium, but this is not critical for the use of the medium. In fermentation media, precipitation will be more extensive due to higher concentrations of phosphate, ammonium, and magnesium. An alternative is to add 1/3 of the ammonium chloride and trace mineral solution from the start. The rest is added as one or two batches during the fermentation. The need for ammonium and trace minerals is closely connected to cell growth. Thus, when the cell concentration in the fermenter reaches 5–7 g d.w./L, more ammonium chloride and trace minerals should be added. If phytosterol is present in the medium, cell concentration can be difficult to determine as optical density or dry weight, but the cell concentration can be estimated from the accumulated

CO_2 expired in the off gas by assuming that 50% of the metabolized carbon goes to CO_2, the rest to cell mass, and that carbon constitutes 50% of the cell mass. Thus, an accumulated expiration of 210 mmol CO_2 from a fermenter with 1 L medium corresponds to roughly 5 g dry weight cells/L. The estimate may be refined by correcting for CO_2 in the air (approx. 0.04%), but unless the aeration rate is very high or the fermentation time very long (several days), the influence of CO_2 in the air on this crude estimate is negligible. It is preferable to add more nutrients before they limit cell growth.

4. The shake flask medium is designed to be C-source (phytosterol) limited and more phytosterol (1–2 g/L) and corresponding amounts of Me-β-CD may be added if desired. The molar ratio between phytosterol and Me-β-CD should be at least 1:1. A too high concentration (>50 g/L) Me-β-CD may have negative effects on cell growth. A ratio between phytosterol and Me-β-CD of 1:4 by weight corresponds to a molar ratio of approx. 1:1.3. More phytosterol and/or Me-β-CD may be added during the fermentation. Based on observations it appears that when Me-β-CD is complexed with steroids, it affects the cells less.
5. The fermenter medium is designed to yield approx. 15 g/L cells (d.w.). If a higher final cell density is required, the concentration of the components should be increased correspondingly, but oxygen transfer may become a challenge if the final cell density becomes too high. The problems with precipitation of medium components (*see* **Note 3**) increase with increasing concentration of medium components.
6. The selected antifoam is chosen based on previous experience. Other antifoams may work equally well.
7. The bioconversion fermenter medium is designed to be C-source (phytosterol) limited and more phytosterol (20–50 g/L) and corresponding amounts of Me-β-CD may be added if desired. However, because a too high concentration of Me-β-CD (>50 g/L from start) may inhibit cell growth, more phytosterol and Me-β-CD should be added at intervals during the bioconversion process.
8. The trace mineral solution can be stored for at least a year at room temperature in a closed bottle. It is composed so that 1 mL solution provides roughly the amount of trace minerals required by 1 g (d.w.) bacterial cells.
9. Prepare Eluent A (25 mM formic acid in water) fresh prior to analysis.
10. All reagents should be of analytical or HPLC grade unless otherwise stated. All solvents used for the HPLC eluents should be of mass spectrometric grade. Water used should be deionized, e.g., on a Milli-Q system.

11. Solubility of especially the phytosterols is limited in the eluents used in the LC–MS analysis. Therefore, always prepare the stock solutions at high concentrations in a good solvent (such as chloroform for phytosterols and ethyl acetate for steroids) and prepare only the diluted standards in ethyl acetate or methanol.

12. Strains from culture collections are usually received as freeze-dried samples, and the culture collection will usually provide recommended resurrection procedures and initial cultivation medium. In other cases, the strain may be received as colonies on agar medium or dispersed in buffer or medium. It is advisable to first cultivate the strain in a rich or semi-defined medium.

13. Freeze-dried bacteria often require some time in liquid medium before they start to grow. If, in addition, the survival rate in the freeze-dried sample is low, it may take some time before visible growth is observed. The necessary incubation time is therefore difficult to predict. Observe the culture daily. Visible growth should normally be obtained within a week. If not, the backup suspension of freeze-dried cells plated on agar medium may have yielded colonies.

14. With some mycobacteria we have experienced that it is difficult to obtain growth if the bacterium is transferred directly from a rich or semi-defined medium to a defined medium. In such cases an adaption process may be employed. The culture is transferred from the rich or semi-defined medium to a defined medium containing a small amount of, e.g., 33% of the complex components from the former medium, and then in a second step to a medium with 5–10% of the complex medium components before final transfers to the completely defined medium.

15. Some mycobacteria produce considerably amounts of acid when growing on defined medium and the pH in shake flask cultures can become very low (pH <4). It may then be preferable to harvest the culture at an early stage, i.e., in mid-logarithmic growth phase rather than late logarithmic growth phase to avoid the most extreme pH reduction. The disadvantage is that the standard inoculum will be thinner, and it will take longer before a dense culture is obtained from the inoculum.

16. The idea behind standard inoculum is to minimize inoculum effects in further studies. The standard inoculum should preferably be prepared on a medium resembling the medium to be used in later studies. Different "schools" recommend from 15% to 25% final concentration of glycerol, but in our experience concentrations within this range are not very critical when

freezing bacteria. It is advisable to incubate the cryotubes in a refrigerator for an hour or two before transferring them to the −80 °C freezer to give the glycerol time to penetrate the cells. By this method, 30–100 cryotubes with the same frozen culture are easily prepared. An alternative to freezing at −80 °C is to maintain the organisms on agar slants. The culture is inoculated onto agar slants in a test tube and cultivated at 30 °C until small colonies are visible. Then the test tube is put in a refrigerator. The disadvantage with this method is that new slant cultures must be prepared at intervals (weeks to months) and over time genetic drift in the organisms may make it slightly different from the organisms you started with. The risk of contamination is also higher.

17. Glycerol-preserved bacteria frozen at −80 °C remain viable for several years, often tens of years. However, freezers do not last that long. To avoid losing all the cryotubes with standard inoculum if the freezer breaks down, it is recommended, if possible, to divide the cryotubes between two freezers.
18. Every time a new experiment is started, take a cryotube with frozen culture from the freezer and inoculate into a shake flask with defined medium. The frozen culture should be thawed rapidly, either in lukewarm water or by clutching it in your hand. However, remember that the culture temperature should not exceed 30 °C.
19. Microscopy of the cultures along the purity test may also be used to ensure that the culture is pure and has the added advantage of making the experimenter familiar with the appearance of the organism in culture. This is a great advantage if, at a later stage in the studies, contamination is suspected. If contamination of the standard inoculum is found or suspected, pick 3–4 separate colonies from the agar plates that appear to be of the right type and use these to prepare a new standard inoculum. The old standard inoculum is discarded. Alternatively, if it is suspected that the contamination may have occurred during the purity test, repeat the purity test with three new cryotubes.
20. A challenge when cultivating mycobacteria in shake flasks is wall growth. Many mycobacteria like to cling to the glass wall at or above the water line. This tendency is amplified when the cells are cultivated on phytosterols as the insoluble phytosterols often adhere to the glass wall above the water line. Commercially available shake flasks with baffles vary in their baffle design. We have found that shake flasks with the baffles relatively high upon the wall are better at keeping the cells in the water phase (less splashing), especially if combined with a low shaking rate (100 rpm). As the bacterium grows slowly,

especially when growing on phytosterols, the oxygen transfer rate does not need to be very high, and a low shaking rate can be employed.

21. The inoculum size in bioconversion studies depend on whether you want the cells to grow and multiply on the phytosterol substrate or use a system where the bulk of the cell mass has been produced on another substrate, here glycerol. To avoid changing the concentration of phytosterol significantly and transferring significant amounts of glycerol to the conversion medium with the inoculum, the cells from the inoculum culture are centrifuged down and re-dispersed in a small volume of defined bioconversion medium without phytosterol before transfer to the bioconversion shake flask or fermenter.

22. Many *Mycolicibacterium* spp. tend to adhere to the wall above the water line so "splashing" inside the fermenter should be minimized. This can be achieved by keeping the aeration rate low and the volume in the fermenter well above or well below the uppermost stirrer in the fermenter. Because the cell growth is slow, oxygen transfer is seldom a challenge. In fermentations with water-soluble carbon sources such as glycerol, the problem with wall growth can be kept reasonably under control. However, when water-insoluble phytosterols are added, the problem increases considerably. Often, large lumps of cells and phytosterol accumulate above the water line in the fermenter. Small laboratory fermenters (1–3 L) can often, after the stirring is turned off, be loosened, and manually shaken, and this has often been an efficient method to bring most of the phytosterol back into the water phase. As the bioconversion process proceeds and more and more phytosterol is converted into AD(D) that is solubilized by Me-β-CD, the problem with phytosterol/cell aggregates diminishes.

23. We have found 5 mL to be a suitable sample volume, but this will depend on the system and the number of samples planned from the culture. Representative sampling is a challenge. In cultures containing enough Me-β-CD (≥1 mol Me-β-CD/mol AD(D)), almost all steroids are solubilized, but the phytosterols are mainly present as insoluble particles and clumps, usually intermixed with bacterial cells.

24. We use glass centrifuge tubes (approx. 13 mL) with short conical bottom (Assistant) closed with plastic caps (Ulstrup Plast), but other types of tubes and caps may also be suitable. The volumes in the extraction procedure are based on a sample volume of 5 mL. If a much larger sample volume, i.e., 20 mL or more, is used, the amount of cell pellet will increase, and the volumes may need to be adjusted.

Acknowledgments

This work was supported by grants from the European Union program ERA IB for the MySteri project and ERA CoBioTech for the project Syntheroids. MySteri funding was received from the European Union's 7th Research Framework Programme (EIB.12.010) and financed through the Research Council of Norway (project number 230037/O30). Syntheroids funding was received from the European Union's Horizon 2020 research and innovation programme under grant agreement No [722361] and financed through the Research Council of Norway (project number 285849). The authors thank Gunn Broli, Therese Myhre Harbak, and Kai Vernstad for their technical support. They are grateful to have had the MySteri and Syntheroids consortium members as valuable discussions partners during the work.

References

1. Moreau RA, Whitaker BD, Hicks KB (2002) Phytosterols, phytostanols, and their conjugates in foods: structural diversity, quantitative analysis, and health-promoting uses. Prog Lipid Res 41:457–500
2. Cantrill R, Kawamura Y (2008) Phytosterols, phytostanols and their esters, chemical and technical assessments. Proceedings of 69th JEFCA, 1–13. Available at http://www.fao.org/fileadmin/templates/agns/pdf/jecfa/cta/69/Phytosterols.pdf
3. Al Jasem Y, Khan M, Taha A, Thiemann T (2014) Preparation of steroidal hormones with an emphasis on transformations of phytosterols and cholesterol - a review. Mediterranean J Chem 3:796–830
4. Szentirmai A (1990) Microbial physiology of sidechain degradation of sterols. J Ind Microbiol 6:101–116
5. Malaviya A, Gomes J (2008) Androstenedione production by biotransformation of phytosterols. Bioresour Technol 99:6725–6737
6. The Human metabolome database, http://www.hmdb.ca/metabolites/hmdb00852
7. Saad HY, Higuchi WI (1965) Water solubility of cholesterol. J Pharm Sci 54:1205–1206
8. Yuan JJ, Guan YX, Wang YT, Wang HQ, Yao SJ (2016) Side-chain cleavage of phytosterols by *Mycolicibacterium sp.* MB 3683 in a biphasic ionic liquid/aqueous system. J Chem Technol Biotechnol. Published online in Wiley Online Library: (wileyonlinelibrary.com). https://doi.org/10.1002/jctb.4865
9. Pendharkar GB, Anjum SD, Patil S (2014) Enhanced biotransformation of phytosterols, a byproduct of soybean refineries, to a key intermediate used for synthesis of steroidal drugs. Asian J Pharm Clin Res 7:178–180
10. Yalkowski SH, He Y (2003) Handbook of Aqueous Solubility Data. CRC Press, Boca Raton
11. NIH PubChem open chemistry database, https://pubchem.ncbi.nlm.nih.gov/
12. Khomutov SM, Dovbnya DV, Donova MV (2001) Dissolution of a mixture of steroids in cyclodextrin solutions: A model description. Pharmaceut Chem J 55:627–629
13. Lee CY, Liu WH (1992) Production of androsta-l,4-diene-3,17-dione from cholesterol using immobilized growing cells of *Mycolicibacterium sp.* NRRL B-3683 adsorbed on solid carriers. Appl Microbiol Biotechnol 36: 598–603
14. Smith M, Zahnley J, Pfeifer D, Goff D (1993) Growth and cholesterol oxidation by *Mycolicibacterium* species in Tween 80 medium. Appl Environ Microbiol 59:1425–1429
15. Pérez C, Falero A, Llanes N, Hung BR, Herve ME, Palmero A, Marti E (2003) Resistance to androstanes as an approach for androstandienedione yield enhancement in industrial mycobacteria. J Ind Microbiol Biotechnol 30:623–626
16. Donova MV (2007) Transformation of steroids by actinobacteria: a review. Appl Biochem Microbiol 43:1–14
17. Xu YG, Guan YX, Wang HQ, Yao SJ (2014) Microbial side-chain cleavage of phytosterols by mycobacteria in vegetable oil/aqueous

two-phase system. Appl Biochem Biotechnol 174:522–533

18. Stefanov S, Yankov D, Beschkov V (2006) Biotransformation of phytosterols to androstenedione in two phase water-oil system. Chem Biochem Eng Q 20:421–427
19. Carvalho F, Marques MPC, de Carvalho CCCR, Cabral JMS, Fernandes P (2009) Sitosterol bioconversion with resting cells in liquid polymer based systems. Bioresour Technol 100:4050–4053
20. Marques MPC, Carvalho F, de Carvalho CCCR, Cabral JMS, Fernandes P (2010) Steroid bioconversion: towards green processes. Food Bioprod Process 88:12–20
21. Cruz A, Fernandes P, Cabral JMS, Pinheiro HM (2002) Effect of phase composition on the whole-cell bioconversion of β-sitosterol in biphasic media. J Mol Catal B Enzym 19-20: 371–375
22. Cruz A, Angelova B, Fernandes P, Cabral JMS, Pinheiro HM (2004) Study of key operational parameters for the side-chain cleavage of sitosterol by free mycobacterial cells in bis-(2-ethylhexyl) phthalate. Biocatal Biotransformation 22:189–194
23. Del Valle EM (2004) Cyclodextrins and their uses: a review. Process Biochem 39: 1033–1046
24. Brewster ME, Loftsson T (2007) Cyclodextrins as pharmaceutical solubilizers. Adv Drug Deliv Rev 59:645–666
25. Gao XQ, Feng JX, Wang XD, Hua Q, Wei DZ (2015) Enhanced steroid metabolites production by resting cell phytosterol bioconversion. Chem Biochem Eng Q 29:567–573
26. Donova MV, Nikolayeva VM, Dovbnya DV, Gulevskaya SA, Suzina NE (2007) Methyl-β-cyclodextrin alters growth, activity and cell envelope features of sterol-transforming mycobacteria. Microbiology 153:1981–1992
27. Shen Y, Wang M, Zhang L, Ma Y, Ma B, Zheng Y et al (2011) Effects of hydroxypropyl-β-cyclodextrin on cell growth, activity, and integrity of steroid-transforming Arthrobacter simplex and Mycolicibacterium sp. Appl Microbiol Biotechnol 90:1995–2003
28. Gulla V, Banerjee T, Patil S (2008) Quantitative TLC analysis of steroid drug intermediates formed during bioconversion of soysterols. Chromatographia 68:663–667
29. Kollerov VV, Shutov AA, Fokina VV, Sukhodolskaya GV, Donova MV (2008) Biotransformation of 3-keto-androstanes by Gongronella butleri VKM F-1033. J Mol Catal B Enzym 55: 61–68
30. Gomez C, Fabregat A, Pozo OJ, Marcos J, Segura J, Ventura R (2014) Analytical strategies based on mass spectrometric techniques for the study of steroid metabolism. Trends Anal Chem 53:106–116
31. Marwah A, Marwah P, Lardy H (2001) High-performance liquid chromatographic analysis of dehydroepiandrosterone. J Chromatogr A 935:279–296
32. Sánchez-Machado DI, López-Hernández J, Paseiro-Losada P, López-Cervantes J (2004) An HPLC method for the quantification of sterols in edible seaweeds. Biomed Chromatogr 18:183–190
33. Rocco A, Fanali S (2009) Analysis of phytosterols in extra-virgin olive oil by nano-liquid chromatography. J Chromatogr A 1216: 7173–7178
34. Ahonen L, Keski-Rahkonen P, Saarelainen T, Paviala J, Ketola RA, Auriola S et al (2012) Comparison of liquid chromatography-microchip/mass spectrometry to conventional liquid chromatography–mass spectrometry for the analysis of steroids. Anal Chim Acta 721:115–121
35. Mo S, Dong L, Hurst WJ, van Breemen RB (2013) Quantitative analysis of phytosterols in edible oils using APCI liquid chromatography–tandem mass spectrometry. Lipids 48:949–956
36. Palmgren JJ, Töyräs A, Mauriala T, Mönkkönen J, Auriola S (2005) Quantitative determination of cholesterol, sitosterol, and sitostanol in cultured Caco-2 cells by liquid chromatography–atmospheric pressure chemical ionization mass spectrometry. J Chromatogr B 821:144–152
37. theLabRat.com - Biotech Jobs and Research Resources http://www.thelabrat.com/protocols/WolfesVitaminSolution.shtmL

Chapter 16

Selective Microbial Conversion of DHEA into 7α-OH-DHEA

Vyacheslav V. Kollerov, Andrei A. Shutov, and Marina V. Donova

Abstract

7α-Hydroxy dehydroepiandrosterone (7α-OH-prasterone, 7α-OH-DHEA) is a key steroid intermediate in the synthesis of valuable pharmaceuticals widely used in the treatment of autoimmune illness, rheumatoid arthritis, colitis, and other severe diseases. The steroid can be produced using a filamentous fungus, which is capable of regio- and stereospecific hydroxylation of the steroid 3β-alcohol (DHEA) in the allylic position C7. Here, we describe a method for highly selective microbial production of 7α-OH-DHEA from DHEA using the zygomycete *Backusella lamprospora* VKM F-944. The method ensures high yield of 7-α-OH-DHEA (up to 89%, mol/mol) even at high concentration of the substrate DHEA (15 g/L).

Key words Filamentous fungi, *Backusella lamprospora*, Biocatalysis, Steroids, Transformation, DHEA, 7α-Hydroxylation, 7α-OH-DHEA

1 Introduction

The regio- and stereospecific hydroxylation plays a key role in the synthesis of valuable steroid pharmaceuticals, and it is a challenge that is difficult or even impossible to solve using traditional chemical methods [1].

3β,7α-Dihydroxyandrost-5-en-17-one (7α-OH-DHEA) is a well-known biologically active steroid with immunomodulatory property which is widely used in medicine to treat rheumatoid arthritis and other autoimmune diseases [2–4]. The ability to catalyze 7α-hydroxylation of DHEA to form 7α-OH-DHEA was reported for the wild-type filamentous fungi representing *Actinomucor*, *Gongronella*, *Fusarium*, *Cunninghamella*, and some other genera [3, 5, 6]. However, when incubating with DHEA, most of the strains also produce, along with 7α-OH-DHEA, the epimeric 7β-OH-, as well as 11α- and 15α-hydroxylated metabolites [4, 5, 7–9]. The soil-dwelling zygomycete *Backusella lamprospora* VKM F-944 selected in our lab is characterized by high regio- and stereospecific 7α-hydroxylase activity toward DHEA (Fig. 1) [10].

Carlos Barreiro and José-Luis Barredo (eds.), *Microbial Steroids: Methods and Protocols*, Methods in Molecular Biology, vol. 2704, https://doi.org/10.1007/978-1-0716-3385-4_16,

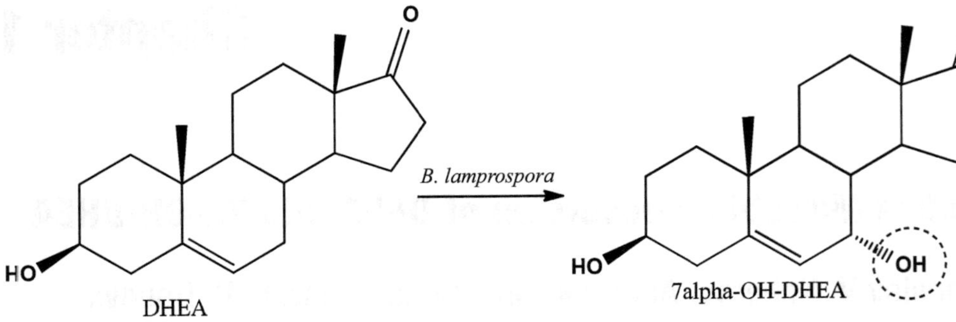

Fig. 1 Transformation of DHEA by *Backusella lamprospora* VKM F-944

The 7α-hydroxylase activity of the fungus was shown to depend on the nutrient medium composition, mode, and time of the substrate addition [10].

Here we describe a method for effective microbial production of 7α-OH-DHEA from DHEA by *B. lamprospora* under the revealed optimal bioconversion conditions. The method ensures the yield of the target 7α-OH-DHEA up to 89% (mol/mol) or 94% (w/w) even at high DHEA concentration (15 g/L). The amount of residual (unconverted) substrate does not exceed 2% (mol/mol).

2 Materials

2.1 Cultivation and Transformation

1. *Backusella lamprospora* VKM F-944 (All-Russian Collection of Microorganisms, Institute of Biochemistry and Physiology of Microorganisms, Russian Academy of Sciences; VKM IBPM RAS).
2. 10% KOH solution in distilled water.
3. Slant medium: 40 g/L of potato starch, 10 g/L of yeast extract (Difco), 10 g/L of corn steep solids (Sigma-Aldrich), and 15 g/L of agar. Adjust to pH 6.0 with 10% KOH if necessary (*see* **Note 1**).
4. Seed medium: 20 g/L of glucose, 5 g/L of corn steep solids (Sigma-Aldrich), 2.5 g/L of yeast extract (Difco), 2 g/L of K_2HPO_4, and 16 g/L of KH_2PO_4. Adjust to pH 6.0 with 10% KOH if necessary (*see* **Note 2**).
5. Bioconversion medium: 30.0 g/L of fructose, 5 g/L of corn steep solids (Sigma-Aldrich), 5 g/L of yeast extract (Difco), 10 g/L of tryptone (Sigma-Aldrich), and 0.5% (v/v) of PEG 400. Adjust to pH 6.0 with 10% KOH if necessary (*see* **Note 3**).
6. 3β-Hydroxyandrost-5-en-17-one (dehydroepiandrosterone, prasterone, DHEA) with a purity of 99% (Sigma-Aldrich).

7. 3β,7α-Dihydroxyandrost-5-en-17-one (7α-hydroxy-DHEA, 7α-OH-DHEA) with a purity of 99% (Sigma-Aldrich).
8. Dimethylformamide.
9. Erlenmeyer flasks (750 mL).
10. Rotary shaker.
11. Benchtop centrifuge.
12. Autoclave.
13. Vacuum rotary evaporator.
14. pH meter.

2.2 TLC and HPLC Analysis

1. Centrifuge tubes.
2. Plastic container.
3. Cyclomixer.
4. Ethyl acetate (EtOAc).
5. Acetonitrile (ACN) (HPLC grade).
6. Benzene.
7. Acetone.
8. $MnCl_2 \times 4H_2O$.
9. $MnCl_2$ reagent: 0.2 g of $MnCl_2 \times 4H_2O$, 30 mL of H_2O, 30 mL of methanol, and 2 mL of H_2SO_4 (97%).
10. Methanol.
11. H_2SO_4.
12. Pre-coated TLC sheets ALUGRAM SIL G/UV254 (Macherey-Nagel GmbH and Co. KG).
13. TLC developing chamber (Merck).
14. CN-15MC UV Darkroom (Vilber Lourmat).
15. Reverse-phase column Symmetry C18, 5 μm, 4.6 mm × 250 mm (Waters).
16. Guard column Symmetry C18, 5 μm, 3.9 mm × 20 mm (Waters).
17. Conditioned 50% ACN: mix equal volumes of ACN (HPLC grade) and deionized water (ASTM Type I), warm to an ambient temperature, and degas by vacuum filtration through 0.45 μm pore size nylon filtration disk (47 mm).
18. Conditioned HPLC mobile phase: mix 52 volumes of ACN (HPLC grade), 48 volumes of deionized water (ASTM Type I), and 0.01 volumes of acetic acid (HPLC grade), warm to an ambient temperature, and degas by vacuum filtration as described above.

19. Standard solution of DHEA (for TLC analysis): weight a precisely known amount (about 10 mg) of analytical standard of 3β-hydroxyandrost-5-en-17-one using analytical balance and dissolve it in 10 mL EtOAc. Calculate the accurate concentration of DHEA (about 1 mg/mL).
20. Standard solution of 7α-OH-DHEA (for TLC analysis): prepare the solution of analytical standard of 3β,7α-dihydroxyandrost-5-en-17-one and calculate its accurate concentration as described above for DHEA (about 1 mg/mL).
21. Standard solution of DHEA (for HPLC analysis): weight a precisely known amount (about 10 mg) of analytical standard of 3β-hydroxyandrost-5-en-17-one using analytical balance and dissolve it in the conditioned 50% ACN in a 100 mL class A volumetric flask. Calculate the accurate concentration of DHEA (about 0.1 mg/mL).
22. Standard solution of 7α-OH-DHEA (for HPLC analysis): prepare the solution of analytical standard of 3β,7-α-dihydroxyandrost-5-en-17-one and calculate its accurate concentration as described above for DHEA (about 0.1 mg/mL).

3 Methods

3.1 Generation of First Inoculum

1. Transfer the spore suspension from one agar slant (2–4 weeks old) into a 750 mL Erlenmeyer flask, which contains 50 mL seed medium.
2. Incubate on a rotary shaker (200 rpm) at 28 °C for 48 h (*see* **Note 4**).

3.2 DHEA Bioconversion

1. Inoculate the aliquot (10%, v/v or 5 mL) of the first-generation mycelium into bioconversion medium (45 mL) in a 750 mL Erlenmeyer flask.
2. Add steroid substrate DHEA as DMF (1.75%, v/v) solution (*see* **Note 5**) into bioconversion medium simultaneously with 10% (v/v) aliquot of the first inoculum.
3. Carry out the transformation reaction under the same conditions as described above for the first inoculum generation for a maximum of 168 h.

3.3 TLC Analysis

1. Take samples of cultivation broth (1 mL) every 12 h of bioconversion.
2. Add 7.5 volumes of EtOAc to the bioconversion sample, mix thoroughly on a shaker for 1 min, and then centrifuge at 2000 × *g* for 2 min.

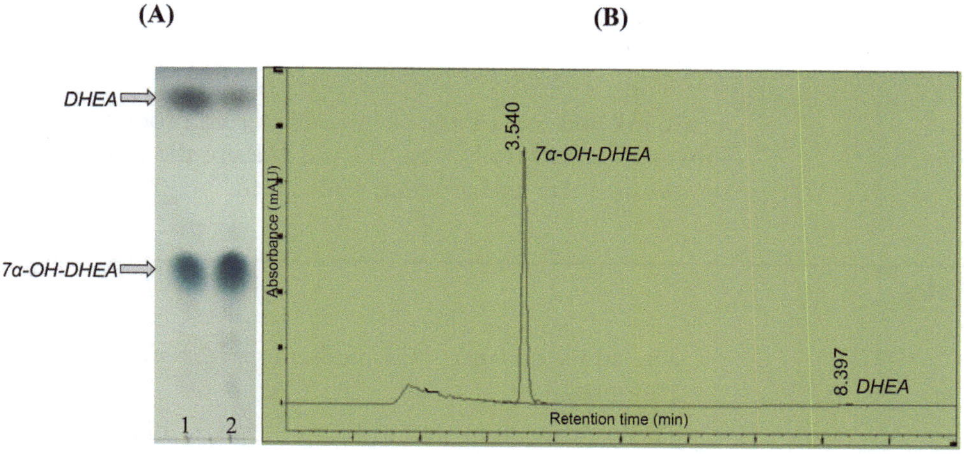

Fig. 2 TLC profile of standard steroids (1) (top to bottom): DHEA (7 μg), 7α-OH-DHEA (5 μg), and culture broth sample (2) of DHEA (15 g/L) transformation by *B. lamprospora* (144 h) (~10 μg of steroids in the spot) (**a**); HPLC profile of 168 h culture broth sample (**b**)

3. Load 5 μL of EtOAc extract (~10 μg of steroids in the spot) and 1–10 μL of standard solutions of DHEA and 7α-OH-DHEA to TLC sheet and develop in a mixture of benzene–acetone (3: 1.5, v/v).
4. Stain TLC sheet with $MnCl_2$ reagent and heat it to 105 °C until color develops (Fig. 2a).

3.4 HPLC Analysis

1. Run and equilibrate the HPLC system in the following conditions: column temperature 50 °C, mobile phase flow rate 1 mL/min, and detection wavelength 200 nm; acquisition times are 10 min for the standards and 30 min for broth samples.
2. Consecutively inject 5, 10, and 20 μL of the DHEA and 7-α-OH-DHEA standard solutions, register stable peak signals (*see* **Note 6**), and built linear calibration curves considering the precise concentration of compounds in the standard solutions.
3. Prepare bioconversion broth samples as follows: shake the sample on cyclomixer, take precise aliquot immediately, and dilute 1:50 with the conditioned 50% ACN. Spin the dilution at ≥1000 × *g* for 15 min in a tightly closed centrifuge tube. Take out and keep the tube in vertical position.
4. Apply the supernatant for injection to the HPLC system (20 μL). Register and integrate the chromatogram. Calculate DHEA and 7α-OH-DHEA titers using the calibration curves obtained and the sample dilution factor (Fig. 2b).
5. Calculate the molar yield (MY) of 7α-OH-DHEA according to the following equation:

$$MY = \frac{(mP/MP)}{(mS/MS)} \times 100\%,$$

where mP and mS are the weights of steroid products and steroid substrates, respectively, and MS and MP are the molar weights of steroid substrate and product, respectively.

4 Notes

1. The strain on agar slant medium can be stored at 4 °C for 1 month.
2. The seed medium should be sterilized at 121 °C for 30 min before using.
3. The bioconversion medium should be sterilized at 121 °C for 30 min before using.
4. The use of seed medium usually provides homogeneous, fine-fibrous growth of *B. lamprospora* mycelium of the first generation.
5. Steroid substrate should be dissolved in dimethylformamide (DMF) under heating prior adding into bioconversion medium. The final DMF concentration in the medium should not exceed 1.75% (v/v).
6. HPLC reference retention times (Rt) are ~8.46 min for DHEA and ~3.53 min for 7α-OH-DHEA.

Acknowledgments

This work is financially supported by the Russian Science Foundation under Grant No. 21-64-00024.

References

1. Nassiri-Koopaei N, Faramarzi MA (2015) Recent developments in the fungal transformation of steroids. Biocatal Biotransformation 33: 1–28. https://doi.org/10.3109/10242422.2015.1022533
2. Traish AM, Kang HP, Saad F, Guay AT (2011) Dehydroepiandrosterone (DHEA) - a precursor steroid or an active hormone in human physiology. J Sex Med 8:2960–2982. https://doi.org/10.1111/j.1743-6109.2011.02523.x
3. Kozłowska E, Urbaniak M, Kancelista A, Dymarska M, Kostrzewa-Susłow E, Ste˛pie´n Ł, Janeczko T (2017) Biotransformation of dehydroepiandrosterone (DHEA) by environmental strains of filamentous fungi. RSC Adv 7:31493–31501. https://doi.org/10.1039/C7RA04608A
4. Kozłowska E, Hoc N, Sycz J, Urbaniak M, Dymarska M, Grzeszczuk J, Kostrzewa-Susłow E, Stępień Ł, Pląskowska E, Janeczko T (2018) Biotransformation of steroids by entomopathogenic strains of *Isaria farinosa*. Microb Cell Factories 17:71. https://doi.org/10.1186/s12934-018-0920-0
5. Lobastova TG, Gulevskaya SA, Sukhodolskaya GV, Turchin KF, Donova MV (2007) Screening of mycelial fungi for 7α- and

7β-hydroxylase activity towards dehydroepiandrosterone. Biocatal Biotransformation 25: 434–442. https://doi.org/10.1080/10242420701568492

6. Janeczko T, S´wizdor A, Dmochowska-Gładysza J, Białonska A, Ciunik Z, Kostrzewa-Susłow E (2012) Novel metabolites of dehydroepiandrosterone and progesterone obtained in Didymosphearia igniaria KCH 6670 culture. J Mol Catal B Enzym 82:24–31. https://doi.org/10.1016/j.molcatb.2012.05.009
7. Milecka-Tronina N, Kołek T, Swizdor A, Panek A (2014) Hydroxylation of DHEA and its analogues by *Absidia coerulea* AM93. Can an inducible microbial hydroxylase catalyze 7α- and 7β-hydroxylation of 5-ene and 5α-dihydro C19-steroids? Bioorg Med Chem 15:883–891. https://doi.org/10.1016/j.bmc.2013.11.050
8. Kołek T, Milecka N, Świzdor A, Panek A, Białońska A (2011) Hydroxylation of DHEA, androstenediol and epiandrosterone by *Mortierella isabellina* AM212. Evidence indicating that both constitutive and inducible hydroxylases catalyze 7α- as well as 7β-hydroxylations of 5-ene substrates. Biomol Chem 9:5414–5422. https://doi.org/10.1039/c1ob05350g
9. Kozłowska E, Urbaniak M, Hoc N, Grzeszczuk J, Dymarska M, Stępień Ł, Pląskowska, Kostrzewa-Susłow E, Janeczko T (2018) Cascade biotransformation of dehydroepiandrosterone (DHEA) by *Beauveria species.* Sci Rep 8:13449. https://doi.org/10.1038/s41598-018-31665-2
10. Kollerov V, Shutov A, Kazantsev A, Donova M (2022) Hydroxylation of pregnenolone and dehydroepiandrosterone by zygomycete *Backusella lamprospora* VKM F-944: selective production of 7α-OH-DHEA. Appl Microbiol Biotechnol 106:535–554. https://doi.org/10.1007/s00253-021-11737-6

Chapter 17

Production of 11α-Hydroxysteroid Derivatives by *Corynebacterium glutamicum* Expressing the *Rhizopus oryzae* Hydroxylating System

Beatriz Galán, Carmen Felpeto-Santero, and José Luis García

Abstract

Hydroxylation of steroids has acquired special relevance for the pharmaceutical industry. Particularly, the 11α-hydroxylation of steroids is a process of biotechnological importance currently carried out at industrial scale for the production of contraceptive drugs and glucocorticoids. This process is performed by several fungal species including *Rhizopus nigricans*, *Aspergillus ochraceus*, *Aspergillus niger*, and *Rhizopus oryzae* that are used to produce by biotransformation hydroxylated steroids for pharmaceutical purposes (Wang et al., J Steroid Biochem Mol Biol 171:254–261, 2017). However, the development of more efficient biotransformation processes is essential since the steroidal derivatives obtained by the in vivo hydroxylation are often a mixture of hydroxylated compounds in different positions of the steroid molecule. This phenomenon is due to the large number of different CYPs contained in the fungal strains.

The genes responsible for the 11α-hydroxylase activity in *R. oryzae* consisting in the cytochrome CYP509C12 and its redox partner, the reductase RoCPR1, have been chemically synthetized forming a synthetic operon named FUN optimized to be expressed in bacteria. To express this operon, we have selected the strain *Corynebacterium glutamicum* R31 that is a robust and GRAS bacterial strain widely used for industrial purposes. The synthetic operon has been cloned in the pECXK-99E vector, yielding pXKFUN plasmid, and transformed *C. glutamicum* R31 to generate *C. glutamicum* R31 (pXKFUN) strain. This strain is not a steroid degrader and can efficiently transport C19 and C21 steroids across the cytoplasmic membrane (García-Fernandez et al. Catalysts 316:1–12, 2017). *C. glutamicum* can be used as a clean host for steroid biotransformation, because it does not introduce additional undesired side reactions on the steroids, thus reducing the contamination of the final products (Felpeto-Santero et al., Microbiol Biotechnol 12:856–868, 2019). Here we show a proof of concept that *C. glutamicum* can be used as a suitable chassis to perform steroid biotransformation expressing eukaryotic cytochromes. The protocol below provides detailed information on steroid 11*α*-hydroxylations by *Corynebacterium* recombinant strain.

Key words *Corynebacterium glutamicum*, Steroids, Biotransformation, 11α-Hydroxylation, Fungal CYP450

Carlos Barreiro and José-Luis Barredo (eds.), *Microbial Steroids: Methods and Protocols*, Methods in Molecular Biology, vol. 2704, https://doi.org/10.1007/978-1-0716-3385-4_17,

1 Introduction

Steroids represent one of the most widely used drugs for multiple clinical purposes (e.g., anti-inflammatory, immunosuppressive, anti-allergic, anticancer) [3]. The oxidation state of steroid rings and the presence of different attached functional groups determine the specific physiological function of each steroid [4], and therefore, structural modifications of steroids, such as hydroxylation, dehydrogenation, or esterification, highly affect their biological activity [5, 6]. Among them, hydroxylation is one of the most important steroid modifications introduced by the pharmaceutical industry, since it introduces deep changes not only in their clinical activities but also in their physicochemical properties (e.g., solubility, adsorption). The steroid oxyfunctionalization is mainly catalyzed by fungal cytochromes P450 (CYPs) acting as monooxygenases that insert a single oxygen atom into a nonactivated C–H bond of the substrate with a concomitant reduction of other oxygen atom to water [7]. Fungi have been reported to carry out hydroxylations at almost centers of the steroid molecule [8], and they have been widely applied at industrial scale for the production of hydroxylated steroids with a broad range of biological activities [9]. Fungal hydroxylations are usually carried out by two-component enzyme systems consisting of a CYP monooxygenase and a NAD(P)H CYP-reductase (CPR) [8]. The 11-α-hydroxylation of steroids by fungi is the synthetic procedure currently used by the steroid industry [10]. The 11α-hydroxylating system from *R. oryzae* has been used for the production of hydroxyprogesterone [10, 11]. However, expression of CYP509C12 in yeast has demonstrated that this system hydroxylates predominantly at 11α and 6β positions of steroids [10].

After cloning and expressing both genes in *C. glutamicum* R31, we have selected *C. glutamicum* as a chassis since the publication of its complete genome [12, 13] has provided the basis for an enormous progress in the use of this microorganism for different biotechnological applications, placing it as an ideal chassis as cell factory [14]. The resulting recombinant strain is able to synthesize 11β-hydroxysteroids from different steroid substrates. Following the same strategy, in order to produce 11α-hydroxylated steroids, we have rationally designed a new recombinant bacterium carrying the synthetic operon named FUN that contains the 11-α-hydroxylating enzymes (CYP509C12 and CPR59830) from *R. oryzae*. We have cloned and expressed both genes in *C. glutamicum* R31, and the resulting recombinant strain is able to synthesize 11α-hydroxysteroids from different steroid substrates [15, 16].

2 Materials

2.1 Bacterial Strains

1. *C. glutamicum* R31 [17].
2. *C. glutamicum* R31 (pXKFUN) carrying the 11α-hydroxylase genes from *R. oryzae* [18, 19] (*see* **Note 1**).

2.2 Culture and Biotransformation Media

Chemical substances for preparing bacterial culture media can be purchased individually from any common supplier such as Sigma or as prepared media from Difco.

1. Tryptic Soy Agar (TSA): 15 g of Bacto Tryptone, 5 g of Bacto Soytone, 5 g of sodium chloride, and 15 g of agar. Dissolve all the components, adjust pH to 7.3 ± 0.2, bring the volume up to 900 mL, and autoclave at 121 °C for 20 min.
2. TSA-kanamycin: After autoclaving TSA, cool to 45 °C and add 20 μg/L of kanamycin. Mix and pour into Petri plates.
3. Tryptic Soy Broth (TSB): 15 g of Bacto Tryptone, 5 g of Bacto Soytone, and 5 g of sodium chloride. Dissolve all the ingredients, adjust pH to 7.3 ± 0.2, bring the volume up to 1 L, and autoclave at 121 °C for 20 min.
4. TSB-kanamycin: After autoclaving TSB, cool to 45 °C, add 20 μg/L of kanamycin, and mix.
5. 1 mM IPTG Dioxane-Free: Add high purity IPTG (e.g., Calbiochem) to the bacterial growing medium.
6. 0.5 mM δ-aminolevulinic acid (ALA): Add ALA to the bacterial growing medium (*see* **Note 2**).
7. NaCl–Tween 80 solution: 0.9% NaCl and 0.05% Tween 80 (*see* **Note 3**).
8. 50 mM sodium phosphate buffer pH 7.4.
9. Steroid stock solution: 5 mM solution stock prepared in 10% (v/v) tyloxapol (*see* **Note 4**).
10. Biotransformation medium: Prepare 50 mM potassium phosphate buffer (pH 7.4) and filter throughout 0.22 μm pore size to sterilize. Add the steroid to a final concentration of 0.5 mM (*see* **Note 5**).

2.3 Steroids

Prepare the following steroids as 5 mM stock solution in 10% (v/v) tyloxapol:

1. Deoxycortisone (DOC; Sigma).
2. Deoxycortisone (DOC; Sigma).
3. 1,4-Androstene-3,17-dione (ADD; Sigma).
4. 4-Androstene-3,17-dione (AD; Sigma).
5. Testosterone (TES; Sigma).

6. Progesterone (PROG; Sigma).
7. Dehydroepiandrosterone (DHEA; Sigma).
8. Tyloxapol (Sigma) (*see* **Note 4**).

2.4 Bacterial Preculture

1. TSA-Km plate: After autoclaving TSA, cool to 45 °C and add 20 μg/L of kanamycin. Mix and pour into Petri plate.
2. TSB-Km: Add 20 μg/L of kanamycin to TSB.
3. 50 mL sterile tube.
4. Orbital shaker.

2.5 Steroid Extraction

1. Chloroform for analysis (Supelco).
2. Vortex.
3. Centrifuge Mini Spin.
4. Acetonitrile for HPLC gradient grade (VWR Chemicals).
5. Thermoblock (80 °C).

2.6 Monitoring Steroid Biotransformation by TLC Analysis

1. Flat Bottom Chamber for 20 × 20 cm plates with stainless steel lid (CAMAG).
2. TLC Silica gel 60 F254 aluminum sheets 20 × 20 cm (Merck).
3. Mobile phase: Hexane/ethyl acetate (10/4).
4. TLC/HPTLC Sprayer (CAMAG).
5. Development solution: 20% H_2SO4 (v/v) prepared in water.

2.7 Steroid Quantification: HPLC–DAD Chromatography–Mass Spectrometry

1. Steroids: DOC, TES, AD, ADD, PROG, and DHEA stock solutions dissolved in 10% tyloxapol.
2. Formic acid 0.1% dissolved in water.
3. Acetonitrile for HPLC gradient grade (VWR Chemicals).
4. DAD detector and LXQ Ion Trap Mass Spectrometer equipped with an atmospheric pressure chemical ionization source and interfaced to a Surveyor Plus LC system. Surveyor Autosampler and MS Pump. Xcalibur Software for data analysis (Thermo Electron).
5. C18 column (4.6 × 150 mm, 5 μm particle size).

2.8 General Equipment

1. Spectrophotometer.
2. Filter 0.22 μm pore size.
3. Vacuum pump device.
4. Centrifuge.
5. Bullet Blender 50-DX-CE (Next Advance).

3 Methods

3.1 Shaken Flask Biotransformation

3.1.1 Preculture Preparation

C. glutamicum R31 (pXKFUN) preculture is necessary to obtain the biomass required to inoculate the culture medium at an appropriate OD_{600} in order to achieve an efficient phytosterol biotransformation.

1. Inoculate an isolated colony of *C. glutamicum* R31 (pXKFUN) from a TSA plate in 10 mL of TSB-kanamycin placed in a sterile 50 mL tube.
2. Place it in an orbital shaker at 30 °C and 250 rpm to allow bacterial growth for 24–30 h.

3.1.2 Culture Preparation

1. Take a well-grown ($OD_{600} = 2–2.5$) preculture of *C. glutamicum* R31 (pXKFUN) and use it to inoculate flasks 200 mL of TSB-kanamycin and 0.5 mM δ-aminolevulinic acid.
2. Incubate at 30 °C and 250 rpm, and when OD_{600} reaches 1.5, induce expression with 1 mM IPTG during 24 h.

3.1.3 Steroid Biotransformation by Resting Cells

1. Harvest *C. glutamicum* R31 (pXKFUN) cells by centrifugation (10 min at 4000 × *g*).
2. Wash twice with 0.85% (v/v) NaCl by centrifugation (10 min at 4000 × *g*).
3. Resuspend the cells in 100 mL flasks containing 25 mL of 50 mM potassium phosphate buffer (pH 7.4) and add the steroid solubilized in tyloxapol (*see* Subheading 2.3) at a final concentration of 0.5 mM. Shake (250 rpm) at 30 °C for 72–96 h.
4. Take 1 mL aliquots every 24 h to monitor steroid biotransformation: OD_{600}, TLC, and DAD–HPLC–MS.

3.2 Steroid Extraction of Shaken Flask Biotransformation

Before TLC and DAD–HPLC analysis, steroids must be extracted in an organic phase by duplicates.

1. Take duplicate samples of 1 mL from each culture (*see* Subheading 3.1.3).
2. Only for DAD–HPLC analysis, add 50 μL of 5 mM of each steroid in 10% (v/v) tyloxapol as an internal standard (ISTD). The ISTD is different depending on the biotransformation as it is shown in Table 1.
3. Add 10 mL of chloroform to each sample.
4. Homogenize samples using a Bullet Blender 50-DX-CE at speed 12 for 3 min.

Table 1
The steroidal ISTD used for each biotransformation is indicated

Substrate	Product	ISTD
PROG	11α-OH-PROG	TEST
DOC	11α-OH-DOC	ADD
AD	11α-OH-AD	DOC
TEST	11α-OH-TEST	PROG
DHEA	11α-OH-DHEA	PROG
ADD	11α-OH-ADD	DOC

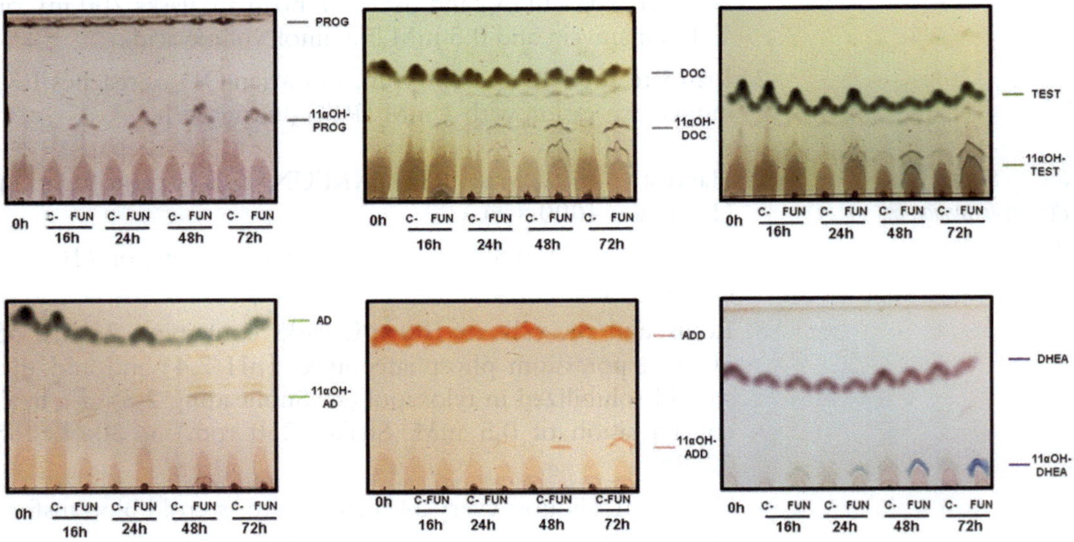

Fig. 1 TLC analysis of biotransformation using *C. glutamicum* R31 (pXKFUN) strain. The substrate used in each biotransformation assay and the obtained 11α-hydroxylate derivative are indicated to the right of each picture

5. Centrifugate the lysate at 4000 × *g* during 10 min.
6. Freeze the tubes at −80 °C for 30 min to separate the organic phase from the aqueous phase.
7. Discard the aqueous phase (*see* **Note 6**).
8. Place the tubes in a thermoblock at 70 °C until the samples are dry.

3.3 Monitoring Sterol Biotransformation by TLC Analysis

The biotransformation of steroids is monitored by TLC every 24 h during the process (72 h) (Fig. 1).

1. Extract the steroid fraction as described in Subheading 3.2.
2. Cut TLC plates leaving 1 cm margin on each side, 8 cm in length to allow the sample running, and 1 cm separation to upload the different samples.
3. Dissolve the samples in 20–100 μL of acetonitrile depending on the expected steroid concentration and always use the same volume for all the samples.
4. Apply 10 μL of each sample obtained (*see* Subheading 3.2) onto the TLC plates and let it dry for 5 min at room temperature. It is recommended to load standards of the substrate and expected products in the same TLC plate to better interpret the results of the analysis.
5. Place the loaded plate into the TLC chamber containing 50 mL of chloroform:ethanol (95:5) as the mobile phase.
6. Run the TLC until the mobile phase gets to the delimited frame on the TLC plate (approximately 1 cm from the end).
7. Dry the plate for 10 min at room temperature.
8. Develop the plate by spraying the surface with a solution containing 20% H_2SO4 in water (v/v) and heat the developed plate at 100 °C in an oven until color is maximal (approximately 10 min) (*see* **Note 7**).

3.4 Monitoring Sterol Biotransformation by DAD–HPLC–MS Analysis

1. Take a sample (25 μL) for chromatographic analysis by HPLC–DAD–MS.
2. Use the following interface parameters: capillary temperature 350, 60 °C for gas temperature in the vaporizer, capillary voltage 9 V, amplifier 400 Vp, and power source 100 μA.
3. Use high purity nitrogen used as nebulizer, sheath, and auxiliary gas. Perform MS analysis both in full scan and in selected ion monitoring (SIM) mode by scanning all the daughter ions of the products in positive ionization mode. Perform the quantification from parent mass of compounds, and obtain the specificity by following the specific fragmentations of all compounds. Perform chromatographic separation on a C18 column. Run the chromatography using water containing 0.1% (v/v) of formic acid and acetonitrile containing 0.1% (v/v) of formic acid as mobile phases A and B, respectively (flow 1 mL min^{-1}). Gradient is as follows: 40% B for 5 min, reaching to 80% B in 30 min, hold for 5 min, and return 40% B in 5 min. Re-equilibrate the HPLC column for 15 min in initial conditions. Set the valve to direct LC flow to the mass spectrometer from 1 to 45 min, with the remaining LC eluent diverted to waste.

Follow the next steps below to identify and quantify the steroids produced in the biotransformation.

1. Take 1 mL from the culture every 24 h during 72–96 h in duplicate.
2. Construct a standard curve from 500 μM to 15.625 μM DOC, TES, AD, ADD, PROG, and DHEA. Add the internal standard prior to extraction (*see* Table 1) (*see* **Note 8**).
3. Extract the steroid fraction as described in Subheading 3.2.
4. Resuspend the dry fraction in 0.5 mL of acetonitrile LC-MaScan.
5. Perform the chromatography using water containing 0.1% (v/v) of formic acid and acetonitrile containing 0.1% (v/v) of formic acid as mobile phases A and B, respectively (flow 1 mL/min). Gradient is as follows: 40% B for 5 min, reaching to 80% B in 30 min, hold for 5 min, and return 40% B in 5 min. Re-equilibrate the HPLC column for 15 min in initial conditions. Set the valve to direct LC flow to the mass spectrometer from 1 to 45 min, with the remaining LC eluent diverted to waste.
6. Perform MS analysis both in full scan and in selected ion monitoring (SIM) mode by scanning all the daughter ions of the products in positive ionization mode. Perform the quantification from parent mass of compounds, and obtain the specificity by following the specific fragmentations of all compounds. For quantification of the peak area, use the following quantitative masses: 289 m/z for testosterone; 331 m/z for DOC; 331 m/z for 11α-OH-DOC; 287 m/z for AD; 303 m/z for 11α-OH-AD; 285 m/z for ADD; 301 m/z for 11α-OH-ADD; 315 m/z for PROG; 331 m/z for 11-α-OH-PROG; 271 m/z for DHEA; and 287 for 11-α-OH-DHEA (Figs. 2, 3, 4, 5, 6, and 7).

4 Notes

1. The recombinant strains of *C. glutamicum* R31 can be stored at −80 °C in 50% glycerol (v/v) (viable >5 years) or at 4 °C on TSA plates (viables for 1 month).
2. 0.5 mM δ-aminolevulinic acid (ALA) is added to the bacterial growing medium to enhance the biosynthesis of heme group.

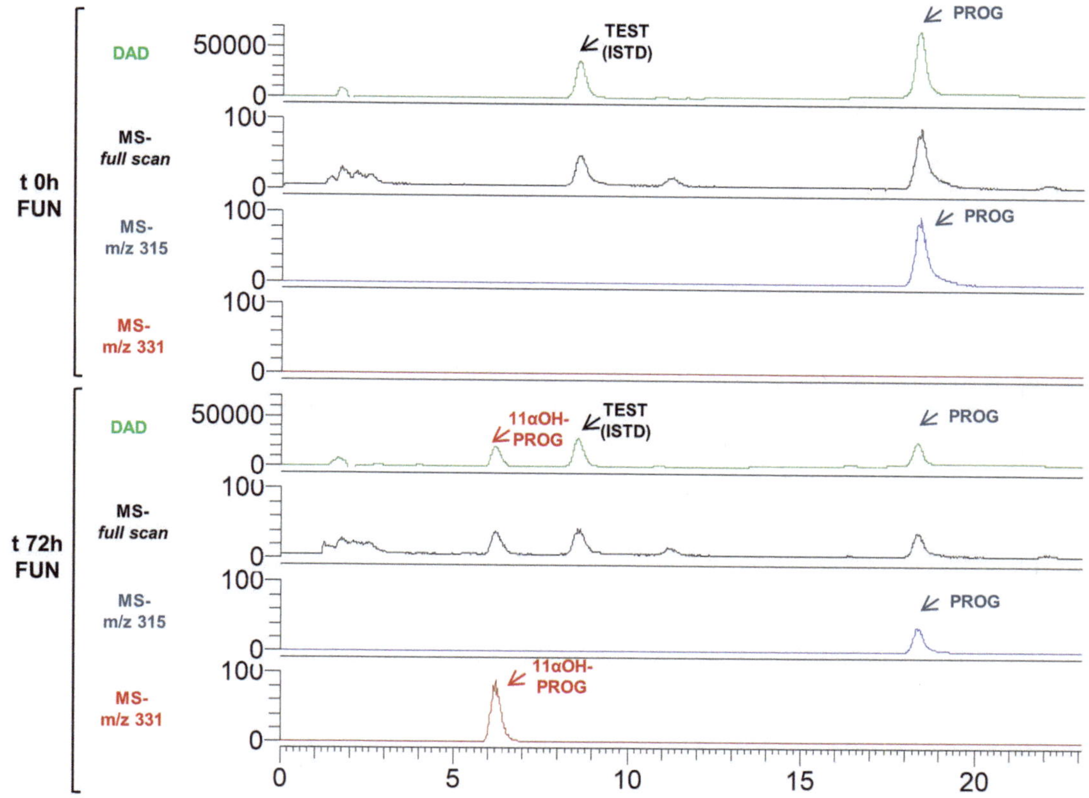

Fig. 2 PROG biotransformation by *C. glutamicum* R31 (pXKFUN) resting cells. HPLC–DAD chromatogram (50–600 nm) (green) and full scan mass spectra (m/z 150–400) at 0 h and 72 h of growth are shown. Mass spectra of m/z 315 and m/z 331 corresponding to progesterone and 11αOH-progesterone are shown

3. Add 0.05% Tween 80 (v/v) from a 20% sterile stock to avoid the *C. glutamicum* R31 clumping in the TSB medium when tyloxapol is not present.
4. Tyloxapol is a nonionic polymeric surfactant not used as a carbon source by *C. glutamicum* R31 that allows the dissolution of all kinds of steroids. Solubilizing the steroids enables the monitoring of the cultures following its DO at 600 nm. To pipette the tyloxapol correctly, preheat it using a microwave to reduce the thickness.
5. Place the biotransformation medium in a sonication bath at 80 °C for 1–4 h to enhance the steroid solubilization before autoclaving.

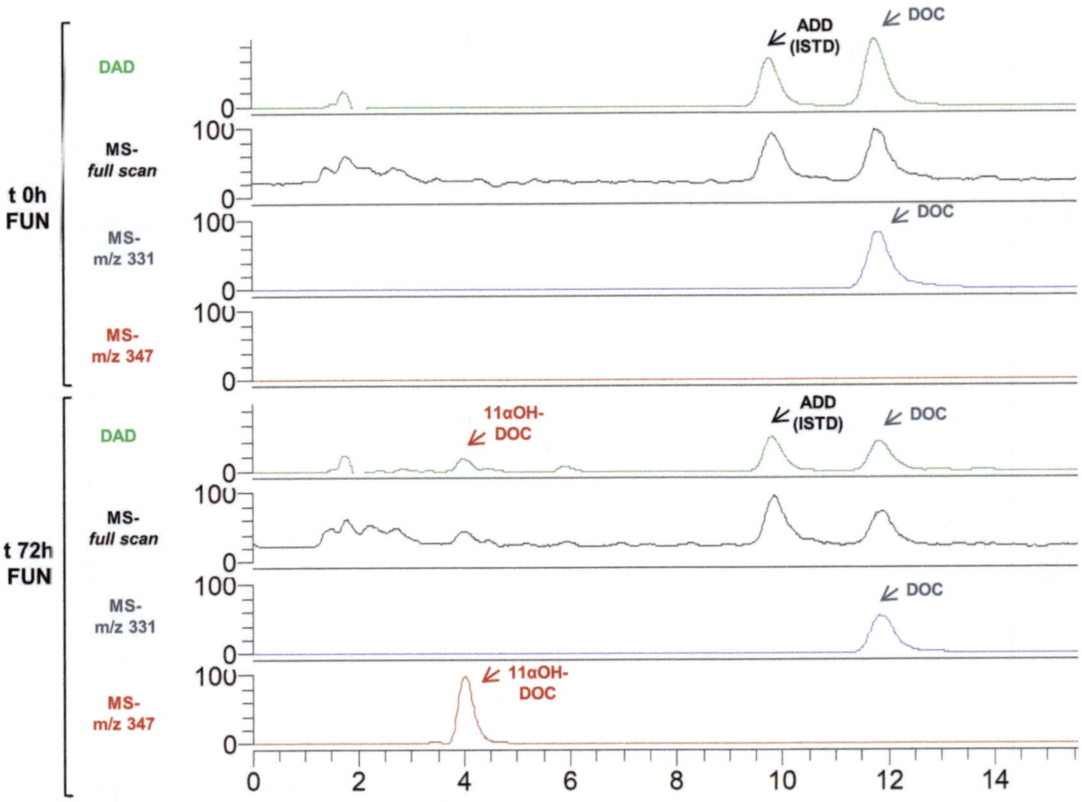

Fig. 3 DOC biotransformation by *C. glutamicum* R31 (pXKFUN) resting cells. HPLC–DAD chromatogram (50–600 nm) (green) and full scan mass spectra (m/z 150–400) at 0 h and 72 h of growth are shown. Mass spectra of m/z 315 and m/z 331 corresponding to DOC and 11αOH-DOC are shown

6. Place the samples at room temperature for 5 min and recover the organic phase that is not frozen with a pipette before the ice melts.
7. Before spraying the TLC with 20% H_2SO_4, those compounds that contain conjugated double bonds like ADD can be observed under UV light.
8. A calibration curve should be generated for each compound in the sample, and the concentration of the standards should be chosen on the basis of the concentration range expected in the study. To prepare a good calibration curve, a blank and at least six samples covering the expected range must be prepared and analyzed.

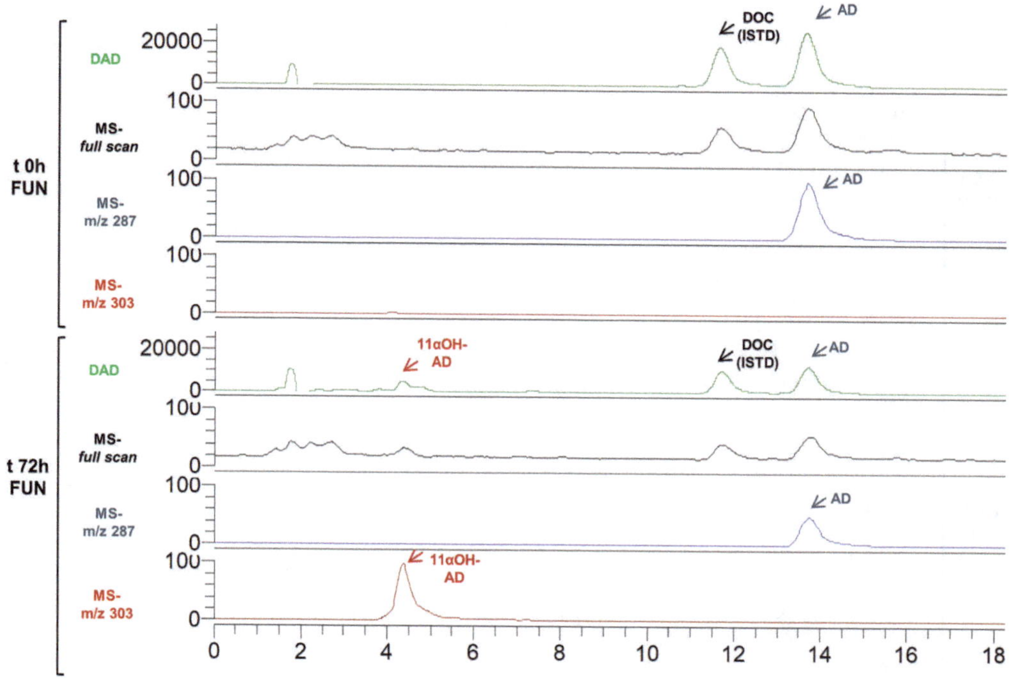

Fig. 4 AD biotransformation by *C. glutamicum* R31 (pXKFUN) resting cells. HPLC–DAD chromatogram (50–600 nm) (green) and full scan mass spectra (m/z 150–400) at 0 h and 72 h of growth are shown. Mass spectra of m/z 287 and m/z 303 corresponding to AD and 11αOH-AD are shown

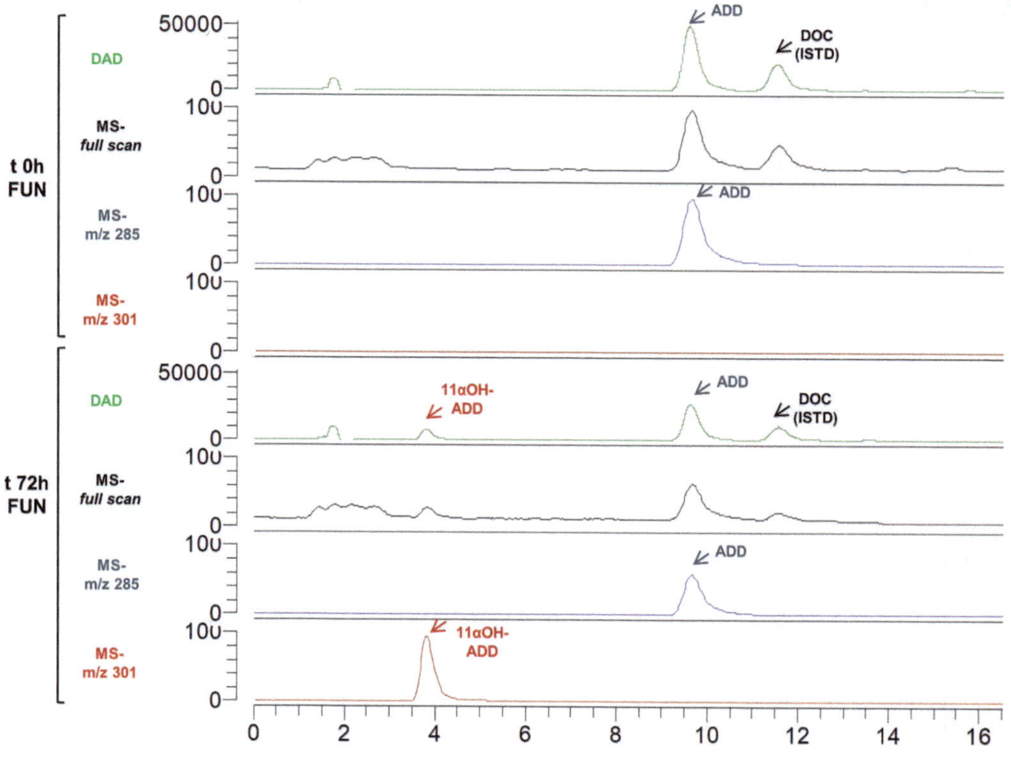

Fig. 5 ADD biotransformation by *C. glutamicum* R31 (pXKFUN) resting cells. HPLC–DAD chromatogram (50–600 nm) (green) and full scan mass spectra (m/z 150–400) at 0 h and 72 h of growth are shown. Mass spectra of m/z 285 and m/z 301 corresponding to ADD and 11αOH-ADD are shown

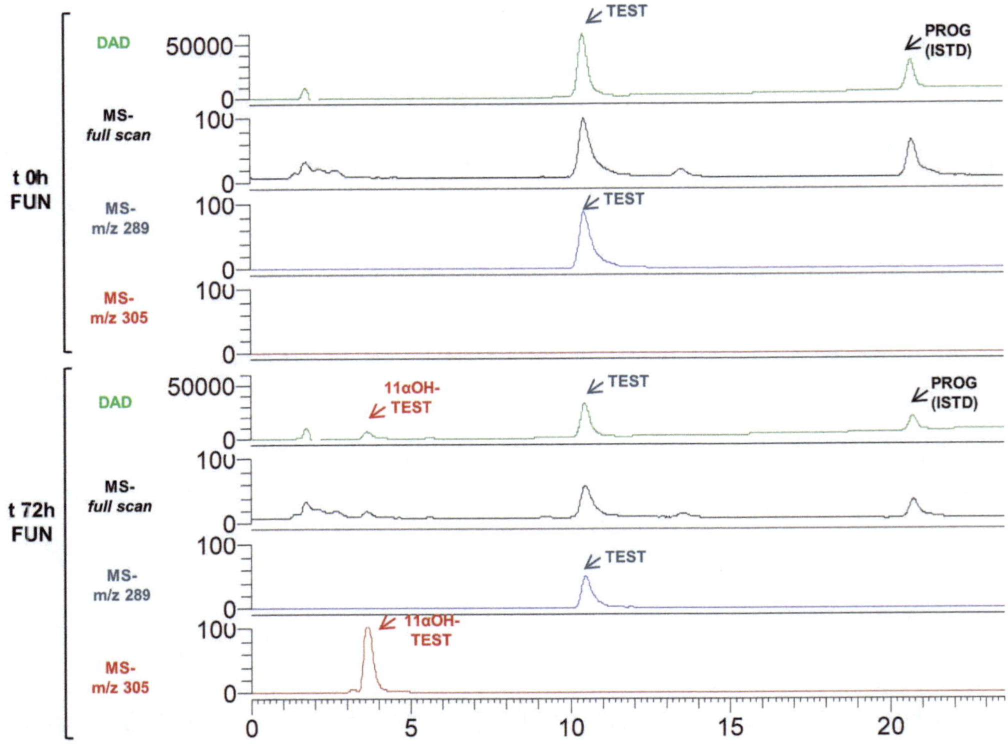

Fig. 6 Testosterone biotransformation by *C. glutamicum* R31 (pXKFUN) resting cells. HPLC–DAD chromatogram (50–600 nm) (green) and full scan mass spectra (m/z 150–400) at 0 h and 72 h of growth are shown. Mass spectra of m/z 289 and m/z 305 corresponding to TES and 11αOH-TES are shown

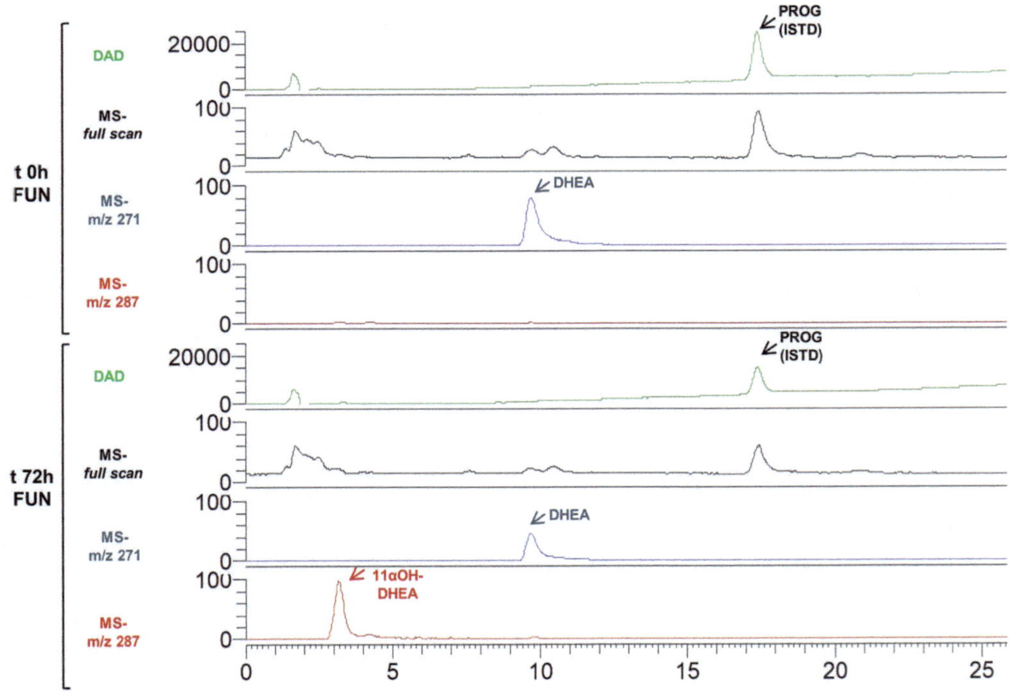

Fig. 7 DHEA biotransformation by *C. glutamicum* R31 (pXKFUN) resting cells. HPLC–DAD chromatogram (50–600 nm) (green) and full scan mass spectra (m/z 150–400) at 0 h and 72 h of growth are shown. Mass spectra of m/z 271 and m/z 287 corresponding to DHEA and 11αOH-DHEA are shown

References

1. Kollerov V, Shutov A, Fokina V et al (2010) Bioconversion of C19- and C21-steroids with parent and mutant strains of *Curvularia lunata*. Prikl Biokhim Mikrobiol 46:212–220
2. Wang R, Sui P, Hou X et al (2017) Cloning and identification of a novel steroid 11-α-hydroxylase gene from *Absidia coerulea*. J Steroid Biochem Mol Biol 171:254–261
3. García-Fernandez J, Galán B, Felpeto-Santero C et al (2017) Production of 4-Ene-3-ketosteroids in *Corynebacterium glutamicum*. Catalysts 316:1–12
4. Felpeto-Santero C, Galán B, Luengo JM et al (2019) Identification and expression of the 11-β-steroid hydroxylase from *Cochliobolus lunatus* in *Corynebacterium glutamicum*. Microb Biotechnol 12:856–868
5. Fernández-Cabezón L, Galán B, García JL (2018) New insights on steroid biotechnology. Front Microbiol 9:958
6. Lednicer D (2011) Steroid chemistry at a glance. Wiley, New York
7. Donova MV, Egorova OV (2012) Microbial steroid transformations: current state and prospects. Appl Microbiol Biotechnol 94:1423–1447
8. Szaleniec M, Wojtkiewicz AM, Bernhardt R et al (2018) Bacterial steroid hydroxylases: enzyme classes, their functions and comparison of their catalytic mechanisms. Appl Microbio Biotechno 102:8153–8171
9. Bernhardt R (2006) Cytochromes P450 as versatile biocatalysts. J Biotechnol 124:128–145
10. Kristan K, Rižner TL (2012) Steroid-transforming enzymes in fungi. J Steroid Biochem Mol Biol 129:79–91
11. Donova MV (2017) Steroid bioconversions. Methods Mol Biol 1645:1–13
12. Petrič Š, Hakki T, Bernhardt R et al (2010) Discovery of a steroid 11α-hydroxylase from *Rhizopus oryzae* and its biotechnological application. J Bacteriol 150:428–437
13. Fernandes P, Cruz A, Angelova B et al (2003) Microbial conversion of steroid compounds: recent developments. Enzym Microb Technol 32:688–705
14. Ikeda M, Nakagawa S (2003) The *Corynebacterium glutamicum* genome: features and impacts on biotechnological processes. Appl Microbiol Biotechnol 62:99–109
15. Felpeto-Santero C, Galán B, García JL (2021) Production of 11α-hydroxysteroids from sterols in a single fermentation step by *Mycolicibacterium smegmatis*. Microbl Biotechnol 14: 2514–2524
16. Felpeto-Santero C, Thesis D (2016) Producción de esteroides mediante ingeniería de citocromos. Facultad de Ciencias Químicas, Universidad Complutense de Madrid
17. Kalinowski J, Bathe B, Bartels D et al (2003) The complete *Corynebacterium glutamicum* ATCC 13032 genome sequence and its impact on the production of L-aspartate-derived amino acids and vitamins. J Biotechnol 104: 5–25
18. De Lorenzo V (2015) Chassis organism from *Corynebacterium glutamicum*: the way towards biotechnological domestication of Corynebacteria. Biotechnol J 10:244–245
19. Santamaría RI, Gil JA, Martin JF et al (1985) High-frequency transformation of *Brevibacterium lactofermentum* protoplasts by plasmid DNA. J Bacteriol 162:463–467

Chapter 18

Obtaining of 24-Norchol-4-ene-3,22-dione from Phytosterol with Mutants of *Mycolicibacterium neoaurum*

Dmitry V. Dovbnya, Tanya V. Ivashina, Sergey M. Khomutov, Andrei A. Shutov, Natalia O. Deshcherevskaya, and Marina V. Donova

Abstract

Engineered mutants of *Mycolicibacterium* spp. are known producers of valuable steroid synthons with C_{19} or C_{22} skeleton. Here we describe a method for site-directed mutagenesis of *Mycolicibacterium neoaurum* strains, bioconversion from phytosterol, and selective purification of C_{23} steroid 24-norchol-4-ene-3,22-dione (24-NCED) and C_{22} steroid 20-hydroxymethylpregn-4-ene-3-one (20-HMP). The yields of crystalline products with 95% purity by the method here described are 2.74 ± 0.085 g for 24-NCED and 1.42 ± 0.085 g for 20-HMP from 10 g/L phytosterol. 20-HMP is recognized as the key precursor in chemical syntheses of pharmaceutical corticosteroids and 24-NCED is a promising synthon for the synthesis of valuable steroids and own potent biological activity.

Key words Phytosterol, Cholesterol, Bioconversion, Steroid synthon, 24-Norchol-4-ene-3,22-dione, 20-Hydroxymethylpregn-4-en-3-one, *Mycobacterium neoaurum*, *Mycolicibacterium*, *fadA5*

1 Introduction

Modern production of precursors of synthetic steroid hormones (steroid synthons) is based on the microbial transformation of natural sterols. The most accessible and cost-effective raw materials for this industry are phytosterols (PS) produced from tall pitch, tall oil, or vegetable oils that are a mixture of plant sterols with a skeleton consisting of 28–29 carbon atoms [1].

The possibility of sterol bioconversion to useful steroids is due to the functioning of modified biochemical pathways of oxidative degradation of sterols. Representatives of many taxa of prokaryotes have the ability to oxidize sterols; however, the largest number of known non-pathogenic and industrially significant strains belongs to actinobacteria, in particular, *Mycolicibacterium* (basonym *Mycobacterium*) [2].

Carlos Barreiro and José-Luis Barredo (eds.), *Microbial Steroids: Methods and Protocols*, Methods in Molecular Biology, vol. 2704, https://doi.org/10.1007/978-1-0716-3385-4_18,

Multiple statistical mutagenesis (chemical, ultraviolet, or transposon) of natural strains has been used earlier to breed strains able to produce steroid precursors [3, 4]. Intensive development of genomics and molecular biology of actinobacterial strains that has taken place over the past decade provided the basis for the construction of recombinant strains producing steroid precursors by site-directed mutagenesis and metabolic engineering [5–7].

Traditional steroid synthons with fully oxidized side chain of sterols at C17 (3,17-diketo-androstanes, e.g., AD and ADD) (Fig. 1) are large-scale biotechnological products and the basic compounds for industrial syntheses of all groups of steroid hormones. At the same time, their use for preparation of synthetic hormones of the pregnane series requires the reconstruction of the side chain at C17 [7]. An actual task is to develop effective methods for a single-step biotechnological production of the precursors bearing oxygen function in the side chain at the C22 position, to shorten production chains while obtaining corticosteroids, progestogens, and vitamin D3 derivatives. For example, the steroid C22 alcohols can be converted into steroids of the pregnane series in two chemical steps through C22 aldehyde [8].

Producer strains and methods for the biotechnological production of steroid C22 alcohols, such as 20-hydroxymethyl-pregn-4-ene-3-one (20-HMP, Fig. 1) and its 1,2-dehydro-analogue (20-HMPD), their 17,20-dehydro-analogs [7, 9–12], as well as methyl esters of the corresponding C22 carboxylic acids [10, 13], have been reported previously.

Additional advantages for the synthesis of pregnane hormones can be provided by using a steroid precursor containing a carboxyl or carbonyl group at the C22 position. For instance, 9α-hydroxy-20-carboxypregna-4,17(20)-diene-3-one (HCPD, Fig. 1) can be converted in two or one step to 9α-hydroxy-progesterone, corresponding to 16-dehydro-20-oxo- or 17α-hydroxy-20-oxo-derivatives [14, 15]. In addition, the presence of a C22 carbonyl group may be useful for the synthesis of other steroids bearing various substituents at C22.

The 24-norchol-4-ene-3,22-dione (24-NCED, Fig. 1) is naturally occurring steroid with 23 carbon atoms in its skeleton that was found in the skin of some marine invertebrates and presumably performs a protective function [16]. The compound can be obtained in a multistage chemical synthesis through chlorine-substituted steroid intermediate or in the 4-stage chemical synthesis with 63% yield [16], both requiring purified stigmasterol as a starting material which complicates their practical implementation.

The mutant strains *Mycobacterium* sp. CCM 3528, *Mycobacterium parafortuitum complex* MCI 1412 and MCI 1414, and some others have been reported to convert cholesterol to 24-NCDED (Fig. 1) with 24-NCED being formed as a minor product [17, 18]. The strains provided extremely low productivity for

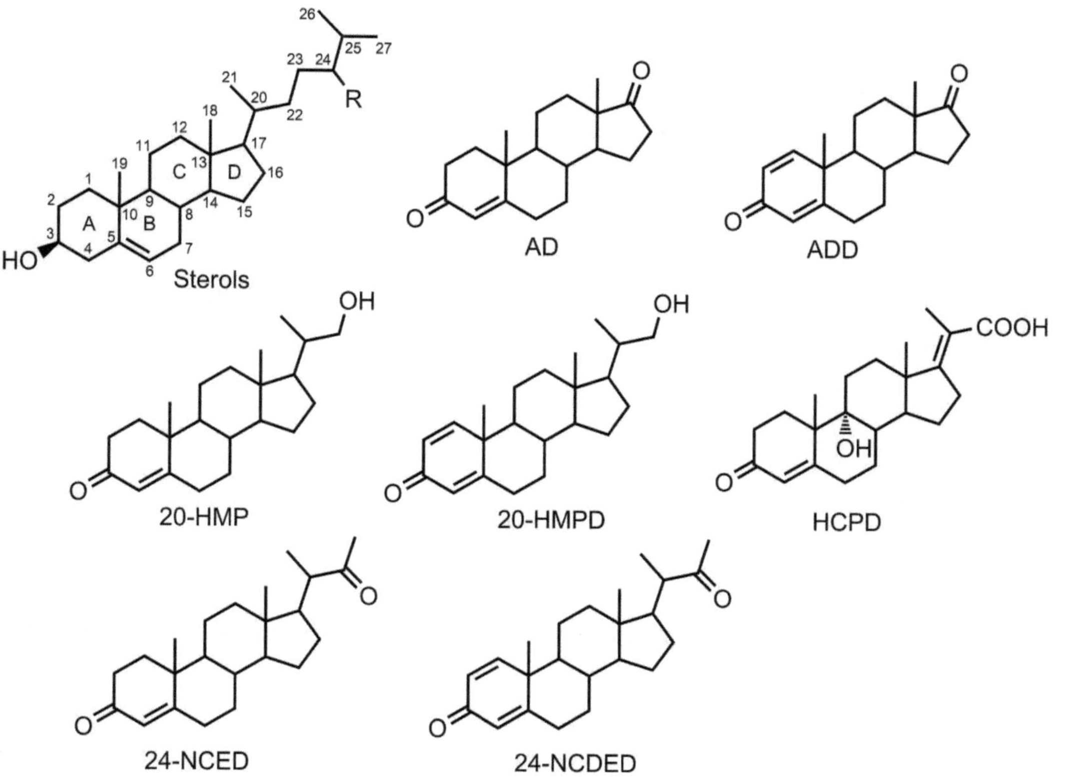

Fig. 1 Chemical structure of natural sterols, some valuable steroid synthons produced by microbial bioconversion, and C_{23}-steroids. Sterols: R = H, cholesterol; C_2H_5, β-sitosterol; CH_3, campesterol; C_2H_5 (Δ^{22}), stigmasterol; CH_3 (Δ^{22}), brassicasterol. Steroid synthons: AD, androst-4-ene-3,17-dione; ADD, androsta-1,4-ene-3,17-dione; 20-HMP, 20-hydroxymethylpregn-4-ene-3-one; 20-HMPD, 20-hydroxymethylpregna-1,4-diene-3-one; HCPD, 9α-hydroxy-20-carboxypregna-4,17(20)-diene-3-one. C_{23}-steroids: 24-NCED, 24-norchol-4-ene-3,22-dione; 24-NCDED, 24-norchol-1,4-diene-3,22-dione

24-NCED in optimized conditions. Even lower yields (0.75 g for 24-NCDED and 0.05 g for 24-NCED from 8 g substrate) were observed from PS. The low productivity and the complex nutrient additives used complicated isolation and purification of the target steroids.

The formation of C_{23}-steroids by actinobacterial strains is non-trivial as the sub-pathway of sterol side chain degradation normally goes through C_{24} and C_{22} intermediates [19]; however, it was suggested that 24-NCED could be formed by decarboxylation of a C_{24}-intermediate [16].

The β-ketoacyl-CoA thiolase FadA5 catalyzes the side chain shortening steps in sterol degradation pathway—at least two reactions in the case of cholesterol and one or more in the case of plant sterols, including the cleavage of acetyl-CoA residue from the C24 CoA thioester of chol-4-ene-3,22-dione-24-oic acid (CHEDA, Fig. 2). Targeted disruption of the *fadA5* gene in *Mycobacterium tuberculosis* H37Rv resulted in the loss of the strain ability to

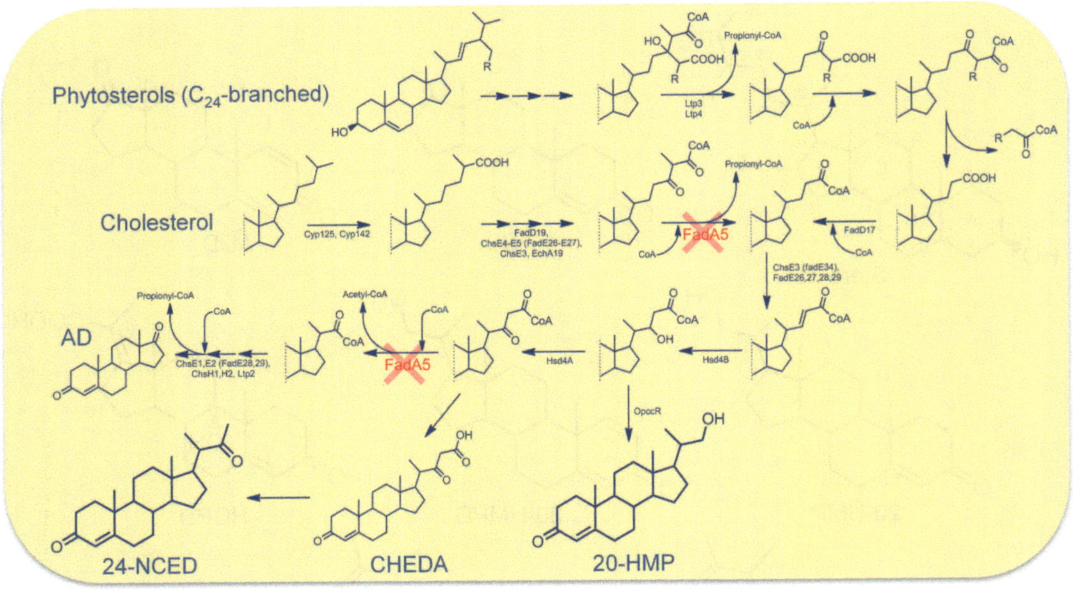

Fig. 2 Sterol side chain degradation sub-pathway in *Mycolicibacterium neoaurum* and formation of 24-NCED by the NRLL B-3805Δ*kstD*Δ*fadA5* mutant strain

convert cholesterol into AD(D) [20]. Knockout of *fadA5* in *Mycolicibacterium neoaurum* ATCC 25795 resulted in some increase in the production of steroidal primary C22-alcohols 20-HMP and 20-HMPD from PS. Inactivation of the bi-functional *hsd4A* gene (homologous to genes of the *fabG* family) in the same strain led to a significant increase in the production of 20-HMP [7], especially in combination with *fadA5* overexpression and inactivation of 3-ketosteroid 1,2-dehydrogenases KstD (18–49%, molar yield) [21]. No 22-oxo-functionalized products were accumulated by the constructed strains.

Here we describe a method for *fadA5* knockout in *Mycolicibacterium neoaurum* resulting in the recombinant strains capable of producing 24-NCED as a major product from phytosterol, an optimized laboratory bioconversion procedure resulting in preferential accumulation of 24-NCED, and the method for selective isolation and purification of 24-NCED and 20-HMP from the conversion broth. The method can be exploited for production of valuable 24-NCED for further practical applications in biotechnology, chemical synthesis, and biomedicine.

1.1 Parent Bacterial Strains and Gene Target

Analysis of the genomes of VKM Ac-1815D [5] and NRRL B-3805 [6] (GenBank accession nos. CP006936.2 and CP011022.1) revealed 18 open reading frames (ORFs) presumably encoding acetyl-CoA acetyltransferases or thiolases. Transcriptomic study of *M. neoaurum* VKM Ac-1815D grown revealed the only gene (D174_RS24025, referred to below as *fadA5*) showed a

75–100-fold increase of expression in the presence of sterols. The *fadA5* genes of *Mycolicibacterium neoaurum* strains NRRL B-3805 or VKM Ac-1815D are identical and show high level of the encoded protein homology (99% identical amino acids) with FadA5 protein from *M. neoaurum* ATCC 25795.

Both the VKM Ac-1815D and NRRL B-3805 strains form limited amounts of 1-dehydrosteroids from PS (up to 10% of the total yield) due to the residual activity of 3-ketosteroid-1-dehydrogenase encoded by *kstD*. To prevent possible 1-dehydrogenation of 24-NCED and 20-HMP, the strain lacking KstD activity should be used [22]. The suitable genetic techniques for unmarked gene knockout in NRRL B-3805 have been described [23].

1.2 Overview of the Gene Knockout Method

The principal scheme of obtaining *M. neoaurum* Δ*fadA5* mutants is shown in Fig. 3. For the *fadA5* knockout, a 2602 bp DNA fragment containing the ORF (1102 bp) along with adjacent genomic DNA sequences (1500 bp) of the parent strain genomic DNA should be amplified and cloned into pEX18Tc plasmid vector (or another vector that does not replicate in mycobacteria). Then a mutant *fadA5* allele should be obtained by deletion of the 282 bp fragment from the coding sequence of the gene that flanked with two *SmaI* sites that are absent in the vector DNA and *fadA5* flanking sequences. The deletion does not lead to a frameshift and corresponds to the amino acid residues 68–162 in the FadA5 protein. Then the deleted fragment should be substituted with appropriate selective marker, e.g., the kanamycin resistance (Km^R) cassette (1.252 kb) from pUC4K plasmid.

The resulting plasmid pEX::fadA5-Km is designed to generate the marked *fadA5* deletion mutations via the homologous recombination between the mutant allele of the gene localized in the

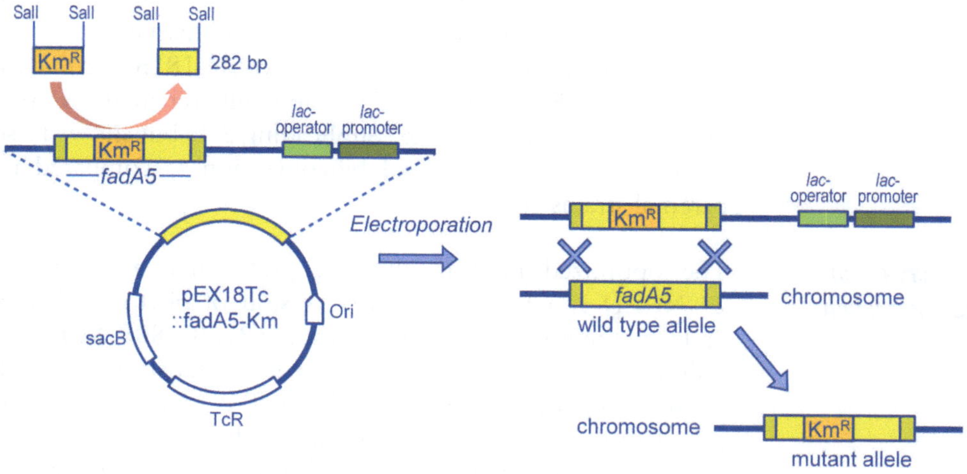

Fig. 3 Scheme of pEX18Tc::fadA5-Km plasmid construction and *fadA5* knockout in *Mycolicibacterium neoaurum* strains

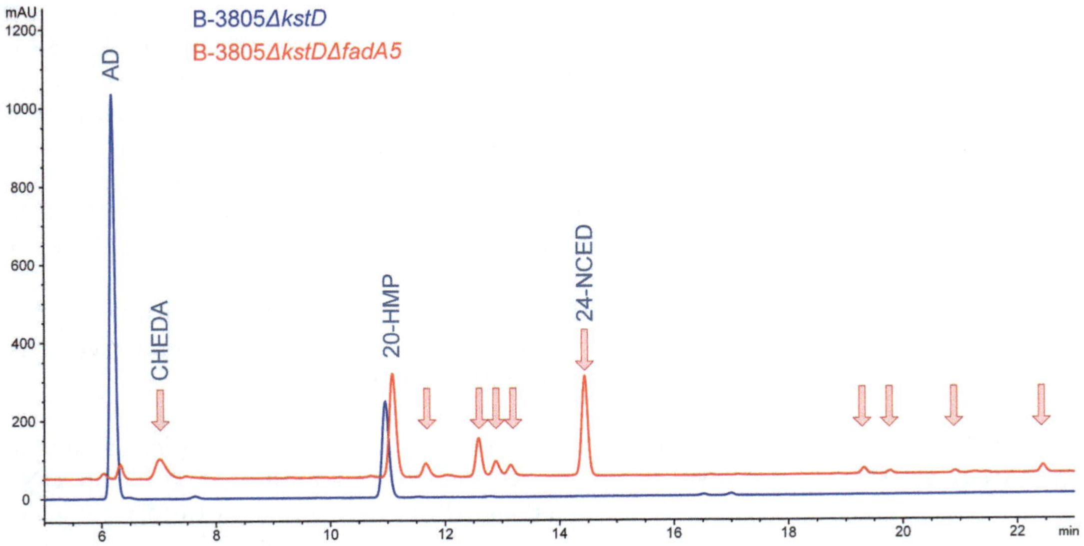

Fig. 4 Comparative HPLC profiles of steroid products at the biotransformation of phytosterol (5 g/L) by *M. neoaurum* NRRL B-3805Δ*kstD* and the Δ*kstD*Δ*fadA5* mutant strain. Red arrows point at steroid products accumulated by the mutant but not the parent strain

plasmid and the wild-type allele in the chromosome of Ac-1815D, NRRL B-3805, and other closely related strains having an identical nucleotide sequence in a similar genome fragment.

The parent strain is then transformed with the recombinant plasmid by electroporation according to the previously optimized method [24]. The recombinant clones were then selected for resistance to Km (cassette marker), which indicated the presence of a mutant *fadA5* allele in the bacterial chromosome. To determine whether Km^R colonies are the desired mutants (resulting from double crossover) or merodiploids (resulting from plasmid integration into the chromosome due to the single crossover), the colonies are analyzed for the presence of the plasmid markers.

The obtained *M. neoaurum* VKM Ac-1815D Δ*fadA5* and NRRL B-3805 Δ*kstD*Δ*fadA5* mutants fully retained the ability to grow on MYCA medium (*see* Subheading 2.1) but showed a somewhat suppressed phytosterol bioconversion as compared to the parental strains (Fig. 4).

1.3 Overview of Biotransformation Process

The optimized method allowing preferential accumulation of 24-NCED is a fed-batch bioconversion of phytosterol in shake flasks by the growing *M. neoaurum* NRRL B-3805 Δ*kstD*Δ*fadA5* cells in the medium supplemented with statistically substituted hydroxypropyl-β-cyclodextrin (HPbCD). During the initial 2–3 days of a bioconversion batch a wide spectrum of C_{29}, C_{24}, and C_{22} steroid products appeared in the broth. Finally, most of them are converted to 24-NCED, whereas 20-HMP, minor amounts of AD, and some other steroids formed as by-products (Fig. 5).

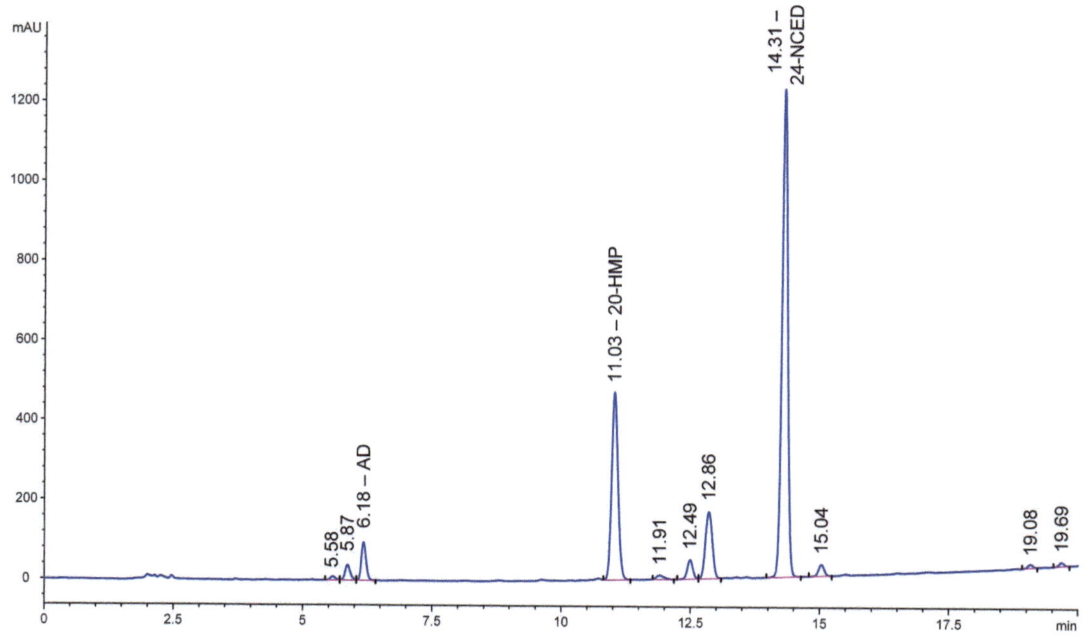

Fig. 5 The final HPLC profile of steroid products in biotransformation broth by *M. neoaurum* NRRL B-3805Δ*kstD*Δ*fadA5* mutant strain in the optimized conditions

The application of β-cyclodextrin (bCD) derivatives to enhance the efficiency of biotransformation of various steroids, including PS conversion to 3,17-diketosteroids or 20-HMP, is widely applied [12, 25, 26]. Recent analysis has demonstrated that from over 70 steroid-producing *Mycolicibacterium* strains obtained in the last two decades, 31 strains were characterized in a cyclodextrin-containing medium [27]. It is worthy to note in the present method the application of HPbCD provided more selective final ratio 2.45:1 of 24-NCED to 20-HMP (molar yields 52.0% and 21.2%, respectively) in comparison with 1.51:1 for the statistically methylated β-cyclodextrin (MbCD). When cholesterol is applied instead of plant sterols with C24-branched side chain (Fig. 2), the side chain degradation and the production of C_{22} and C_{23} steroids by the *M. neoaurum* Δ*fadA5* mutants slowed down.

1.4 Overview of Steroid Product Isolation and Purification

The general scheme of the isolation and purification of 24-NCED and 20-HMP is presented in Fig. 6. The steroid products are extracted from the supernatant of aqueous broth with water-immiscible organic solvent and purified by stepwise crystallization and re-crystallizations to the desired quality (90–95%). In order to avoid cross-contamination of 24-NCED and 20-HMP, the latter is selectively precipitated with calcium chloride [28]. The biotransformation and isolation methods allow obtaining 2.74 ± 0.085 g of

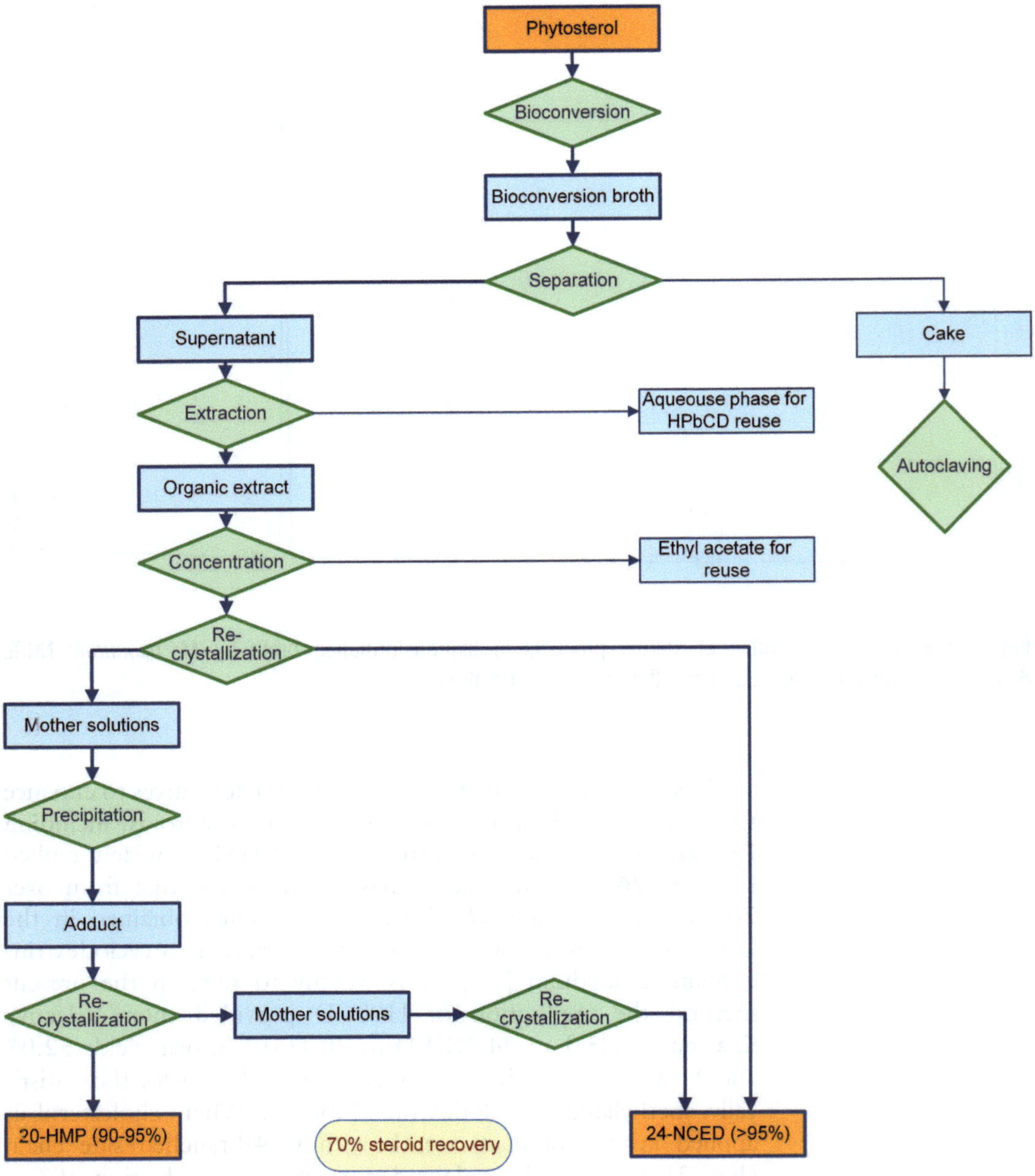

Fig. 6 Flowchart of 24-NCED and 20-HMP selective purification from biotransformation broth

24-NCED and 1.42 ± 0.085 g of 20-HMP crystalline products from 1 L of bioconversion broth (10 g phytosterol). This is the highest productivity for 24-NCED based on the phytosterol biotransformation ever reported.

2 Materials

2.1 Bacterial Strains and Nutrient Media

1. Parent bacterial strain *Mycolicibacterium* sp. NRRL B-3805 [3, 6].
2. *Mycolicibacterium* sp. NRRL B-3805Δ*kstD* (inactivated *kstD*) [23].
3. *Mycolicibacterium neoaurum* VKM Ac-1815D [5, 29].
4. *E. coli* DH5α (Novagen).
5. Nutrient medium MYCA: 10 g/L of glycerol, 1.5 g/L of $(NH_4)_2HPO_4$, 0.55 g/L of $K_2HPO_4 \times 3H_2O$, 0.5 g/L of KH_2PO_4, 0.2 g/L of $MgSO_4 \times 7H_2O$, 0.005 g/L of $FeSO_4 \times 7H_2O$, 0.002 g/L of $ZnSO_4 \times 7H_2O$, and 20 g/L of agar. Adjust to pH 7.1 with 2N KOH if necessary. Sterilize in autoclave at 120 °C for 30 min. (*see* **Note 1**).
6. Nutrient medium MYCB: the same as MYCA without agar, supplemented with 1.0 g/L of Tween 80 (*see* **Note 1**).
7. Nutrient medium MYC-EC: the same as MYCB, but supplemented with 15 g/L of Tween 80 and 15 g/L of glycine (*see* **Note 1**).
8. LB agar: 10 g/L of tryptone, 5 g/L yeast extract, 10 g/L of NaCl, and 15 g/L of agar. Adjust the pH to 7.0 with 5 N NaOH. Sterilize in autoclave at 120 °C for 30 min [30].

2.2 Genomic DNA Isolation, Recombinant Plasmid Construction, and Gene Knockout

1. Primer fp1: 5′-TTAGGTACCACAAAGCACCACTGCGC-3′.
2. Primer rp1: 5′-TTCGAGCTCCAGATCTTCTGTCCG-3′.
3. Primer fp2: 5′- TTATTTCATATGCAGAGATCTCCTTGCT CAAATC-3′.
4. Primer rp2: 5′-TTAAGCTTGCGGTCTCAGATCCTTTCG-3′.
5. pEX18Tc suicide plasmid [31].
6. Km^R cassette from pUC4K plasmid [32, 33].
7. Q5® High-Fidelity DNA Polymerase (New England Biolabs).
8. *SmaI* restriction endonuclease.
9. SalI restriction endonuclease.
10. Klenow DNA polymerase.
11. T4 DNA ligase.
12. QIAprep Spin Miniprep Kit (Qiagen).
13. QIAquick gel extraction kit (Qiagen).
14. QIAquick PCR purification kits (Qiagen).
15. Glycerol solution in deionized water (10% v/v), sterilize in autoclave at 120 °C for 30 min.
16. Kanamycin stock solution 50 mg/mL.

17. Tetracycline stock solution 10 mg/mL.
18. Sucrose stock solution 400 g/L.
19. Lysozyme stock solution 20 mg/mL.
20. SDS stock solution 10%.
21. Proteinase K stock solution 20 mg/mL.
22. DNase-free RNase A stock solution 10 mg/mL.
23. TE buffer: 10 mM Tris–HCl and 1 mM EDTA, pH 8.0.
24. Phenol–chloroform–isoamyl alcohol mixture (25:24:1, v/v).
25. Chloroform–isoamyl alcohol mixture (24:1, v/v).
26. Automatic pipettes set.
27. Spectrophotometer for 600 nm wavelength and 10 mm cuvettes.
28. Microvolume UV-Vis spectrophotometer, e.g., NanoDrop OneC (Thermo Scientific).
29. Centrifuge for microtubes, 13,000 × *g*.
30. Cyclomixer (vortex).
31. PCR amplificator.
32. Thermoblock for microtubes.
33. Electroporation apparatus Gene Pulser (Bio-Rad) with 2 mm electrode gap cuvettes.
34. Horizontal electrophoresis chamber and power supply.
35. Low temperature freezer for −70 °C.

2.3 Phytosterol Biotransformation

1. Biotransformation medium BT: 1 g/L of KH_2PO_4, 5.0 g/L of $K_2HPO_4 \times 3H_2O$, 3 g/L of $(NH_4)_2SO_4$, 5 g/L of glycerol, 0.25 g/L of urea, 0.2 g/L of $MgSO_4 \times 7H_2O$, 0.01 g/L of $FeSO_4 \times 7H_2O$, 0.002 g/L of $ZnSO_4 \times 7H_2O$, 0.1 g/L of Tween 80, 62 g/L of HPbCD (Cavasol W7 HP, Wacker Chemie AG, Germany), and 10 g/L of PS, where PS is a high-grade commercial phytosterol containing ≥95% of convertible sterols (e.g., Arboris). Sterilize in autoclave at 110 °C for 30 min. (*see* **Note 1**).
2. 2N KOH sterile solution in distilled water.
3. 2N HCl sterile solution in distilled water.
4. Sterile D-fructose feed solution 500 g/L in distilled water (sterilize in autoclave at 110 °C for 30 min); approx. 25 mL of the solution is required for 1 L BT medium.
5. Sterile distilled water.
6. pH indicator stripes.
7. Shake flasks (1 L) with well-aerating plugs.
8. Orbital shaker incubator with 50 mm stroke for 1 L flasks.

9. pH meter.
10. Analytical balance.
11. HPLC equipment.
12. Laboratory centrifuge for 15 mL centrifuge tubes, 2000–10,000 × *g*.
13. Laboratory centrifuge for 100–250 mL bottles, 6000–10,000 × *g*.

2.4 Steroid Product Isolation and Purification

1. 1N HCl.
2. Ethyl acetate, research grade.
3. Isopropanol, research grade.
4. Distilled water.
5. Sodium sulfate, anhydrous.
6. Hot solution of $CaCl_2$ in isopropanol (prepared shortly before use): Dissolve 2.8 g of anhydrous $CaCl_2$ in 16 mL of isopropyl alcohol under agitation and heating at 60–70 °C. The given amount is for 1 L of bioconversion broth. The solution can be stored for a few hours.
7. pH meter.
8. Analytical balance.
9. HPLC equipment.
10. Extraction reactor: stirring flask or reactor with back cooler for extraction with organic solvents.
11. Vacuum rotary evaporator.
12. Hot plate magnetic stirrer.
13. Separation funnel.
14. Fritted glass (fine) filtration funnel and Bunsen flask.
15. Vacuum pump.
16. Drying oven or vacuum drying oven.

2.5 Steroid Analyses

1. HPLC equipment: binary gradient pump, column oven, UV absorbance detector, autosampler, computer, and integration software (Agilent Infinity 1260 or equivalent).
2. HPLC column Symmetry C_{18}, 5 μm, 250 × 4.6 mm (Waters or equivalent).
3. Guard column Symmetry C_{18}, 20 × 3.9 mm (Waters or equivalent).
4. Ultrasonic bath or 2 L vacuum filtration assembly with 0.45 μm nylon membranes.
5. Diluent solution: mix equal volumes of ACN (HPLC grade) and HPLC-grade deionized water (ASTM Type I), add 0.05

volume of trifluoroacetic acid (TFA), warm to an ambient temperature, and degas by ether vacuum filtration through 0.45 μm filter or sonication.

6. Mobile phase A: mix 60 volumes of ACN (HPLC gradient grade), 40 volumes of deionized water (ASTM Type I), and 0.05 volumes of HPLC grade TFA, warm to an ambient temperature, and degas by sonication or vacuum filtration.
7. Mobile phase B: add 0.05 volumes of TFA (HPLC grade) to 100 volumes of ACN (HPLC gradient grade), mix, and degas.
8. Standard solutions of steroids: weigh a precisely known amount (about 10 mg) of analytical standards of AD and 20-HMP (e.g., Steraloids) using analytical balance, dissolve each separately in the diluent solution in a 100 mL class A volumetric flask, and bring volume to the mark. Calculate the accurate concentrations of steroids in the solutions (about 0.1 mg/mL).

3 Methods

3.1 Maintenance of M. neoaurum Stock Cultures

Subculture *M. neoaurum* on agar slants or Petri dishes with MYCA at 30 °C during 3–4 days (*see* **Note 2**).

3.2 Genomic DNA Isolation from M. neoaurum

Purify the total DNA from *M. neoaurum* parental strain using one of the previously published specialized protocols [23, 34] or as follows.

1. Prepare sterile (120 °C, 30 min) MYCB and MYC-EC dispensed by 5 mL on 50 mL test tubes.
2. Prepare the first-generation culture: inoculate 50 mL MYCB with one loop of *M. neoaurum* from fresh culture grown on MYCA. Cultivate in orbital shaker incubator at 200 rpm, 30 °C for 48 h.
3. Prepare the second-generation culture in early stationary phase: inoculate 0.5 mL of the first-generation culture to 5 mL of MYC-EC and cultivate in the same conditions for 24 h.
4. Harvest the cells from 5 mL broth at 8000 × *g* for 10 min and suspend in 720 μL of TE buffer pH 8.0.
5. Add lysozyme stock solution to a final concentration of 1 mg/mL and incubate for 30 min at 37 °C.
6. Add SDS (final concentration 1%) and proteinase K (final concentration 400 μg/mL). Incubate for 1 h at 56 °C.
7. Add DNase-free RNase A (final concentration 50 μg/mL) and incubate at 37 °C for 30 min.

8. Extract the obtained lysate sequentially with equal volumes of phenol–chloroform–isoamyl alcohol and chloroform–isoamyl alcohol mixtures until the precipitate formation stops. After each extraction transfer the upper aqueous phase containing DNA to new test tube.
9. Precipitate DNA from the extracted aqueous phase by adding 0.1 volume of 3.0 M potassium acetate pH 5.3 and 0.7 volume of isopropanol at a room temperature for 15 min.
10. Wash the precipitate twice by 70% ethanol, air-dry, and re-suspend in TE buffer pH 8.0.
11. Determine the concentration and purity of DNA on a micro-volume UV-Vis spectrophotometer (by measuring the absorbance at 260 and 280 nm).

3.3 Recombinant Plasmid Construction

1. Amplify by PCR the DNA fragment (2.58 kb) containing the *fadA5* ORF flanked with upstream and downstream genome sequences using the specific primers fp1 and rp1, and total DNA of the parental strain as a template.
2. Clone the resulting DNA fragment into the pEX18Tc suicide vector.
3. To obtain a deletion within the *fadA5* ORF, digest the recombinant plasmid with *SmaI* restriction endonuclease according to the manufacturer's instructions. Ligate blunt ends under conditions for circular molecule formation.
4. Transform competent *E. coli* DH5α cells with the ligation mixture and select plasmid-containing clones on plates with LB agar supplemented with 10 μg/mL tetracycline.
5. Analyze the selected transformants by PCR, and detect the presence of a 282 bp deletion by agarose gel electrophoresis.
6. Isolate and purify the desired pEX::fadA5 plasmid.
7. Digest pUC4K plasmid with *SalI* and purify a DNA fragment containing 1.25 kb Km^R cassette by agarose gel electrophoresis and gel extraction kit. Blunt the sticky ends by Klenow polymerase and ligate with *SmaI*-digested pEX::fadA5 plasmid.
8. Transform competent *E. coli* DH5α cells with the ligation mixture and select plasmid-containing clones on the LB agar plates supplemented with 20 μg/mL kanamycin.
9. Purify the desired pEX::fadA5-Km plasmid with the plasmid purification kit, using deionized water for DNA elution.

3.4 M. neoaurum Electrocompetent Cell Preparation

1. Prepare sterile (120 °C, 30 min) MYCB dispensed by 5 mL on 50 mL test tubes.
2. Prepare sterile (120 °C, 30 min) MYC-EC dispensed by 50 mL on 1 liter shake flasks.

3. Inoculate 5 mL MYCB with one loop of *M. neoaurum* from agar slant. Cultivate in orbital shaker incubator at 200 rpm, 30 °C for 48 h.
4. Transfer 1 mL of the resulting culture to 5 mL of MYC-EC and cultivate in the same conditions until OD_{600} 0.8–1.0.
5. Cool down the culture in an ice bath for 30 min and harvest on centrifuge at 5000 × *g* for 20 min at 4 °C. Wash the cell pellet twice with 25 mL of chilled deionized water (centrifuge in the same conditions), once with a 10% solution of glycerol in deionized water (centrifuge 5000 × *g* for 30 min at 4 °C), and suspend in 1 mL of the same solution.
6. Dispense 100–200 μL on sterile cryovials or microtubes and store at −70 °C.

3.5 Electroporation and Gene Knockout

1. Add 1–5 μg of the pEX::fadA5-Km plasmid DNA in deionized water (not more than 5 μL) to 200 μL of *M. neoaurum* electro-competent cells, incubate in an ice bath for 5 min, and transfer into pre-chilled 2 mm electroporation cuvettes.
2. Undergo the mixture to an electrical impulse (2.5 kV) twice. Then add 2 mL of MYCB medium and incubate for 7–8 h for recovery at 28 °C with aeration.
3. Spread out the mixture on MYCA agar supplemented with 10 μg/mL kanamycin and incubate at 30 °C for 5–7 days. Most of the clones are merodiploids resulting from a single crossover ($Km^RTc^RSuc^S$ phenotype).
4. Pick two to five merodiploid clones, transfer each to test tube with 3 mL MYCB without antibiotics and grow for 2 days with aeration on shaker.
5. Prepare serial dilutions of the cultures and plate on MYCA agar with 10 μg/mL kanamycin to get individual colonies.
6. Make replica plates Km^R clones on MYCA supplemented with 5 μg/mL tetracycline or 6% sucrose and perform a counter-selection of the mutant clones with $Km^RTc^SSuc^R$ phenotype resulting from double crossover.
7. Confirm allelic exchange by PCR using the primers fp2 and rp2. Approximately one of ten clones is sensitive to Tc and grew on a medium with sucrose, indicating the presence of a mutation (Fig. 7).

3.6 Preparation of M. neoaurum Inoculum for Biotransformation

1. Prepare sterile (120 °C, 30 min) MYCB dispensed by 100 mL to 1 liter shake flasks. The total volume of the medium should consist at least 11% of the desired bioconversion scale.
2. Prepare the first-generation seed culture: use one slant to inoculate 75 mL of the MYCB in one 1 liter shake flask. Suspend well whole *Mycobacterium* sp. growth from one agar slant in

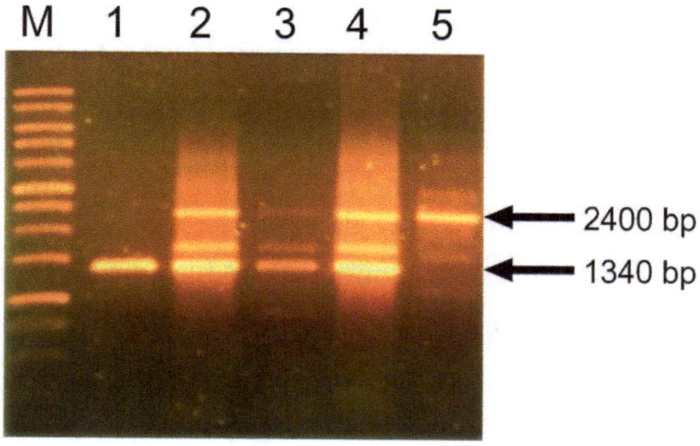

Fig. 7 PCR analysis of clones obtained after electroporation of *M. neoaurum* NRRL B-3805 cells with pEX::fadA5-Km plasmid. From merodiploid clones two fragments amplified, one corresponding to wild-type *fadA5* allele (~1340 bp) and the other (~2400 bp) to the gene with a deletion and insertion of the Km^R cassette. Lanes: M, DNA markers; 1, 1340 bp amplified fragment from total DNA of the wild-type strain; 2–4, amplified fragments obtained from merodiploid cells; 5, ~2400 bp amplified fragment obtained from the mutant Δ*fadA5* strain

3–4 mL of the sterile medium by a pipette and transfer to a shake flask. Incubate on orbital shaker (200 rpm) at 30 °C for 40–48 h.

3. Prepare the second-generation liquid seed culture: inoculate MYCB fresh medium with 10% (v/v) of the first-generation culture and incubate on orbital shaker (200 rpm) at 30 °C for 24–26 h. Keep in mind that amount of the second-generation liquid seed culture should be at least 10% (v/v) from the desirable bioconversion scale (*see* **Note 3**).

3.7 Phytosterol Biotransformation

The biotransformation and product purification operations described in Subheadings 3.3, 3.4, 3.5, and 3.6 are based on the total amount of 1 L of biotransformation broth in 10 shake flasks. For other scale, change the amounts of materials proportionally.

1. Prepare the necessary amount of BT in 90% of desired working volume (bioconversion scale, e.g., 1 L) and dispense by 90 mL on 1 liter shake flasks. Add weighted portions of PS individually to each flask, based on 1 g of pure convertible sterols. Autoclave at 110 °C for 30 min and cool down. Shake well and sonicate the sterile BT medium on ultrasonic bath (~1 min) before the inoculation.
2. Inoculate the BC in each shake flask with 10 mL of fresh second-generation *Mycobacterium* sp. inoculum.
3. Incubate shake flasks aerobically on an orbital shaker at 30 °C and 200 rpm.

4. Every 24 h add to each flask 0.6 mL of 500 g/L D-fructose solution until 120 h of incubation.
5. Maintain pH of the broth in the range of 6.5–7.4 until 120 h of incubation. Every 24 h check pH with precise indicator strips and correct to 7.2–7.4 with 2N KOH (600–200 μL per flask per day).
6. Continue the biotransformation process for 11–12 days until 24-NCED accumulation stops. Take samples of bioconversion broth at 24–72 h intervals and follow the steroid titers by HPLC (*see* Subheading 3.9; *see* **Note 4**).

3.8 Separation of Biomass

1. Distribute all the broth to a proper scale centrifuge bottles and spin on a laboratory centrifuge at 6000–10,000 × *g* and 20–25 °C during 1 h.
2. Decant the supernatant immediately and collect.
3. Inactivate the biomass in autoclave (120 °C, 30–45 min) and discard.

3.9 Extraction of Steroids

1. Load the supernatant of bioconversion broth to a proper scale extraction reactor. Titrate with 1N HCl to pH 2.5–3.0 under vigorous stirring (*see* **Note 5**).
2. Add 0.6 volumes of ethyl acetate and agitate vigorously during 1 h for organic solvent extraction.
3. Keep the mixture without agitation until a good separation of the phases is obtained.
4. Collect the upper (organic) phase I. Place the aqueous phase back to the reactor.
5. Repeat the extraction two times more as described, combining the organic phases I to III. Discard the aqueous phase (*see* **Note 6**).

3.10 Purification of 24-NCED and 20-HMP

1. Concentrate the organic phases I–III on a vacuum rotary evaporator at water bath temperature 40–50 °C until ~10% of initial volume (approx. 170–190 mL by extracting supernatant from 1 L broth).
2. Wash the concentrate into an extraction funnel twice with equal volumes of warm distilled water under periodical shaking for 30 min. Keep the mixture still until a complete separation of the phases. Collect the upper (organic) phase IV. Discard the aqueous phase.
3. Dry the organic phase IV over anhydrous sodium sulfate and vacuum-filter it through fine fritted glass filter. Wash the drying agent on the filter with ethyl acetate.
4. Concentrate the organic phase IV on a vacuum rotary evaporator until approx. 10% volume of the biotransformation broth.

5. Pour the hot concentrate to a glass beaker immediately, cool down to room temperature, cover, and place in a refrigerator for crystallization overnight (≥8 h) at 4 °C.
6. Suspend the formed steroid crystals with glass rod and separate them on fritted glass filter under vacuum. Collect the mother solution I (filtrate).
7. Keep the crystals on the filter under fume hood for drying for 2–3 h. Gather 4.1–4.4 g of yellowish steroid crystals I, containing 70–74% of 24-NCED and 16–18% of 20-HMP.
8. Dissolve the steroid crystals I in a minimal volume of hot ethyl acetate and recrystallize at 4 °C (≥2 h), and gather the crystals II by vacuum filtration and re-crystallization once more. Dry the crystals obtained in a drying oven at 60 °C until constant weight (about 2 h). Get 1.9–2.1 g of colorless crystals III. Collect mother solutions II and III.
9. Check purity of the 24-NCED crystals III by HPLC (*see* Subheading 3.7). Normally it should be ≥95%. If necessary, repeat re-crystallization from a minimal amount of ethyl acetate as described in **step 8**.
10. Combine the mother solutions I–III for isolation of 20-HMP. Dry over anhydrous sodium sulfate and vacuum-filter them through fritted glass filter.
11. Gradually add the $CaCl_2$ solution in isopropanol to the mother solutions I–III under agitation on a magnetic stirrer. Continue agitation for 2 h at 20–25 °C.
12. Separate the white amorphous $CaCl_2$–20-HMP precipitate formed by filtration, wash on the filter with ethyl acetate, and dry. Collect the filtrate I.
13. Suspend the $CaCl_2$–20-HMP precipitate in 50 mL of hot distilled water and agitate on a magnetic stirrer near the boiling point until the adduct will hydrolyze. Collect the formed sediment of 20-HMP by filtration, wash it with distilled water, and dry at 80 °C. Get 2.01–2.10 g of a colorless 20-HMP crystals IV (*see* **Note 7**).
14. Check purity of the 20-HMP preparation (crystals IV) by HPLC (*see* Subheading 3.7). Normally it should be ≥90%.
15. Dissolve the 20-HMP crystals IV in the minimal amount of hot isopropanol and re-crystallize at 4 °C (≥4 h). Gather and dry the crystals obtained on the filter in a drying oven at 60 °C until constant weight (about 2 h). Get 1.68–1.84 g of 20-HMP crystals V.
16. Check purity of the 20-HMP preparation (crystals V) by HPLC (*see* Subheading 3.7). Normally it should be ≥95%.
17. Pour the filtrate I obtained at **step 12** after 20-HMP precipitation to a separation funnel and wash twice with equal volumes

of distilled water. Collect and dry the organic phase over anhydrous sodium sulfate, and clear it by vacuum filtration. Get the filtrate II.

18. Evaporate the filtrate II under vacuum at 50 °C until crystallization begins (~90–100 mL). Continue crystallization overnight (≥8 h) at 4 °C. Gather the crystals by vacuum filtration and dry on filter. Get 24-NCED crystals VI with a purity of 85–90%.
19. Dissolve the 24-NCED crystals VI in the minimal amount of hot ethyl acetate and re-crystallize at 4 °C (≥4 h). Gather and dry the crystals obtained on the filter in a drying oven at 60 °C until constant weight (about 2 h). Get 0.64–0.72 g of an additional portion of 24-NCED crystals VII.
20. Check purity of the 24-NCED crystals VII by HPLC (*see* Subheading 3.7). Normally it should be ≥95%.
21. Combine the 24-NCED crystals III, obtained at **step 8** with 24-NCED crystals VII to get a total of 2.65–2.82 g of 24-NCED of approx. 95% purity. If necessary, repeat re-crystallization from a minimal amount of ethyl acetate as described in **step 19**.

3.11 HPLC Analysis of Steroids

1. Prepare bioconversion broth samples as follows: shake the sample on cyclomixer (in the case of frozen samples unfreeze and sonicate first), take precise aliquot immediately and dilute 25 times with the diluent solution, close tightly in a centrifuge tube, mix and incubate for 3 h at 50 °C, and cool down. Centrifuge the dilution for 15 min at room temperature and 4000–7000 × *g*. Gently take out the tubes and keep them in vertical position. Take 1.5 mL of the supernatant with a pipette and transfer to an HPLC vial.
2. Run and equilibrate the HPLC system in the following conditions: column temperature 50 °C, mobile phase flow rate 1 mL/min, 100% mobile phase A, detection wavelength 240 nm, and acquisition time 26 min. Set the following conditions for sample elution:

Segment	Time, min	Solvent ratio (acetonitrile), % v/v	Channel ratio, % v/v	
			A	B
1 Isocratic	0–5	60	100	0
2 Linear gradient	5–15	60→84	100→40	0→60
3 Linear gradient	15–20	84→100	40→0	60→100
4 Isocratic	20–21	100	0	100
5 Equilibration	21–26	0	100	0

3. For quantification of steroids, perform multilevel calibration of the steroid signals (peak area) according to the external standard technique. Inject sequentially 3–5 increasing portions (e.g., 5, 10, and 20 μL) of each standard solution, build the calibration curves considering the concentrations of pure steroid, and calculate the calibration factors.
4. Inject 20 μL of the prepared broth sample to the HPLC system. Register and integrate the chromatogram. Calculate the steroid titers using the appropriate calibration and sample dilution factors. Use the calibration for 20-HMP to calculate 24-NCED weight titer with correction factor × 1.036. *See* **Note 8** for the estimated reference retention times.
5. For purity estimation of the crystalline steroid product, make a test solution of the obtained preparation in the same way as described for the standard solutions (*see* Subheading 2.5), in duplicate or triplicate if necessary, and inject (20 μL). Calculate the steroid content in the preparation as described above.

4 Notes

1. For the media MYCA, MYCB and BT, dissolve the sulfates separately from other compounds in small amount of water and then mix.
2. *M. neoaurum* slants can be stored at +4 °C within 1 month. The first-generation liquid cultures can be stored at +4 °C for 24 h (the second cultivation should be prolonged by ~2–4 h in this case).
3. OD_{600} of the second-generation *M. neoaurum* culture should be 8–10 units and wet biomass 20–30 g/L.
4. Normally the main product yields in the broth comprised 4.13 ± 0.35 g/L (molar yield ~52%) for 24-NCED and 1.69 ± 0.12 g/L (molar yield ~21%) for 20-HMP with approx. 0.12 g/L AD. *See* Fig. 3 for reference of the HPLC profile.
5. Operate with organic solvents under fume hood and comply with fire safety rules.
6. Normally the steroid residue in the aqueous phase after extraction is less than 0.2–0.4%. The HPbCD from aqueous phases can be re-used after vacuum concentration in large-scale operations.
7. As 20-HMP is a commercially available steroid, **steps 12–15** in Subheading 3.6 can be omitted and $CaCl_2$–20-HMP precipitate can be discarded.

8. Estimated reference retention times (Rt) are as follows:

Steroid	Rt, min
ADD	5.0
Testosterone (T)	5.33
AD	6.19
20-HMPD	8.22
20-HMP	10.95
24-NCED	14.3
Methyl-3-oxopregna-4,17(20)-diene-20-carboxylate (MOPDC)	16.4

Acknowledgments

This work was supported by Russian Science Foundation under Grant No. 21-64-00024.

References

1. Donova MV (2007) Transformation of steroids by actinobacteria: a review. Appl Biochem Microbiol 43:1–14. https://doi.org/10.1134/S0003683807010012
2. Fernández-Cabezón L, Galán B, García JL (2018) New insights on steroid biotechnology. Front Microbiol 9:958. https://doi.org/10.3389/fmicb.2018.00958
3. Marsheck WJ, Kraychy S, Muir RD (1972) Microbial degradation of sterols. Appl Microbiol 23:72–77. https://doi.org/10.1128/am.23.1.72-77.1972
4. Andor A, Jekkel A, Hopwood DA, Jeanplong F, Ilkőy É, Kónya A, Kurucz I, Ambrus G (2006) Generation of useful insertionally blocked sterol degradation pathway mutants of fast-growing *Mycobacteria* and cloning, characterization, and expression of the terminal oxygenase of the 3-ketosteroid 9-α-hydroxylase in *Mycobacterium smegmatis* mc 2 155. Appl Environ Microbiol 72:6554–6559. https://doi.org/10.1128/AEM.00941-06
5. Shtratnikova VY, Bragin EY, Dovbnya DV, Pekov YA, Schelkunov MI, Strizhov N, Ivashina TV, Ashapkin VV, Donova MV (2014) Complete genome sequence of sterol-transforming *Mycobacterium neoaurum* strain VKM Ac-1815D. Genome Announc 2: e01177–e01113. https://doi.org/10.1128/genomeA.01177-13
6. Rodríguez-García A, Fernández-Alegre E, Morales A, Sola-Landa A, Lorraine J, Macdonald S, Dovbnya D, Smith MCM, Donova M, Barreiro C (2016) Complete genome sequence of '*Mycobacterium neoaurum*' NRRL B-3805, an androstenedione (AD) producer for industrial biotransformation of sterols. J Biotechnol 224:64–65. https://doi.org/10.1016/j.jbiotec.2016.03.021
7. Xu L-Q, Liu Y-J, Yao K, Liu H-H, Tao X-Y, Wang F-Q, Wei D-Z (2016) Unraveling and engineering the production of 23,24-bisnorcholenic steroids in sterol metabolism. Sci Rep 6:21928. https://doi.org/10.1038/srep21928
8. Peng H, Wang Y, Jiang K, Chen X, Zhang W, Zhang Y, Deng Z, Qu X (2021) A dual role reductase from phytosterols catabolism enables the efficient production of valuable steroid precursors. Angew Chem Int Ed 60:5414–5420. https://doi.org/10.1002/anie.202015462
9. Imada Y, Takahashi K (1978) A process for producing steroidal alcohols. Patent Eur 0001622

10. Wowcha MG, Antosz FJ, Beaton JM, Garcia AB, Kominek LA (1980) Process for preparing 9α-OH BN acid methyl ester. US Patent 4214051
11. Wang Y, Li X, Feng J, Wu Q, Zhu D, Ma Y. (2017) The mycobacteria of one plant of ketone of 4 alkene of biosynthesis 20 hydroxyl, 23,24 dinorchola 3 and synthetic method. Patent CN 106854630
12. Hu Y, Wang D, Wang X, Wei D (2020) A recycled batch biotransformation strategy for 22-hydroxy-23,24-bisnorchol-4-ene-3-one production from high concentration of phytosterols by mycobacterial resting cells. Biotechnol Lett 42:2589–2594. https://doi.org/10.1007/s10529-020-02991-1
13. Jiu J, Marsheck WJ (1976) 3-Oxo-pregna-4,17(20)-dien-20-carboxylic acid and esters. US Patent 3994933
14. Toró A, Ambrus G (1990) Oxidative decarboxylation of 17(20)-dehydro-23,24-dinorcholanoic acids. Tetrahedron Lett 31:3475–3476. https://doi.org/10.1016/S0040-4039(00)97426-4
15. Toró A, Ambrus G (1992) Synthesis of 17-α-hydroxy-20-oxo-pregnanes from 17(20)-dehydro-23,24-dinorcholan-22-oic acids. Tetrahedron Lett 33:5265–5266. https://doi.org/10.1016/S0040-4039(00)79150-7
16. Gavagnin M, Ungur N, Mollo E, Templado J, Cimino G (2002) Structure and synthesis of a progesterone homologue from the skin of the dorid nudibranch *Aldisa smaragdina*. Eur J Org Chem 2002:1500–1504. https://doi.org/10.1002/1099-0690(200205)2002:9<1500::AID-EJOC1500>3.0.CO;2-D
17. Schwarz V, Pihera P, Protiva J, Mickova R (1984) Biodegradation of cholesterol by a mutant of the *Mycobacterium* species. Collect Czechoslov Chem Commun 49:2713–2719
18. Imada Y, Takahashi K (1981) Steroid and its preparation. Patent Japan JPS56120698
19. Shtratnikova VY, Schelkunov MI, Fokina VV, Pekov YA, Ivashina T, Donova MV (2016) Genome-wide bioinformatics analysis of steroid metabolism-associated genes in *Nocardioides simplex* VKM Ac-2033D. Curr Genet 62:643–656. https://doi.org/10.1007/s00294-016-0568-4
20. Nesbitt NM, Yang X, Fontán P, Kolesnikova I, Smith I, Sampson NS, Dubnau E (2010) A thiolase of *Mycobacterium tuberculosis* is required for virulence and production of androstenedione and androstadienedione from cholesterol. Infect Immun 78:275–282. https://doi.org/10.1128/IAI.00893-09
21. Xiong L-B, Liu H-H, Xu L-Q, Sun W-J, Wang F-Q, Wei D-Z (2017) Improving the production of 22-hydroxy-23,24-bisnorchol-4-ene-3-one from sterols in *Mycobacterium neoaurum* by increasing cell permeability and modifying multiple genes. Microb Cell Factories 16:89. https://doi.org/10.1186/s12934-017-0705-x
22. Dovbnya DV, Ivashina TV, Khomutov SM, Shutov AA, Donova MV (2022) Recombinant mycobacterial producer strains and their use to produce 22-functionalised steroids. Patent RU 2767610
23. Loraine JK, Smith MCM (2017) Genetic techniques for manipulation of the phytosterol biotransformation strain *Mycobacterium neoaurum* NRRL B-3805. In: Barredo J-L, Herráiz I (eds) Microbial Steroids. Springer New York, New York, pp 93–108. https://doi.org/10.1007/978-1-4939-7183-1_7
24. Ivashina TV, Nikolayeva VM, Dovbnya DV, Donova MV (2012) Cholesterol oxidase ChoD is not a critical enzyme accounting for oxidation of sterols to 3-keto-4-ene steroids in fast-growing *Mycobacterium* sp. VKM Ac-1815D. J Steroid Biochem Mol Biol 129:47–53. https://doi.org/10.1016/j.jsbmb.2011.09.008
25. Hesselink PGM, van Vliet S, de Vries H, Witholt B (1989) Optimization of steroid side chain cleavage by *Mycobacterium* sp. in the presence of cyclodextrins. Enz Microb Technol 11:398–404. https://doi.org/10.1016/0141-0229(89)90133-6
26. Zhang L, Wang M, Shen Y, Ma Y, Luo J (2009) Improvement of steroid biotransformation with hydroxypropyl-β-cyclodextrin induced complexation. Appl Biochem Biotechnol 159:642–654. https://doi.org/10.1007/s12010-008-8499-2
27. Zhao A, Zhang X, Li Y, Wang Z, Lv Y, Liu J, Alam MA, Xiong W, Xu J (2021) *Mycolicibacterium* cell factory for the production of steroid-based drug intermediates. Biotechnol Adv 53:107860. https://doi.org/10.1016/j.biotechadv.2021.107860
28. Weber A, Muller R, Kurzidim J (1976) Method for isolation of 3-hydroxy steroids and 3-keto steroids. US Patent 4057541
29. Donova MV, Dovbnya DV, Kalinichenko AN, Arinbasarova AYu, Vagabova LM, Morozova ZV, Koshcheyenko KA (1997) Method of androst-4-ene-3,17-dione preparing. Patent RU 2079258
30. Sambrook J, Fritsch EF, Maniatis T (1989) Molecular cloning: a laboratory manual, 2nd edn. Cold Spring Harbor Laboratory Press, Cold Spring Harbor

31. Hoang TT, Karkhoff-Schweizer RR, Kutchma AJ, Schweizer HP (1998) A broad-host-range Flp-FRT recombination system for site-specific excision of chromosomally-located DNA sequences: application for isolation of unmarked *Pseudomonas aeruginosa* mutants. Gene 212:77–86. https://doi.org/10.1016/S0378-1119(98)00130-9
32. Vieira J, Messing J (1982) The pUC plasmids, an M13mp7-derived system for insertion mutagenesis and sequencing with synthetic universal primers. Gene 19:259–268. https://doi.org/10.1016/0378-1119(82)90015-4
33. Oka A, Sugisaki H, Takanami M (1981) Nucleotide sequence of the kanamycin resistance transposon Tn903. J Mol Biol 147:217–226. https://doi.org/10.1016/0022-2836(81)90438-1
34. Bragin EY, Shtratnikova VY, Dovbnya DV, Schelkunov MI, Pekov YA, Malakho SG, Egorova OV, Ivashina TV, Sokolov SL, Ashapkin VV, Donova MV (2013) Comparative analysis of genes encoding key steroid core oxidation enzymes in fast-growing *Mycobacterium* spp. strains. J Steroid Biochem Mol Biol 138:41–53. https://doi.org/10.1016/j.jsbmb.2013.02.016

Chapter 19

Stabilization of Enzymes by Using Thermophiles

Ana-Luisa Ribeiro, Mercedes Sánchez, Sandra Bosch, José Berenguer, and Aurelio Hidalgo

Abstract

Manufactured steroid compounds have many applications in the pharmaceutical industry. Due to the chemical complexity and chirality of steroids, there is an increasing demand for enzyme-based bioconversion processes to replace those based on chemical synthesis. In this context, thermostability of the involved enzymes is a highly desirable property as both the increased half-life of the enzyme and the enhanced solubility of substrates and products will improve the yield of the reactions. Metagenomic libraries from thermal environments are potential sources of thermostable enzymes of prokaryotic origin, but the number of expected hits could be quite low for enzymes handling substrates such as steroids, rarely found in prokaryotes. An alternative to metagenome screening is the selection of thermostable variants of well-known steroid-processing enzymes. Here we review and detail a protocol for such selection, where error-prone PCR (epPCR) is used to introduce random mutations into a gene to create a variants library for further selection of thermostable variants in the thermophile *Thermus thermophilus*. The method involves the use of folding interference vectors where the proper folding of the enzyme of interest at high temperature is linked to the folding of a reporter encoding a selectable property such as thermostable resistance to kanamycin, leading to a life-or-death selection of variants of reinforced folding independently of the activity of the enzyme.

Key words *Thermus thermophilus*, Enzyme variants library, Thermostabilization, Folding-interfering reporter

1 Introduction

Bioactive steroids are biologically and commercially relevant compounds. Manufactured steroid compounds have a vast therapeutic applicability as anti-inflammatory, diuretic, contraceptive agents, breast and prostate cancer and anti-obesity agents, and others [1]. Chemical synthesis of steroids is often difficult or not advantageous due to process complexity and low yields. Recently, bioconversion processes involving engineered microorganisms and their enzymes have emerged as efficient and cost-effective alternatives to chemical production and modification of steroid compounds

Carlos Barreiro and José-Luis Barredo (eds.), *Microbial Steroids: Methods and Protocols*, Methods in Molecular Biology, vol. 2704,
https://doi.org/10.1007/978-1-0716-3385-4_19,

[2]. Although some of these processes are well established, increasing the efficiency of the processes and identifying new potentially useful bioconversions are two goals in steroid drug production [3].

Thermozymes—thermostable enzymes that function optimally between 60 and 125 °C—have attracted a great industrial and biotechnological interest because they are better suited for harsh industrial processes. In addition to sustained catalytic activity at high temperatures, they are resistant to organic solvents, detergents, low and high pH, and other denaturants and have a lower susceptibility to proteolysis [4–7]. The low solubility of sterols and steroids in water [8] entails a need for higher temperatures during modification reactions. Therefore, thermostable variants of the steroid bioconverting enzymes used in the industry or new enzymes from thermophilic organisms capable of bioconverting steroids would be of great value.

Thermozymes can be of natural origin or obtained by selection of thermostable variants of normal enzymes. Natural thermozymes are produced by thermophiles, microorganisms that are adapted to grow at temperatures above 50 °C. Access to steroid-modifying enzymes from these microorganisms requires either culture and further selection of a producing strain, or the selection of the appropriate enzyme from a genomic or metagenomic library from a thermophile or a thermophilic environment. As few thermophiles are easily grown under laboratory conditions, metagenomics has evolved as a powerful cultivation-independent approach, which can be applied to gain access to new sterol bioconverting enzymes from difficult to grow thermophilic microorganisms. Over the past decades metagenomic libraries derived from extreme habitats have been constructed and novel enzymes with potential use in industrial applications have been found [9, 10] (Table 1). Two screening methodologies can be used on metagenomic material: the function-driven screening of expression libraries and the sequence-driven analysis based on gene homology. The first approach is the most commonly used for the identification of industrial target proteins and requires the uncovering of positive clones expressing a desired trait by either direct or indirect detection. Direct detection is achieved by the addition of dyes or chromophore-containing chemicals into the growth medium [11, 12]; the indirect approach consists of the use of host strains able to grow under selective conditions only when expressing the targeted gene product in an active form [13].

Despite several successes, metagenomic strategies typically have low rates of target identification, in most cases due to the inability of the host to transcribe and/or translate or fold properly the gene (s) and protein(s) of interest [14]. As an alternative, mesophilic enzymes can be engineered to become more thermostable. Thermal stability is a consequence of the concurrent combination of multiple factors [15–17]. When mesophilic and thermophilic

Table 1
Methods to obtain thermostable proteins

Approach		Method	Technique	Advantages	Disadvantages	References
Search for existing enzymes		Metagenomic library	Cloning, sequencing, functional expression	Potential to discover proteins that would not have been recognized by their sequences; large molecular diversity	Hits do not always meet the required demands in terms of activity, selectivity, and stability	[35]
Protein engineering	Directed evolution	Random mutagenesis	Error-prone PCR (epPCR)	High mutation rates; nearly even mutational spectrum; simplicity	Limited amino acid substitutions; accumulated deleterious mutations; polymerases bias	[36, 37]
			Mutator strains	Easy to use	Low mutation rates; uneven mutational spectrum; limited amino acid substitutions	[36]
			Physical–chemical mutagenesis	Dose-dependent mutation rates, simplicity	Low efficiency; hazardous	[36]
			Saturation mutagenesis	Simplicity; access to all amino acid repertoire; focus on specific genetic locus	Gene sequence required; limited diversity; surplus of inactive clones within simultaneous saturation libraries	[37]
		Homologous recombination	DNA shuffling StEP RACHITT	Robust; flexible	Rely on high sequence homology; low crossover rate	[37]
			Incremental truncation methods	Eliminate recombination bias; structural knowledge not necessary;	Limited to two parents; complex, laborious	

(continued)

Table 1 (continued)

Approach	Method	Technique	Advantages	Disadvantages	References
Rational design	Computational tools	Sequence–function relationship prediction	Prediction of functionally relevant residues; capable of using large volume data	Affected by incorrect annotations in databases	[38, 39]
		Structure-based molecular modeling	Sequence identities as low as 25% are usually enough to predict reliable structure models	Difficulties in accurately predicting stabilizing mutations without affecting activity or selectivity	[38]
	Site-directed mutagenesis	PCR-based site-directed mutagenesis	Allows direct modification of specific residues; simplicity	The information gathered relies entirely on the selection of mutants produced	
		Gene truncations and deletions	Valuable in mapping specific regions of functionally; may improve solubility	Requires structural and sequence knowledge	[40]
De novo design	Computational tools	Fixed backbone methods Flexible backbone methods	De novo design can be extended to 3° and 4° structure	Requires advanced sampling techniques	[41]

counterpart of an enzyme sharing >30% of sequence identity are compared, their three-dimensional structures are easily superimposable, sharing also similar architecture of their respective active centers, thus allowing the study of thermophilic adaptations [7, 17]. In many cases, thermophilic adaptation comes along with a higher packing efficiency through hydrophobic interactions and an increased density of salt bridges or charge-assisted hydrogen bonds. Often, these amino acid changes contribute to an increase in mechanical rigidity of the structure concomitant with a decrease in flexibility at low temperatures [18–20]. Most of these changes are located on surface areas, at the ends of conserved domains and in loops. A predominant accumulation of non-polar alanine and positively charged arginine is frequently observed in thermostable proteins of *Thermus thermophilus* [15]. However, it is important to stress that the general trend is that the local distribution of rigid or functionally important flexible regions in a thermophilic protein is similar to that of a mesophilic homologue, provided that the proteins are compared at their respective optimal temperatures [21, 22].

However, many studies have produced conflicting results on factors and evolutionary mechanisms that lead to increased protein thermostability. They may differ from taxon to taxon, organism to organism, and even from protein to protein within the same organism [23], making it very difficult to assemble a common strategy for enzyme thermostabilization by protein engineering.

The first approach adopted to thermostabilize relevant enzymes was rational design (Table 1), which uses molecular modeling and structural and mechanistic information derived from the comparisons commented above to predict the amino acid (s) substitution required to produce the desired changes in the protein. Actually, the difficulties of making reliable predictions imposed a change of path to the so-called "directed evolution" strategies (Table 1). This approach relies on the generation of thousands or millions of protein variants by random mutagenesis ("variants library") and further screening or selection for those that have the desired properties. In recent years, the line that separated rational design and directed evolution became blurred because most researchers prefer to use combinations of both practices to improve proteins [24].

De novo design (Table 1) is the latest methodology adopted by protein engineers to thermostabilize enzymes and it relies solely on creating any functional enzyme by specifying their amino acid sequence on synthetic genes. This is not impossible, since it is known that the primary sequence of a protein determines its structure [25]. However, despite impressive advances in computational enzyme design, the actual improvements of enzyme performance obtained by the de novo design systems are rather poor [26].

Whichever approach chosen, the main obstacle remains the screening of large mutant libraries or metagenomic libraries for variants with the desired functionality at high temperature. This can be very expensive and time-consuming in the absence of an appropriate high-throughput screening or a growth-dependent selection assay. The folding-interference approach using a thermophile as a host is an activity-independent high-throughput screening technology that selects for enzyme variants with reinforced folding at high temperatures in a life-or-death manner, decreasing the screening efforts needed by orders of magnitude [27]. In this strategy, the non-thermostable enzyme variants are fused to a thermostable protein reporter, and the resulting protein fusion is expressed in a thermophile. This way, the non-thermostable variants will not fold properly at high temperature, leading to protein aggregation and to the "interference" with the folding of the reporter. On the contrary, those variants more thermostable will fold properly at high temperature, allowing the proper folding and the activity of the reporter. As the selection method is independent of the activity of the enzyme under selection, a secondary screening will be needed among those "thermostable" variants to eliminate enzymes without activity. The most commonly used folding interference vector is pNCK (Fig. 1), a bifunctional *Escherichia coli*–*Thermus* spp. vector in which the protein of interest is expressed from the $slpA_p$ promoter fused to a thermostable kanamycin nucleotidyl transferase, conferring resistance to kanamycin a wide temperature range. With this promoter, a life-or-death strategy is

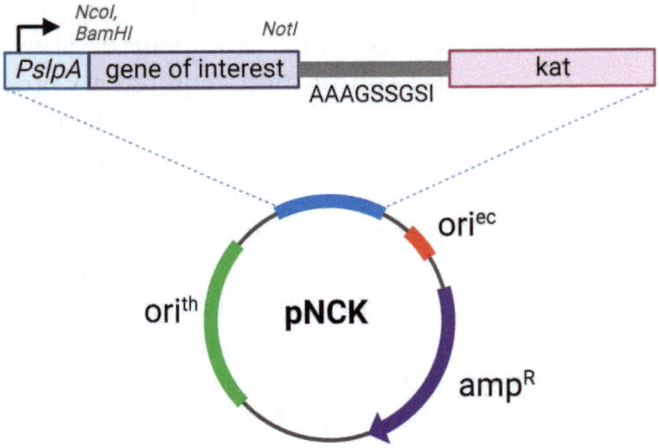

Fig. 1 Schematic structure of plasmid pNCK. The gene of interest is fused to the gene encoding a thermostable kanamycin nucleotidyl transferase variant (*kat*) through a 9-amino-acid linker (AAAGSSGSI). The promoter of the *slpA* gene (*PslpA*) is represented as a yellow bar, upstream of the gene of interest. The plasmid contains the ampicillin resistance gene (amp^R) that codes for the enzyme *β-lactamase*. The *Thermus* spp. origin of replication (ori^{th}), *E. coli* origin of replication (ori^{ec}), and a simplified cloning site for the target genes are shown. (Figure created with Biorender.com)

easily followed, allowing the selection of thermostable variants from large libraries upon transformation in high transformation-efficiency mutants of *T. thermophilus* HB27 (Fig. 2).

However, different conditions should be considered before deciding to use folding interference selection methods. The main requirement is the limitation in the size and structural complexity of the enzyme to thermostabilize, as multi-domain proteins do not easily transmit the unfolding information to the reporter through an intermediate protein domain. Also, limitations due to factors such as different codon usage with respect to the thermophile host, the putative toxicity of the protein in vivo, or the transformation efficiency could be limiting factors to implement this technology [26].

A prime decision to be made for the implementation of this method is the appropriate selection of the screening host. To the best of our knowledge, *Thermus thermophilus* has been the only thermophile for which this strategy has been used successfully for the selection of thermostable proteins and enzyme variants [27, 28]. Here we detail the optimized protocol for the folding-interfering selection of thermostable enzyme variants in this thermophile.

2 Materials

2.1 Bacterial Growth Media

1. *Thermus* Broth (TB): 8 g of tryptone, 4 g of yeast extract, and 3 g of NaCl per liter of mineral water (*see* **Note 1**), pH adjusted to 7.5 with 2 M NaOH.
2. Luria-Bertani (LB) broth: 10 g of tryptone, 5 g of yeast extract, and 5 g of NaCl per liter of water, pH adjusted to 7.0 with 2 M NaOH.
3. SOC liquid medium: 2% (w/v) tryptone, 0.5% (w/v) yeast extract, 10 mM NaCl, 2.5 mM KCl, 10 mM $MgCl_2$, and 20 mM glucose. Prepare a solution containing the first four reagents, sterilize at 121 °C, and then add sterile $MgCl_2$ and glucose.
4. LB or TB solid media: Add 1.5% of agar to the respective liquid medium.
5. If required, after autoclaving and upon cooling down, supplement media with 30 μg/mL kanamycin for LB plates or at the concentration required for selection as defined in Subheading 3.3.

2.2 Bacterial Strains

1. *Escherichia coli* DH5-α: *fhuA2 lac(del)U169 phoA glnV44 Φ80' lacZ(del)M15 gyrA96 recA1 relA1 endA1 thi-1 hsdR17.*

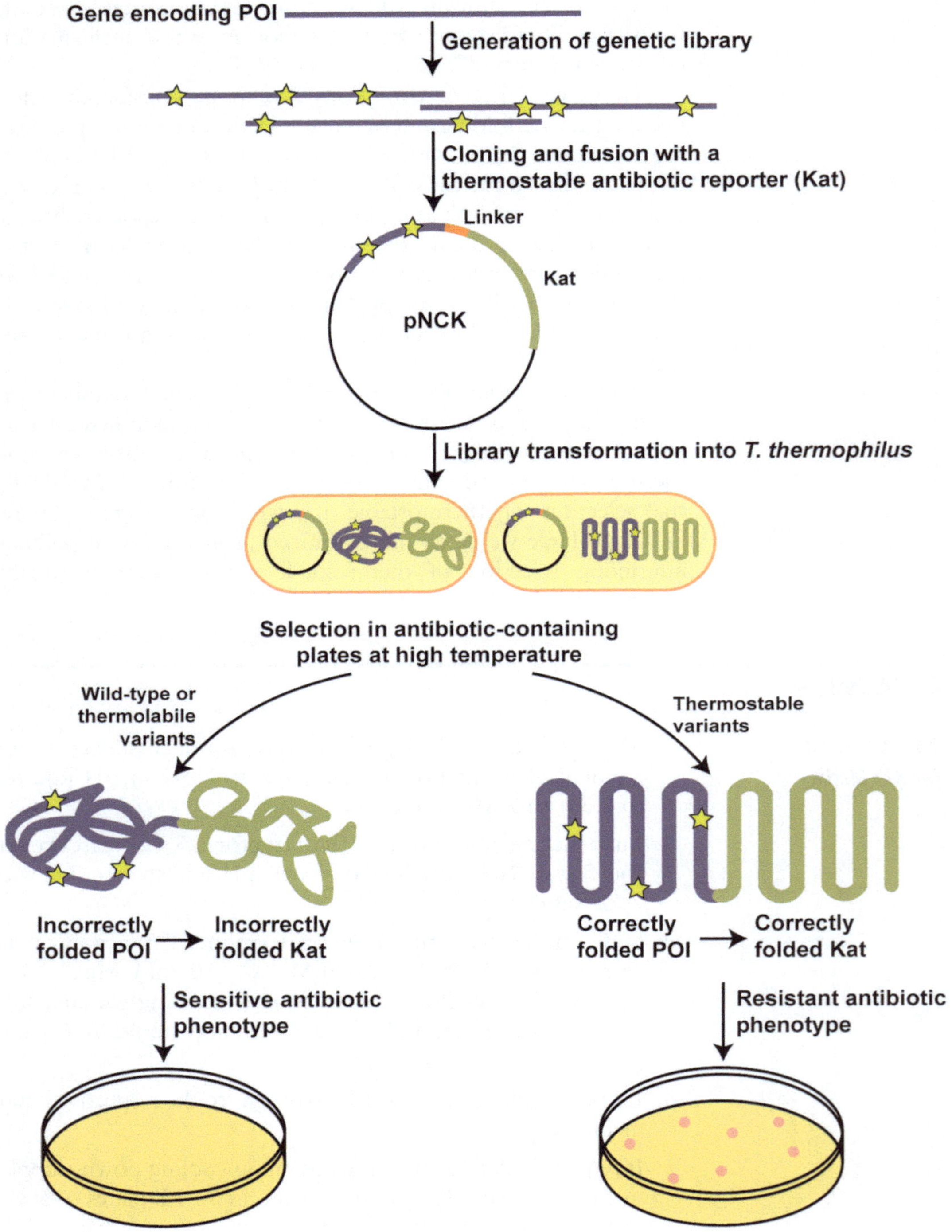

Fig. 2 Selection of thermostable proteins by folding interference. A fusion between a library of mutants of the protein of interest and the thermostable Kat protein is expressed in *T. thermophilus*. When the protein of interest is thermostable, it will fold correctly at high temperatures and as result Kat can fold properly, exhibiting a kanamycin resistance phenotype. Instead, when the protein of interest is thermolabile, its wrong folding interferes with Kat folding and a kanamycin-sensitive phenotype is obtained

2. *Thermus thermophilus* HB27 (ATCC BAA-163) derivative $HB27^{EC}$ showing high transformation efficiency due to the absence of the Argonaute protein [29] (*see* **Note 2**).

2.3 Biological and Chemical Materials

1. DNA primers for PCR and epPCR amplification (*see* **Note 3**).
2. DNA polymerase.
3. DNA template encoding for the enzyme.
4. PCR and epPCR reaction solution: DNA polymerase reaction buffer, dNTP mix, DNA polymerase, $MgCl_2$, and $MnCl_2$ (*see* **Note 4**).
5. pNCK vector (Fig. 1; *see* **Notes 5** and **6**).
6. Agarose gels, stained with 0.5 μg/mL ethidium bromide.
7. 10× TBE electrophoresis buffer: 1 M Tris base, 1 M boric acid, and 20 mM EDTA, pH 8.0.
8. DNA loading buffer and DNA ladder.
9. Gel and PCR purification kit: Wizard SV Gel and PCR Clean-Up System (Promega).
10. Plasmid extraction and purification kit: GeneJet Plasmid Mini-Prep Kit (Thermo Scientific).
11. T4 DNA ligase.
12. FastDigest *NcoI* or *BamHI* and *NotI* (Thermo Scientific).

2.4 Equipment

1. Thermocycler (e.g., Bio-Rad C1000 Touch thermocycler).
2. Agarose gel electrophoresis apparatus.
3. High-precision static incubator (±0.1 °C at 60–80 °C) (e.g., UN55, Memmert).
4. Thermomixer (e.g., Comfort, Eppendorf).
5. Orbital incubator (e.g., MaxQ 420 HP, Thermo Scientific).
6. Nanodrop (e.g., ND-1000, Thermo Scientific).
7. Thermo button for the recording of actual selection temperature (e.g., Thermotrack).

3 Methods

3.1 epPCR

1. Amplify the gene coding for the protein you want to thermostabilize by a PCR reaction. Include in the primers the restriction sites appropriate to get the fusion (*NcoI* and *BamHI* or *NotI*).
2. Purify the PCR products with Wizard SV Gel and PCR Clean-Up System and quantify the DNA concentration with a Nanodrop.

Table 2
Reaction mix for epPCR

Component	Final concentration
Nuclease-free water	Up to 100 μL
10× *Taq* buffer	1×
dNTPs	0.2 mM dATP/dGTP; 1 mM dCTP/dTTP
$MnCl_2$	0.1–0.5 mM (*see* **Note 10**)
Forward primer	0.5 μM
Reverse primer	0.5 μM
Template DNA	5 ng/100 μL
NZY*Taq* DNA polymerase	0.05 U/μL

Table 3
Example of thermocycler program for the randomization by epPCR

Cycle	Step	Temperature (°C)	Time
1	Denaturation	95	3 min
2–31	Denaturation Annealing Elongation	95 65 72	30 s 30 s 75 s
32	Elongation	72	10 min

3. Calculate the amount of DNA that is required for the desired mutation rate (*see* **Note 7**).
4. Prepare the epPCR reagents for ten reactions (*see* **Notes 8** and **9**) (Table 2).
5. Mix the sample and divide it in 10 thin-walled, 0.2 mL tubes.
6. Place the tubes in the thermocycler. Run the epPCR program (Table 3). Select the conditions according to the G+C content and melting temperature of the primers.
7. Verify the quality of the epPCR reaction by agarose gel electrophoresis.
8. Purify the reaction products with a PCR purification kit.
9. Mix the ten the products of the epPCR reactions, digest the mix with the pair of restriction enzymes specified in the forward, and reverse primers (*NcoI/BamHI/NotI*) using Thermo Scientific FastDigest enzymes.

10. Digest pNCK with the aforementioned enzymes and load the digestion product directly onto an agarose electrophoresis gel.
11. Purify both the excised band from the gel and the digested products using Wizard SV Gel and PCR Clean-Up System or any other similar kit (*see* **Note 11**).
12. Ligate pNCK and the PCR products with T4 DNA ligase. A typical reaction setup contains 50 ng of pNCK, 1:3 (plasmid/insert) molar ratio of PCR product, 0.01–0.1 U/ μL of ligase, and 1× ligase buffer. Adjust the final volume to 10 μL with Milli-Q water and incubate overnight at 16 °C.
13. Transform 50 μL of chemically competent *E. coli* DH5α with 5 μL of the ligation. After a 30 min incubation on ice, put the tubes on a thermoblock at 42 °C for 90 s and subsequently incubate again on ice for 5 min. Add 200 μL of SOC medium to the transformed cells and let them at 37 °C and 180 rpm for 3 h. Plate the transformation on LB agar plates with ampicillin (*see* **Note 12**). Other ready-to-use commercial hypercompetent *E. coli* cells can be used also (e.g., One Shot Top10 chemically *E. coli* competent cells; Invitrogen).

3.2 Mutant Library Preparation

1. Select 10 colonies of transformants and inoculate each one onto 5 mL of LB with ampicillin. Let them grow overnight at 37 °C and 180 rpm.
2. Extract the plasmid DNA with GeneJet Plasmid MiniPrep Kit.
3. Determine the concentration and quality of the plasmid preparation with a Nanodrop and sequence the extracted clones to obtain a quick overview of the mutational spectrum.
4. If the mutational spectrum and library size are satisfactory (*see* **Note 13**), pool all the colonies of the plates with LB and extract the plasmids using the MiniPrep Kit. Keep records of the total number of colonies in the library.
5. Grow *T. thermophilus* in TB medium at 65 °C and 180 rpm in an orbital incubator until it reaches exponential phase (OD_{600} 0.3–0.4).
6. Transform *T. thermophilus* HB27 by natural competence or electroporation with the variants library and the wild-type fusions (*see* **Notes 14** and **15**) by adding 0.2 μg of the library to 0.5 mL of cell culture in a 5 mL sterile tube (*see* **Note 16**).
7. Incubate for 4 h at 60 °C under strong aeration.
8. Spread the cell suspension directly onto TB agar selection plates as described below.

3.3 Selection of Thermostable Mutants

The selection of thermostable mutants is carried out using both temperature of selection and kanamycin concentration in the plates as selective pressure (*see* **Note 17**). However, prior to selection the appropriate conditions have to be determined (**steps 1–4**).

1. Prepare TB agar selective plates containing kanamycin at 20, 40, 60, 80, and 100 μg/mL of kanamycin in triplicate (*see* **Note 18**).
2. Spread 500 μL of the suspension of *T. thermophilus* cells transformed with the fusion to the wild-type protein (*see* Subheading 3.2).
3. Each replicate will be incubated at 60 °C, 65 °C, and 70 °C for 48 h in a high-precision static incubator (*see* **Note 19**).
4. Identify the lowest kanamycin concentration and incubation temperature at which the cells transformed with the fusion to the wild-type enzyme do not grow. These are the selective conditions for the primary screening of the variants library.
5. Spread 500 μL of the suspension of *T. thermophilus* transformed with the variants library on TB agar plates supplemented with the selected kanamycin concentration.
6. Incubate the plates at the selective temperature determined in point 4 for 48 h. Register the temperature inside the incubator with a Thermo-track button to control putative temperature changes along the incubation period.
7. Select several colonies for successive rounds of selection at higher temperature (*see* **Notes 20** and **21**).
8. Extract the plasmidic DNA and sequence it to determine which mutations confer thermostability.
9. Combine the most frequent mutations in a single gene and produce the protein in your standard system. Analyze the activity by your standard procedures at different temperatures to confirm the increase in thermostability.

4 Notes

1. Transformation of *T. thermophilus* depends very much on yet undefined components present in mineral water rich in bicarbonate salts. "Evian" water works well, but if not available, test more than one local commercial mineral water to determine which one gives optimum transformability. Test your tap water as well.
2. *T. thermophilus* HB27 grows in TB broth at 60–70 °C under strong aeration and plates are incubated at the same temperature in a closed container in a water-saturated atmosphere.

Colonies of *T. thermophilus* on TB medium are punctiform, show entire margin, and are flat. They have yellow pigmentation due to the presence of the natural carotenoids.

3. PCR primers should be designed to amplify the gene coding for the enzyme that is going to be thermostabilized. The gene is to be cloned in the right translation frame for expression as fusion protein to the thermostable kanamycin nucleotidyl transferase (Kat); therefore, it might be necessary to add nucleotides. The forward primer should contain the *NcoI* or *BamHI* recognition sequence, whereas the reverse primer should have *NotI* restriction site. Both primers have to be non-complementary and have similar melting temperatures. Melting temperatures can be estimated accurately using the OligoAnalyzer tool from Integrated DNA Technologies (http://www.idtdna.com/analyzer/Applications/OligoAnalyzer/) and the following website can be used to design optimal primer sequences: http://bioweb.uwlax.edu/GenWeb/Molecular/seq_anal/primer_design/primer_design.htm.
4. Optimal PCR conditions should be determined for each gene. DNA polymerase without proofreading ability, such as *Taq*, should be used for epPCR.
5. Other vectors could be used but we have more experience in our laboratory with the pNCK *E. coli*/*T. thermophilus* shuttle vector (Fig. 2). It was constructed by inserting a replication origin from *T. thermophilus* and the *slpA*$_p$*-kat* cassette conferring thermostable resistance to kanamycin [30]. The P*slpAp-kat* cassette was further modified by inserting unique cloning sites (*NcoI*, *BamHI*, and *NotI*) downstream the promoter followed by a 9-amino-acid long linker [27].
6. The linker (AAAGSSGSI) was designed to be flexible, hydrophilic, and resistant to any protease.
7. The average rate of mutation by epPCR depends on the presence of Mn^{2+} and the number of cycles of duplication. If the initial amount of template is high, it will undergo few duplications in the epPCR. On the other hand, a low amount of template will result in a greater number of duplications, and more mutations will be introduced. The ideal number of cycles depends also on the size and sequence of the template DNA. It is recommended to evaluate the PCR efficiency before performing an epPCR. In our experience, up to 20 fmol template can be used in an epPCR reaction [31, 32].
8. By running several epPCR reactions in parallel, we are decreasing the bias resulting from the errors introduced during the first cycles.

9. The *Taq* DNA polymerase fidelity is further decreased by the addition of Mn^{2+} and unequal dNTP concentrations. By varying the amount of $MnCl_2$ added to the reaction, the mutation rate can be varied from ~1 to 5 nucleotides per 1 kb [33]. The $MnCl_2$ should be added to the reaction last to prevent dNTP precipitation.
10. It is a good idea to prepare several different error-prone libraries with varying concentrations of $MnCl_2$ (0–0.5 mM) in the PCR reaction.
11. In addition to determining the DNA concentration, the Nanodrop ND-1000 also gives information regarding the 260/280 and 260/230 absorbance ratios. For a successful ligation you need to have a very pure DNA; therefore, the 260/280 ratio should be ≥1.8 and the 260/230 ratio ≥2.0.
12. Electrocompetent cells exhibit a rather reproducible transformation efficiency. However, transformation by electroporation requires expensive, specialized apparatus (the electroporator) that may not be available in every laboratory. Normal rates of transformation are approximately 10^6 to 5.0×10^9 cfu/μg DNA for chemically competent *E. coli* cells and 10^8 to 2.0×10^{10} cfu/μg DNA for electrocompetent cells.
13. A library with an unbiased spectrum of mutations will be maximally diverse to have better chances to contain at least one improved variant. There are three key indicators of bias: (a) the ratio of transition to transversion mutations, (b) the ratio of AT → GC transitions to GC → AT transitions, and (c) the frequency of mutations at A:T base pairs, to mutations at G:C base pairs [34].
14. To reach higher levels of transformation efficiency, libraries can be electrotransformed. A *T. thermophilus* culture is grown in TB at 60 °C until OD_{600} reached 0.3–0.4. The culture is then centrifuged (10,000 × *g*, 5 min), washed twice in ½ volumes of 10% glycerol at 4 °C, and re-suspended in a small volume of the same solution (1% of the volume of the starting culture). The DNA (200 ng) is added to 40 μL of this concentrated cell solution, and the mix is electroporated (12.500 v/cm, 5 ms). Cells are incubated at 60 °C for 4 h in 1 mL TB and then spread on TB plates containing increasing concentrations of kanamycin (20–100 μg/mL). Plates are then incubated at the selective temperature as defined for the wild-type fusion [27].
15. *T. thermophilus* HB27 with this pNCK plasmid reaches 10^5 colonies/μg by natural competence and around 10^6 colonies/μg by electroporation [27].
16. Depending on the manufacturer of the 5 mL sterile tube, sometimes it is necessary to make a small hole with a needle on the lid of the tube to ensure that it does not come out due to broth evaporation at 65 °C.

17. A secondary selection by activity should be made afterwards.
18. At least 30 min before the plating, dry the plates in a laminar flow cabinet and pre-heat them in the static incubator at the desired temperature of selection.
19. The selection temperature usually varies between 60 and 72 °C. Start with the lowest temperature.
20. Increase the temperature 1–2 °C for each selection round until you do not have colonies. Control the real temperature by introducing a thermal sensor (e.g., Thermotrack thermo button) inside the incubator.
21. Plates may be stored afterwards for several days at room temperature or at 4 °C up to 1 month. Avoid picking colonies for subsequent assays after this time of storage.

Acknowledgments

This work has been supported by grant RTC-2014-1439-1 from the Spanish Ministry of Economy and Competitiveness. An institutional grant from Fundación Ramón Areces to CBMSO is also acknowledged.

References

1. García JL, Uhía I, Galán B (2012) Catabolism and biotechnological applications of cholesterol degrading bacteria. Microb Biotechnol 5:679–699
2. Plessis-Rosloniec KZD (2011) Steroid transformation by *Rhodococcus* strains and bacterial cytochrome P450 enzymes. Dissertation, University of Groningen
3. Fernandes P, Cruz A, Angelova B, Pinheiro HM, Cabral JMS (2003) Microbial conversion of steroid compounds: recent developments. Enzym Microb Technol 32:688–705
4. Diane W, Stephan R, Wolfgang Z (1997) Application of thermostable enzymes for carbohydrate modification. In: Praznik W, Huber A (eds) Carbohydrates as raw materials IV : proceedings of the fourth international workshop on carbohydrates as organic raw materials. WUV-Universitatverlag, Vienna
5. Fredrich A, Antrakian G (1996) Keratin degradation by *Fervidobacterium pennavorans*, a novel thermophilic anaerobic species of the order Thermotogales. Appl Environ Microbiol 62:2875–2882
6. Leuschner C, Antranikan G (1995) Heat stable enzymes from extremely thermophilic and hyperthermophilic microorganisms. World J Microbiol Biotechnol 11:95–114
7. Zeikus JG, Vieille C, Savchenko A (1998) Thermozymes: biotechnology and structure-function relationships. Extremophiles 2:179–183
8. Goetschel R, Bar R (1992) Formation of mixed crystals in microbial conversion of sterols and steroids. Enzym Microb Technol 14:462–469
9. Fernández-Arrojo L, Guazzaroni ME, López-Cortés N et al (2010) Metagenomic era for biocatalyst identification. Curr Opin Biotechnol 21:725–733
10. Lorenz P, Liebeton K, Niehaus F, Eck J (2002) Screening for novel enzymes for biocatalytic processes: accessing the metagenome as a resource of novel functional sequence space. Curr Opin Biotechnol 13:572–577
11. Bastien G, Arnal G, Bozonnet S et al (2013) Mining for hemicellulases in the fungus-growing termite *Pseudacanthotermes militaris* using functional metagenomics. Biotechnol Biofuels 6:78–92
12. Rashamuse K, Sanyika W, Ronneburg T et al (2012) A feruloyl esterase derived from a leachate metagenome library. BMB Rep 45:14–19
13. Chen IC, Lin WD, Hsu SK et al (2009) Isolation and characterization of a novel lysine racemase from a soil metagenomic library. Appl Environ Microbiol 75:5161–5166

14. Guazzaroni ME, Silva-Rocha R, Ward RJ (2015) Synthetic biology approaches to improve biocatalyst identification in metagenomic library screening. Microb Biotechnol 8: 52–64
15. Kumwenda B, Litthauer D, Bishop OT et al (2013) Analysis of protein thermostability enhancing factors in industrially important thermus bacteria species. Evol Bioinformatics Online 9:327–342
16. Zhou XX, Wang YB, Pan YJ, Li WF (2008) Differences in amino acids composition and coupling patterns between mesophilic and thermophilic proteins. Amino Acids 34:25–33
17. Vieille C, Zeikus GJ (1996) Thermozymes: identifying molecular determinants of protein structural and functional stability. Trends Biotechnol 14:183–190
18. Robinson-Rechavi M, Alibes A, Godzik A (2006) Contribution of electrostatic interactions, compactness and quaternary structure to protein thermostability: lessons from structural genomics of *Thermotoga maritima*. J Mol Biol 356:547–557
19. Szilágyi A, Závodszky P (2000) Structural differences between mesophilic, moderately thermophilic and extremely thermophilic protein subunits: results of a comprehensive survey. Structure 8:493–504
20. Vogt G, Argos P (1997) Protein thermal stability: hydrogen bonds or internal packing? Fold Des 2:S40–S46
21. Jaenicke R, Bohm G (1998) The stability of proteins in extreme environments. Curr Opin Struct Biol 8:738–748
22. Jaenicke R (1991) Protein stability and molecular adaptation to extreme conditions. Eur J Biochem 202:715–728
23. Trivedi S, Gehlot HS, Rao SR (2006) Protein thermostability in archaea and bacteria. Genetics and molecular research. Genet Mol Res 5: 816–827
24. Bornscheuer U, Kazlauskas RJ (2011) Survey of protein engineering strategies. Curr Protoc Protein Sci Chapter 26(Unit26):7
25. Rubingh DN, Grayling RA (2009) Protein engineering. In: Encyclopedia of life support systems. EOLSS, pp 140–165
26. Van Rossum T, Kengen SW, Van Der Oost J (2013) Reporter-based screening and selection of enzymes. FEBS J 280:2979–2996
27. Chautard H, Blas-Galindo E, Menguy T et al (2007) An activity-independent selection system of thermostable protein variants. Nat Methods 4:919–921
28. Mate DM, Rivera N, Sanchez-Freire E et al (2020) Thermostability enhancement of the *Pseudomonas fluorescens* esterase I by *in vivo* folding selection in *Thermus thermophilus*. Biotechnol Bioeng 117:30–38
29. Swarts DC, Makarova K, Wang Y et al (2014) The evolutionary journey of Argonaute proteins. Nat Struct Mol Biol 21:743–753
30. Lasa I, Castón JR, Fernández-Herrero LA et al (1992) Insertional mutagenesis in the extreme thermophilic eubacteria *Thermus thermophilus* HB8. Mol Microbiol 6:1555–1564
31. Cirino PC, Mayer KM, Umeno D (2003) Generating mutant libraries using error-prone PCR. Methods Mol Biol 231:3–9
32. Wilson DS, Keefe AD (2001) Random mutagenesis by PCR. Curr Protoc Mol Biol Chapter 8:Unit8 3
33. Zhao H, Moore JC, Volkov AA, Arnold FH (1999) Methods for optimizing industrial enzymes by directed evolution. In: Demain AL, Davies JE (eds) Manual of industrial microbiology and biotechnology, 2nd edn. ASM, Washington, DC, pp 597–604
34. Hanson-Manful P, Patrick WM (2013) Construction and analysis of randomized protein-encoding libraries using error-prone PCR. Methods Mol Biol 996:251–267
35. Otte KB, Hauer B (2015) Enzyme engineering in the context of novel pathways and products. Curr Opin Biotechnol 35:16–22
36. Rubin-Pitel SB, Cho CMH, Chen W, Zhao H (2007) Directed evolution tools in bioproduct and bioprocess development. In: Yang ST (ed) Bioprocessing for value-added products from renewable resources. New technologies and applications. Elsevier B.V, pp 49–72
37. Packer MS, Liu DR (2015) Methods for the directed evolution of proteins. Nat Rev Genet 16:379–394
38. Widmann M, Pleiss J, Samland AK (2012) Computational tools for rational protein engineering of aldolases. Comput Struct Biotechnol J 2:e201209016
39. Lee D, Redfern O, Orengo C (2007) Predicting protein function from sequence and structure. Nat Rev Mol Cell Biol 8:995–1005
40. Sobti M, Mabbutt BC (2013) Rational-based protein engineering: tips and tools. Methods Mol Biol 996:233–250
41. Khoury GA, Smadbeck J, Kieslich CA et al (2014) Protein folding and de novo protein design for biotechnological applications. Trends Biotechnol 32:99–109

Chapter 20

Stigmasterol Removal by an Aerobic Treatment System

Soledad Chamorro, Claudia Xavier, and Gladys Vidal

Abstract

Stigmasterol is a phytosterol contained in kraft mill effluent that is able to increase over 100% after aerobic biological treatment. This compound can act as an endocrine disrupter as its structure is similar to that of cholesterol. Furthermore, stigmasterol contained in kraft mill effluent shows high toxicity (25-fold more than β-sitosterol) to aquatic organisms such as *Daphnia magna*. However, the operation of the aerobic treatment and biomass adaptation could be affecting their removal. The performances of activated sludge (AS), aerated lagoon (AL), and moving bed biofilm reactors (MBBR) are compared to remove the stigmasterol contained in kraft mill effluent. The AL operates at a hydraulic retention time of 6 h and removes up to 90% of phytosterols. So, a 96% of stigmasterol is removed by AL when the sterol retention load is 0.6 mg/L · d. However, stigmasterol concentrations increase from 29% to 37% at a low stigmasterol load rate (0.2 mg/L · d). On the other hand, the stigmasterol is removed between 65% and 87% by an AS under a hydraulic retention time of 3 h. Moreover, a 100% of stigmasterol can be removed by the MBBR when the hydraulic retention time is 2 days.

Key words Stigmasterol removal, Kraft mill effluent, Aerated lagoon, Activated sludge, Moving bed biofilm reactors

1 Introduction

Phytosterols and related compounds found in pulp and paper effluents have been identified as potential sources of endocrine disruption [1, 2]. The concentration of these compounds in effluents may vary depending on the raw material and the technology used in kraft pulp production [3, 4]. Reported values range from 0.3 to 3.4 mg/L of phytosterol in effluents, depending on the kind of raw wood used [5]. Several authors [1–7] have noted that the phytosterols most commonly found in kraft mill effluents are β-sitosterol, stigmasterol, stigmastanol, and campesterol. Studies have found that treating effluents with aerobic biological processes significantly reduces phytosterol concentrations [5]. Most bleaching kraft mills in Chile have switched from using aerated lagoons (AL) (Fig. 1) for treating effluent to activated sludge (AS) (Fig. 2) and moving bed

Carlos Barreiro and José-Luis Barredo (eds.), *Microbial Steroids: Methods and Protocols*, Methods in Molecular Biology, vol. 2704, https://doi.org/10.1007/978-1-0716-3385-4_20,

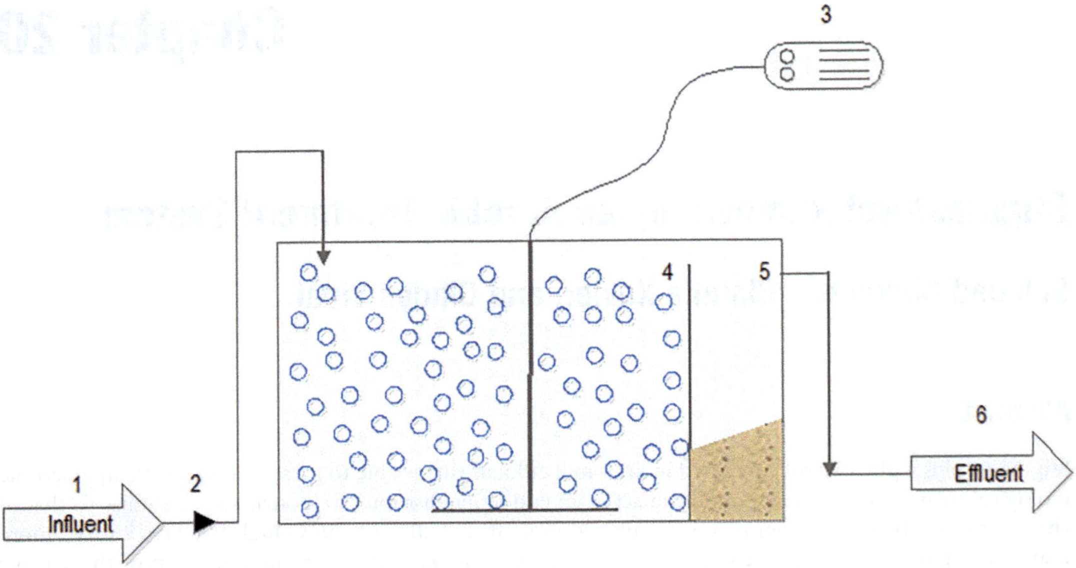

Fig. 1 Aerated lagoon system. Influent (1), feeding pump (2), air compressor (3), aerobic reaction zone (4), settler (5), effluent (6)

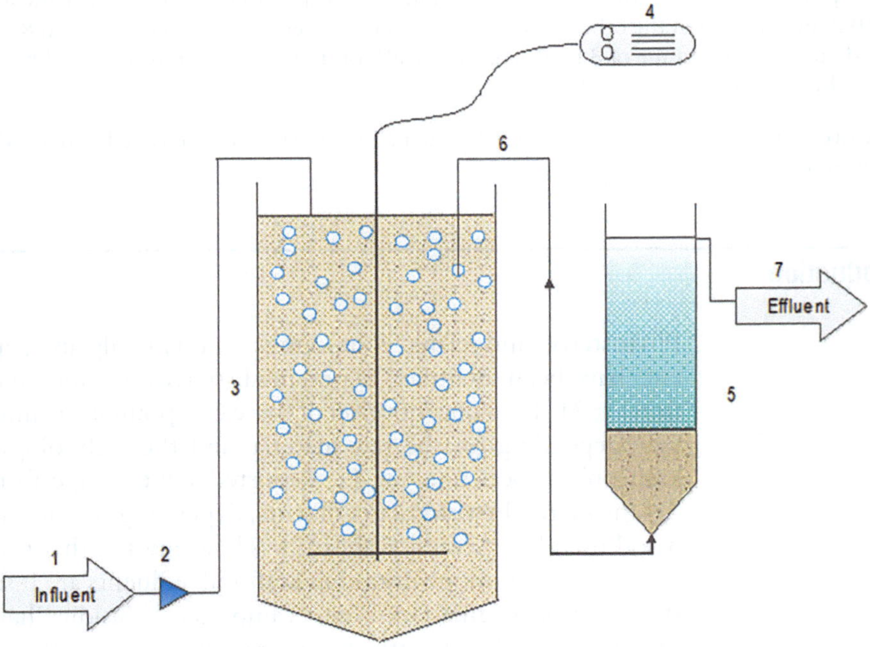

Fig. 2 Activated sludge system. Influent (1), feeding pump (2), aerobic reaction zone (3), air compressor (4), settler (5), biomass recycling (6), effluent (7)

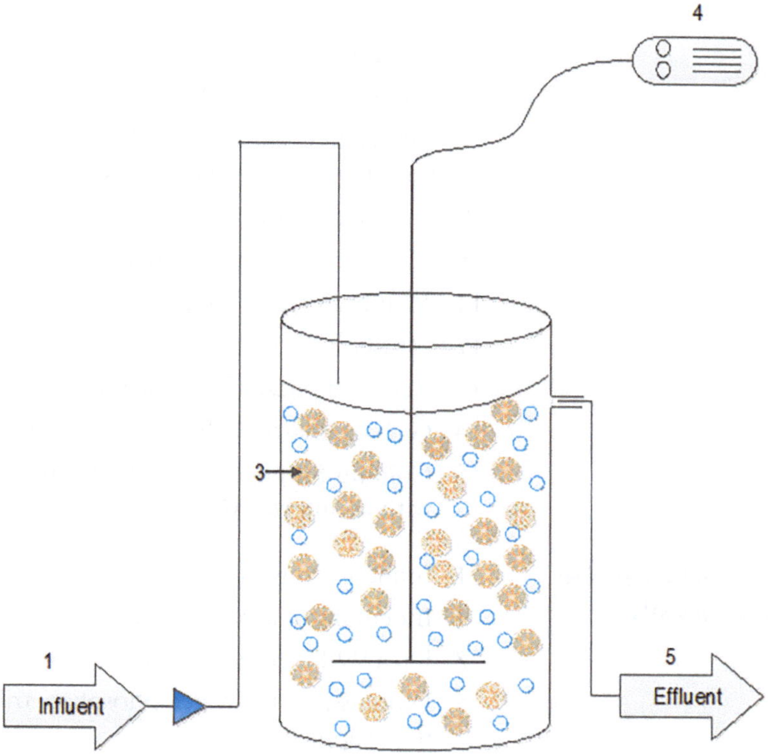

Fig. 3 MBBR system. Influent (1), feeding pump (2), carrier Kaldnes (3), air compressor (4), effluent (5)

biofilm reactors (MBBR) (Fig. 3) [8]. AL and AS can remove 41% of sterols by degrading or transforming them and can eliminate 100% of acute toxicity [9, 10]. However, increases of 192–367% of some phytosterols (stigmasterol) have been observed with AS [11]. Meanwhile, MBBR has been used successfully in the last decade to treat cellulose mill effluent. This approach reduces the probability of bulking during the operation through the use of a carrier for biofilm formation [8]. Nevertheless, little is known about the capacity of MBBR to remove or degrade phytosterol.

2 Materials

2.1 Equipment

1. Aerated lagoon system (AL). Area (0.44 L) and a settling (non-aerated) area (0.22 L). (Fig. 1; *see* **Note 1**).
2. Activated sludge system (AS) composed of an aerobic area (0.90 L) and a settling area (0.45 L) (Fig. 2).
3. Moving bed biofilm reactor (MBBR). Volume 0.509 L (Fig. 3).
4. Polyethylene carrier (Kaldnes K3).

5. Peristaltic pump.
6. Oxygen diffuser system.
7. Membrane filters (0.45 μm).
8. Binder-free glass microfiber filters (4.7 cm in diameter and 0.4 μm particle retention). Grade GF/C (Whatman).
9. 1 × 1 cm quartz cell.
10. UV-VIS spectrophotometer.
11. Membrane filter (0.2 μm).
12. HPLC.
13. Reversed phase C-18 chromatography column.
14. Gas chromatography CG–MS with mass detector.
15. Column 5% phenyl-methylsiloxane, 30 m long, 0.32 mm (id), with an internal diameter of 0.5 μm (MDN-5S; Supelco).

2.2 Raw Materials and Solvents

1. Influent from a local kraft mill.
2. Inoculum of aerobic microorganisms.
3. Urea 99.8% (used as nitrogen supplement).
4. Dipotassium hydrogen phosphate trihydrate (used as phosphorous supplement).
5. NaOH 0.1 M.
6. HCl.
7. Stigmasterol purity >95% (Sigma).
8. β-Sitosterol purity >98% (Sigma).
9. Buffer 0.2 M KH_2PO_4 prepared as follows: Dissolve 27.22 g of KH_2PO_4 in Milli-Q water, adjust the pH to 6.0 with NaOH 30%, and dilute to 1 liter.
10. Phenol purity ≥99% (Merck).
11. Milli-Q water.
12. Methanol MS SupraSolv® purity ≥99.8% (Merck).
13. Ethyl acetate MS SupraSolv® purity ≥99.8% (Merck).
14. n-Hexane SupraSolv® purity ≥98% (Merck).
15. Dichloromethane MS SupraSolv® purity ≥99.8% (Merck).

3 Methods

The method to remove stigmasterol by aerobic treatment has the following steps: (I) Seed an inoculum of aerobic microorganisms to treatment. (II) Feed the treatment (AL or MBBR) with kraft mill influent. (III) Maintain the reactor in optimum conditions, temperature, hydraulic retention time (HRT) between 0.3 and 2 days, and supplement the organic matter, nitrogen, and phosphorous.

(IV) Collect the influent, effluent, and biomass. (V) Extract with solid-phase cartridges previously conditioned. (VI) Assess the phytosterol by gas chromatography with mass detector.

3.1 Inoculum

1. Collect 5 L of aerobic microbial consortium from a kraft mill plant.
2. Determine the volatile suspended solid (VSS) according to APHA as follows. Condition a membrane filter disc (0.45 μm) at constant weight, drying it in an oven at 105 °C overnight. The weight of the membrane disc will be the total suspended solids determination (TSS-1). Take 5 mL of aerobic microbial consortium and filter through the membrane filter (0.45 μm). Place the residual sample in the membrane filter disc in a muffle at 550 °C for 1 h (TSS-2). Record the weights of (TSS-1) and (TSS-2) in a spreadsheet and report the concentration as g/L of the total suspended solids [12].
3. Inoculate the reactor with 5 g of volatile suspended solid per liter of aerobic microbial consortium.
4. Keep 2 weeks at room temperature (about 20 °C) for conditioning in the reactor.

3.2 Primary Treatment System of Kraft Mill Influent

1. Remove the solids from the influent by sedimentation during 30 min at room temperature.
2. Neutralize the influent by adjusting to pH 7.0 with NaOH 0.1 M or HCl, as appropriate. Now the influent is ready to be used in further treatments.

3.3 Influent

1. Collect 30 L of influent after primary treatment.
2. Store the influent in a 30 L PVC tank, in darkness, and refrigerate at 3 °C.
3. Supplement the influent with urea and dipotassium hydrogen phosphate trihydrate to obtain a BOD_5:N:P ratio of 100:5:1.
4. Keep the pH in the range from 6.0 to 7.5.

3.4 Aerated Lagoon (AL) Treatment

1. Feed the system with a peristaltic pump.
2. Keep the dissolved oxygen concentration above 6 mg O_2/L.
3. Conduct the operation at room temperature (10–22 °C).
4. Supplement the influent with stigmasterol (95%) (concentration ranging from 0.2 to 1 mg/L) (*see* **Notes 2** and **3**).

3.5 Activated Sludge (AS) Treatment

1. Conduct the operation at room temperature (10–22 °C).
2. Keep the dissolved oxygen concentration above 6 mg O_2/L.
3. Supplement the influent with stigmasterol and β-sitosterol to reach a concentration of 2 mg/L, respectively (*see* **Note 4**).

Table 1
Conditions of treatment system for stigmasterol removal

Aerobic treatment system	AL			AS		MBBR	
	Step I	Step II		Step 1	Step II	Step 1	Step II
Stigmasterol influent (mg/L)	–	0.2	1.1	0.07	2.0	0.02	0.05
SLR (mg/L·d)	–	0.2	1.1	–	–	–	–
OLR (gCOD/L·d)	0.81	0.96	0.96	0.26–0.66	0.45–9.0	0.155	0.343
HRT (h)	24	12.4	12.4	1.87–1.57	1.49–0.13	2.0	2.0
Removal (%)	–	–37	96	91.1	100	100	100

AL aerated lagoon, *AS* activated sludge, *MBBR* moving bed biofilm reactor, *SLR* stigmasterol load rate, *OLR* organic load rate, *HRT* hydraulic retention time

3.6 Moving Bed Biofilm Reactors (MBBR Treatment): Start-Up and Operation

1. Incorporate the polyethylene carrier (Kaldnes K3) with a density and specific surface area of 100 kg/m^3 in the MBBR reactor (*see* **Note 5**).
2. Maintain the dissolved oxygen concentration above 8 mg O_2/L (*see* **Note 6**).
3. Conduct the operation at room temperature (10–22 °C).
4. Characterize the phytosterol ofinfluent and effluent (*see* **Note 7**).

3.7 Performance of All Reactors

1. The operation for AL, AS, and MBBR is divided into two phases (Table 1) (*see* **Notes 1** and **8**).
2. Filter the samples trough a 0.45 μm membrane.
3. Determine the spectrophotometric measurements of filtered samples performed at the next wavelengths: 280 nm for lignin, 346 nm for the lignosulfonic acid, and 440 nm for the colors.
4. Determine the total phenolic compound concentrations by UV absorbance in a 1 × 1 cm quartz cell at 215 nm, and transform into concentrations using a calibration curve with phenol as a standard solution.

3.8 Extraction Techniques and CG–MS Sample Determinations

1. Filter 500 mL samples of each effluent through binder-free glass microfiber filters (4.7 cm in diameter and 0.4 μm particle retention).
2. Extract the samples by solid-phase extraction (SPE) using reverse phase C-18 cartridges previously conditioned with two volumes (ca 12 mL) of Milli-Q water and one volume each of methanol, ethyl acetate, and n-hexane.
3. Perform the runs at 4 mL min/L (prepare 2 cartridges for each sample).
4. Analyze the influent and effluent composition from one of the cartridges following elution with two volumes each of methanol, ethyl acetate, and n-hexane [1].

3.9 Determination of Phytosterols

1. Filter the kraft mill influent and effluent through a 0.2 μm membrane filter.
2. Extract the compounds from 100 mL of samples with 20 mL dichloromethane at pH 7 [2].
3. Analyze by gas chromatography CG–MS with mass detector using a Supelco MDN-5S capillary column (5% phenyl-methylsiloxane).
4. The detection limit is 1 μg/L (*see* **Note 9**).

4 Notes

1. The AL operation is divided into two phases. In phase I, the system is fed with only kraft mill influent to establish steady-state performance conditions. The HRT of the system is maintained for 24 h, with organic load rates (OLR) of approximately 0.81 g COD/L · d. In phase II, the lagoon is fed with kraft mill effluent supplemented with progressively higher concentrations of stigmasterol. Concentrations ranged from 0.2 to 1 mg/L. In the second phase, the OLR increases to 0.96 g COD/L · d and the HRT is reduced stepwise from 24 to 12 h.
2. AL performance indicates removal efficiencies for BOD_5 and COD ranging from 48% to 96% and 25.0% to 65.2%, respectively. Simultaneously, the average removal efficiency of total phenolic compounds is 20.7% and 28.5% for Phases I and II, respectively. Color and aromatic compounds are polymerized in the aerobic system.
3. The AL removes 96% of β-sitosterol and stigmasterol when the sterol retention load is 0.6 mg/L ·d (Table 1), while stigmasterol concentrations increase by 29–37% at a low stigmasterol load rate (0.2 mg/L· d). This phenomenon was previously described [11], showing that stigmasterol concentrations can increase by more than 300% in the AL system. Khan and Hall [13] also found that aerated systems are not efficient at removing stigmasterol.
4. The removal efficiency of the AS system is higher than AL system. The AS system removes over 95% of BOD_5 and from 57% to 67% of COD. However, phenolic compounds increase by 15% and coloring compounds by 10–40%.
5. The activated sludge system removes from 65% to 87% of phytosterols at an HRT of 3 h.
6. Attached biomass systems like MBBR efficiently reduce organic matter, with average values of 61.5% and 87.4% for COD and BOD_5, respectively. However, the removal efficiency of specific compounds like color and total phenolic compounds are less than 15% [14, 15].

7. The stigmasterol removal by MBBR system is 100% (Table 1) with operational parameters of a 2-day hydraulic retention time and an organic matter load of 0.343 ± 0.001 kg COD/m^3·d.
8. The system operates at biomass concentrations higher than 10 g VSS/L. Excess sludge is withdrawn to keep a 30-day SRT (sludge retention time).
9. These results show that stigmasterol removal is a consequence of the reactor performance, mainly the biomass adaption.

Acknowledgments

This work was supported by ANID/FONDAP/15130015.

References

1. Chamorro S, Pozo G, Jarpa M, Hernández V, Becerra J, Vida G (2010) Monitoring endocrine activity in kraft mill effluent treated by aerobic moving bed bioreactor system. Water Sci Technol 62(1):154–161
2. Chamorro S, Hernández V, Matamoros V, Dominguez C, Becerra J, Vidal J, Piña B, Bayona J (2013) Chemical characterization of organic microcontaminant sources and biological effects in riverine sediments impacted by urban sewage and pulp mill discharges. Chemosphere 90:611–619
3. Chamorro S, Hernández V, Monsalvez E, Becerra J, Mondaca MA, Piña B, Vidal G (2010) Detection of estrogenic activity from kraft mill effluents by the yeast estrogen screen. Bull Environ Contam Toxicol 84:165–169
4. Vidal G, Becerra J, Hernández V, Decap J, Xavier C (2007) Anaerobic biodegradation of sterol contained in Kraft mil effluents. J Biosci Bioeng 104(6):476–480
5. Kostamo A, Holmbom B, Kukkonen JVK (2004) Fate of wood extractives in wastewater treatment plants at kraft pulp mill and mechanical pulp mills. Water Res 37:972–982
6. Dykstra CM, Giles HD, Banerjee S, Pavlostathis SG (2014) Biotransformation of phytosterols under aerobic conditions. Water Res 58: 71–81
7. Fernandes P, Cruz A, Angelova B, Pinheiro HM, Cabral JM (2003) Microbial conversión of steroid compounds: recent developments. Enzyme Microb Technol 32:688–705
8. Diez MC, Castillo G, Aguilar L, Vidal G, Mora ML (2002) Operational factors and nutrients effect on activated sludge treatment for phenolic compounds degradation from *Pinus radiata* kraft mill effluent. Bioresour Technol 83:131–138
9. Odegaard H (2006) Innovations in wastewater treatment: the moving bed biofilm process. Water Sci Technol 53(9):17–33
10. Chamorro S, Xavier C, Hernández V, Becerra J, Vidal G (2009) Aerobic removal of stigmasterol contained in kraft mill effluents. Electron J Biotechnol 12(2):1–7
11. Cook DL, LaFleur L, Parrish A, Jones J, Hoy D (1997) Characterization of plant sterols from 22 US pulp and paper mills. Water Sci Technol 35:297–303
12. APHA-AWWA-WPCF (1985) Standard methods for examination of water and wastewater, 16th edn. American Public Health Association, Washington, D.C.
13. Khan Z, Hall ER (2003) Occurrence and removal of plant sterol in pulp and paper mill effluents. J Environ Eng Sci 2(1):17–26
14. Navia R, Levet L, Mora ML, Vidal G, Diez MC (2003) Allophanic soil adsorption system as a bleached kraft mill aerobic effluent post-treatment. Water Air Soil Poll 148:323–333
15. Villamar C, Jarpa M, Decap J, Vidal G (2009) Aerobic moving bed bioreactor performance: a comparative study of removal efficiencies of kraft mill effluents from *Pinus radiata* and *Eucalyptus globulus* as raw material. Water Sci Technol 59(3):507–514

Part V

Steroids Application

Chapter 21

Glucocorticoid Effect in Cancer Patients

Marta María Blanco-Nistal and Jesús Antonio Fernández-Fernández

Abstract

The use of glucocorticoids is very varied in the context of cancer patients and includes the treatment of symptoms related to cancer, but also the management of the most common side effects of antitumor treatments or adverse events related to the immune system. There is a quantity of experimental evidence demonstrating that cancer cells are immunogenic. However, the effective activation of anticancer T cell responses closely depends on an efficient antigen presentation carried out by professional antigen-presenting cells such as dendritic cells (DCs). The classic strategies to improve the medical management of inflammation are aimed at exacerbating the host's immune response. Although successful in treating a number of diseases, these drugs have limited efficacy and variable responses can lead to unpredictable results. The ideal therapy should reduce inflammation without inducing immunosuppression and remains a challenge for healthcare personnel.

Key words Adjuvant analgesics, Anti-inflammatory, Apoptosis, Cancer patient, Chemotherapy, Glucocorticoid effect, Immune response

1 Introduction

Thanks to their anti-inflammatory, anti-edema, and anti-allergy properties, glucocorticoids (GC) are among the most widely prescribed drugs in patients with cancer. The indications for GC use are very wide and varied in the context of cancer and include the symptomatic management of cancer-related symptoms (compression, pain, edema, altered general state) but also prevention or treatment of common side effects of anticancer therapies (nausea, allergies, etc.) or immune-related adverse events (irAE).

The classic strategies to improve the medical management of inflammation are aimed at exacerbating the host's immune response with aminosalicylates, antibiotics, corticosteroids, thiopurines, methotrexate, and biological antitumor necrosis factor (TNF) agents. Although successful in treating a number of diseases, these drugs have limited efficacy and variable responses can lead to

Carlos Barreiro and José-Luis Barredo (eds.), *Microbial Steroids: Methods and Protocols*, Methods in Molecular Biology, vol. 2704, https://doi.org/10.1007/978-1-0716-3385-4_21,

unpredictable results. The ideal therapy should reduce inflammation without inducing immunosuppression and remains a challenge for healthcare personnel [1].

However, severe even life-threatening side effects can arise from prolonged exposure to synthetic GC, as well as from elevated production of endogenous glucocorticoids, as seen in the rare endocrine cancers known as Cushing's disease. Common side effects include osteoporosis, hypertension, mood disorders, muscle and skin atrophy, and increased susceptibility to infection. Although these side effects are often said to limit the clinical utility of GCs, prolonged treatment with GCs is far from uncommon, particularly among the elderly. Since the discovery of GCs by Philip Hench and others in the 1940s, the uncoupling of their desirable anti-inflammatory or immuno-suppressive effects from their harmful side effects has been pursued with little success [2].

The involvement of GC in protumoral processes, particularly tumor proliferation, cell adhesion, or epithelial–mesenchymal transition, remains controversial. It has been suggested that the use of GC may render tumors resistant or less susceptible to apoptosis after anticancer therapy.

Some cancer types, such as hematologic malignancies, can be effectively treated with GC, whereas responses of epithelial cancers to GC treatment vary, even within cancer subtypes. GCs may have ancillary antitumor effects due to their cytotoxic actions on cancer cells; however, GC can also promote cancer progression, colonization of distant metastatic sites, and metastasis [3].

Indeed, GC treatment was found to downregulate basal and chemotherapy-induced expression of apoptotic effects in human cervical and lung carcinoma cells [4]. Variations in effects have also been observed within tumors, with variable effects shown in breast cancer depending on histologic subtype or tumor microenvironment. For example, some results indicate that the activation of the GC receptor increases heterogeneity and metastasis, which suggests that caution is needed when using GC to treat patients with breast cancer who have developed cancer-related complications [5].

Evidence indicates too that GCs induce apoptosis in hematological cells, thus supporting their use as chemotherapeutic agents for leukemias, lymphomas, and myeloma. GCs are a therapeutic component in their own right in chemotherapy protocols for hemopathic malignancies and have been shown to be very efficacious in association with cytotoxic chemotherapy for the treatment of acute lymphoblastic leukemia, chronic lymphocytic leukemia, Hodgkin and non-Hodgkin lymphoma, as well as multiple myeloma [6]. The mechanisms by which GCs could induce tumor cell apoptosis are still incompletely understood but appear to be multiple and interrelated.

Immune checkpoint inhibitors (ICIs) have revolutionized the management of advanced non-small cell lung cancer. With the intention of generating an antitumor immune response, ICIs can also lead to inflammatory side effects involving a wide variety of organs in the body, termed immune-related adverse events. Although no prospective clinical trial exists to guide recommendations for optimal and more specific immunosuppressive treatments rather than corticosteroids, further studies may lead to a more mechanistic-based approach towards these toxicities in the future. In relation to current practice, adherence to the recent published guidelines is recommend, which emphasize the importance of early recognition and administration of temporary immunosuppressive therapy with corticosteroids in most cases, depending on the organ system involved and the severity of toxicity. Recognition of these toxicities is increasingly important as the use of these agents expands within different indications for patients with lung cancers and to other tumor types [7].

1.1 Definitions

Immune-related adverse events (irAEs) are a unique spectrum of side effects of ICIs that resemble autoimmune responses. irAEs affect almost every organ of the body and are most commonly observed in the skin, gastrointestinal tract, lung, and endocrine, musculoskeletal, and other systems [8].

Biological antitumor necrosis factor agents. Tumor necrosis factor (TNF) plays a central role in the pathogenesis of several inflammatory conditions, including rheumatoid arthritis (RA). TNF is made intracellularly, mainly by activated macrophages. The precursor TNF is converted to soluble TNF after proteolysis by the TNF-converting enzyme. This soluble TNF then oligomerizes and forms the biologically active homotrimer TNF. There are two types of TNF, which are very closely related, TNF-alpha and TNF-beta. The activities of both TNFs are mediated through binding to the TNF receptors I and II (TNFRI and TNFRII), which are present on almost all cell types (except erythrocytes) [9–11].

Immune checkpoint inhibitors (ICIs) are novel therapeutic agents increasingly used in cancer therapy. Tumor cells can evade destruction by the immune system by triggering immune checkpoint receptors, such as cytotoxic T-lymphocyte-associated protein 4 (CTLA-4), programmed cell death protein 1 (PD-1), or programmed death-ligand 1 (PD-L1), that are expressed on T cells and whose engagement inhibits T-lymphocyte function [10]. Immune checkpoint inhibitors are monoclonal antibodies that prevent this immunosuppression by blocking the engagement of these checkpoint molecules, thereby reinvigorating the antitumor immune response [12].

Dendritic cells (DCs) are antigen-presenting cells derived from bone marrow precursors and form a widely distributed cellular system throughout the body. DCs exert immune surveillance for exogenous and endogenous antigens and the later activation of naive T lymphocytes, giving rise to various immunological responses. Different growth factors and cytokines can modulate the differentiation and function of DCs, GM-CSF, M-CSF, Flt3, and TGF-β, resulting in a large variety of DCs with different functional abilities. Functionally, the cDCs may be divided into two states: immature and mature. Immature DCs are specialist in uptaking and processing antigens; in contrast, mature DCs are professional in antigen presentation. It has been observed that immature cDCs can induce immune tolerance, while mature cDCs may induce Th2 or Th1 immune responses. It is worth noting that different subpopulations of DCs have the ability to secrete different cytokine patterns, resulting in the induction of different immunological responses. Furthermore, DCs are involved in the pathophysiology of several diseases such as contact hypersensitivity, autoimmune diseases, or cancer, but they can also be used as therapeutic tools in these conditions [13].

2 Effect of Glucocorticoids (GC) on the Immune Response

Paul Ehrlich at the beginning of the twentieth century presented his immune surveillance theory for cancer. According to Ehrlich, tumor cells appeared spontaneously in the body and were eliminated by the immune system, thus defending the idea of an immune system with anticancer capabilities. In fact, it seems that the efficacy of classical treatments also lies largely in the activation of immune responses after the release of immunostimulatory molecules from necrotic cells [14, 15].

Actually, there is a quantity of experimental evidence demonstrating that cancer cells are immunogenic. However, the effective activation of anticancer T cell responses closely depends on an efficient antigen presentation carried out by professional antigen-presenting cells such as dendritic cells (DCs) [16].

A mathematical model based on a system of ordinary differential equations was developed for the characterization of the regulation of antitumor immune activity.

These mathematical simulations indicated that GC treatment can suppress the antitumor immune response in a dose-dependent manner. The model simulations were in line with previous experimental observations of the inhibitory effects of GCs on T, natural killer (NK) cells, and DCs. Thus, the results of this study could be useful in predicting clinical outcomes in patients receiving GC therapy [17].

NK cells play a crucial role in antitumor immunity because of their innate ability to differentiate between malignant versus normal cells. But these NK cells are severely affected during the period immediately following cancer surgeries. Therefore, an opportunity arises in the aftermath of cancer surgery for residual cancer cells, including distant metastases, to gain a foothold in the absence of NK cell surveillance. In the sense of GC therapy predicting clinical outcomes, it is interesting to know that the release of sympathetic stress-related factors (e.g., cortisol, prostaglandins, catecholamines), anti-inflammatory cytokines (e.g., IL-6, TGF-β), and myeloid derived suppressor cells mediates NK cell dysfunction. Thus, peri-operative therapies to mitigate NK cell suppression in the postoperative period could improve curative outcomes following cancer surgery [18].

Another study investigated the effects of tolerogenic GC-treated DCs on NK and T cell antitumor responses in CD8+ T cells. The effects caused by GC-treated DCs were compared to the responses to immunogenic, CpG-activated DCs. The immunization with CpG and peptide-treated DCs protected against tumor growth by activation of NK cell response. Also, immunogenic DCs induced the expansion of cytotoxic CD8+OT-1 cells. In contrast, the peptide and GC-treated DCs increased the numbers of immature Mac-1+CD27- NK cells as well as Foxp3+ and IL-10 secreting CD8+OT-1 cells with suppressive properties. In conclusion, the generation of tolerogenic DCs is one of many immunosuppressive mechanisms that can be induced by GC [19].

3 The Role of MicroRNA in Glucocorticoid Action

MicroRNAs (miRNAs) are short RNA species, generally 19–22 nucleotides in length, which mediate post-transcriptional down-regulation of protein expression [20].

Although individual miRNAs exert relatively little effect on cortisol biosynthesis, coordinated changes of miRNA abundance may have a more striking impact, for example, in the context of hypoxia or adenocarcinoma [21, 22]. Local GC availability can also be regulated via effects of miRNAs [23].

GCs regulate many aspects of neuronal development and function, having particularly important roles in adaptation to stress. Since pathophysiological point of view, prolonged exposure to stress or elevated GC levels can cause long-lasting reprogramming of neuronal GC responses, which can result in neuropsychiatric disorders such as depression in later life. Such reprogramming at critical developmental stages in utero may contribute to the trans-generational effects of antenatal GC exposure [24, 25]. It has been suggested that miRNA-mediated fine-tuning may contribute to preventing consequences such as GC-induced atrophy of the adrenal glands [26].

Various miRNAs have been implicated in the positive or negative regulation of inflammatory and immune responses, and these mechanisms may be subject to GC modulation. The best characterized example is miR-155, which is generated from the B-cell integrating pool of the precursor transcript [27]. Elevated levels of miR-155 have been described in many types of cancer, leading to its identification as one of the first oncogenic miRNAs (oncomirs).

In experiments with mice, those lacking miR-155 were resistant to synovial inflammation and bone erosion [28]. Dexamethasone GC inhibited lipopolysaccharide (LPS)-induced expression of miR-155 in primary macrophages and macrophage cell lines [29] spleen and liver cells from LPS-injected mice [29, 30] and T lymphocytes from sepsis patients [31].

The regulatory impact of a new miRNA on cancer growth and migration is currently being discussed. The new emerging miRNA-338-3p can regulate the response of cancer cells to chemotherapy and radiotherapy. It appears that miRNA-338-3p has a dual role in cancer chemotherapy, acting as a tumor-promoting or tumor-suppressing factor. The experiments reveal the antitumor activity of miRNA-338-3p in cancer. Therefore, increasing the expression of miRNA-338-3p is important in effective cancer therapy. Long noncoding RNAs, circular RNAs, and hypoxia are possible upstream mediators of miRNA-338-3p in cancer. Antitumor agents including baicalin and arbutin can promote miRNA-338-3p expression to suppress cancer progression [32].

However, successful therapeutic manipulation of miRNA expression is likely to be highly tumor-type specific and mimic viral infection, such as Toll-like receptor 9 (TLR9) ligands. Such agonists render plasmacytoid dendritic cells resistant to GC-induced apoptosis [33].

In summary, as miRNAs exert such pervasive effects on biological processes, it is not surprising that they touch on GC action at several points. Their touch is often light, in the sense that individual miRNAs alter expression of their targets only modestly. Nevertheless, one miRNA can hit several targets, and one target may be hit by several different miRNAs. Therefore, coherent and coordinated changes of miRNA abundance can affect complex biological processes quite profoundly [34].

4 Glucocorticoids and Apoptosis

Clinically, GCs are used for allergies, autoimmunity, and chronic inflammation, because they have strong anti-inflammatory effects and induce the apoptosis of lymphocytes. Past studies have reported the positive and negative effects of GC on the immune system. These opposing properties of GC may regulate the immune balance between the responsiveness to antigens and excessive inflammation in steady-state and stress conditions.

Glucocorticoid receptor (GR) acts as a transcription factor and represses the expression of inflammatory cytokines, chemokines, and prostaglandins by binding to its motif, to GC response element, or to other transcription factors. In mice, GR suppresses the antigen-stimulated inflammation mediated by macrophages, dendritic cells, and epithelial cells and impairs cytotoxic immune responses by downregulating interferon-γ production and inhibiting the development of type-1 helper T cells, CD8+ T cells, and NK. These immune inhibitory effects prevent lethality by excessive inflammation but at the same time increase the susceptibility to infection and cancer. Consistently, stress-induced GCs strongly suppress cell-mediated immunity and cause viral infection and tumor development. They may also enhance the development of pathogenic helper T cells and cause tissue damage through neural and intestinal inflammation [35].

New findings show that GR antagonism enhances tumor cell apoptosis due to cytotoxic agents under physiological GC concentrations in mice. Thus, GR antagonism will help guide the clinical development of relacorilant in combination with cytotoxic agents for the treatment of solid tumors. Anti-proliferative agents, including, but not limited to, paclitaxel, consistently cause tumor cell apoptosis in vitro. Resistance to such therapies is, unfortunately, the norm in clinical practice in patients with solid tumors. The data presented in some studies expand our understanding of the pro-apoptotic effects of GR antagonists described in ovarian and breast cancer cells [36, 37]. Relacorilant improved the activity of diverse cytotoxic agents, and the most pronounced benefits were seen in combination with microtubule-targeted agents [38].

Thus, the optimal clinical application for a GR antagonist would necessarily avoid the use of corticosteroids. For example, paclitaxel formulated with albumin and without cremophor does not require co-administration of corticosteroids and is thus a rational first choice for combination with relacorilant. Additionally, the benefit of relacorilant combined with a diverse set of cytotoxic agents identified further expands the number of potential corticosteroid-sparing regimens with relacorilant [38].

In this way, the aim of a recent study was to identify the effect of ionizing radiation on the structure of prednisolone and cancer cells. Prednisolone is a GC used for the treatment of liver disease and cancer. A new derivative of prednisolone was created by irradiation (prednisolone IR) and its anticancer properties were investigated in liver cancer cells. IR prednisolone promoted apoptosis and arrested the cell cycle at the G0/G1 stage in Huh7 cells. IR prednisolone also altered the mitochondrial membrane potential and activated caspase-associated proteins, thereby activating the intrinsic apoptotic signaling pathway. In conclusion, IR prednisolone promoted anticancer effects in liver cancer cells through activation of apoptosis. This study showed that IR prednisolone may be a potential anticancer agent against liver cancer, although specific molecules have not yet been identified [39].

5 Impact of Glucocorticoid Use on Cancer-Related Symptoms

Cancer remains the leading cause of death worldwide and also carries significant morbidity. Pain resulting, directly or indirectly, from abnormal growth of malignant cells in normal tissue, is the most common and feared symptom associated with cancer [40]. It is estimated that one-third of cancer patients undergoing active treatment, and two-thirds of those with advanced disease, experience pain that requires treatment with analgesic drugs [41].

In addition, the recommendations of the World Health Organization (WHO) and the European Association for Palliative Care for the treatment of pain indicate and establish the adjuvant analgesics that should be used in each step of the WHO analgesic scale, GC being one of the adjuvant analgesics, according to treatment guidelines [42].

In oncology, GCs have long been used to help control cancer-related symptoms, especially those related to inflammation and/or edema caused by the tumor, with most patients receiving GCs followed up in oncology, at some point in their cancer treatment [43].

Due to their anti-inflammatory mechanism of action, corticosteroids are said to provide effective analgesia for pain associated with inflammation and in the treatment of cancer-related complications such as brain metastases and spinal cord compression [41], especially in bone metastases [44, 45], and are also widely used in cancer patients in the palliative phase of the disease [46]. Dexamethasone is often the corticosteroid of choice for specialized palliative care teams [47], as it has a more even tissue distribution and better penetration of the blood–brain barrier [48], and its minimal mineralocorticoid means that dexamethasone causes less retention of liquids, and its long biological half-life of 36–54 h allows once-daily dosing [47].

However, some studies and meta-analyses have shown that the efficacy is lower than expected and should be considered in perspective with the potential adverse effects of this treatment [49].

Recently, a meta-analysis on the use of GC for pain management in adult cancer patients has been carried out, and the authors concluded that more trials are needed, and with a larger number of participants, to evaluate the safety and the effectiveness of corticosteroids for the treatment of cancer pain in adults and to establish an ideal dose, duration of treatment, and route of administration. More randomized controlled clinical trials are indicated, and with adequate statistical power, having pain as the primary outcome and being measured with a standardized and universally accepted tool, using a single GC drug (e.g., dexamethasone, which is the GC most used in this type of patients), at a dose and with a predetermined route of administration, for a short period of time. Likewise, it would be recommended that long-term toxicity be documented during a long follow-up period [50].

Another symptom related to cancer is dyspnea, which has a multifactorial origin and appears in 70% of these patients when they reach the final phase of life and whose treatment with GC can be effective [43]. Currently, and depending on the origin of dyspnea, GCs have several arguments in favor of their use, for example, when dyspnea appears in patients who, due to tumor progression, suffer from bronchospasm, pleural effusion, or superior vena cava syndrome [51], since GCs have an anti-inflammatory component that includes immune infiltration of the wall of the airways, increased proinflammatory cytokines in lung tumors (such as IL-8, IL-6, and C-reactive protein), or increased peripheral activation of neutrophils, which may explain its ability to attenuate dyspnea in view of the high inflammatory response of these patients [52, 53].

However, as dyspnea is known to be multifactorial in cancer patients, GCs may be more effective in some cases, such as lymphangitis carcinomatosa or airway obstruction by tumor, but less so in other circumstances [49].

GCs are also an element used in the treatment of symptomatic cerebral edema (nausea, vomiting, and headache) and are recommended to reduce intracranial pressure related to the progression of a primary brain tumor or intracerebral metastasis [54]. Dexamethasone is the treatment of choice to relieve the symptoms of brain tumor progression with signs of intracranial hypertension and/or neurological deficits [55]. The reason for choosing dexamethasone is due to its limited mineralocorticoid effects and the lower risk of a rebound effect after discontinuation.

The treatment of leptomeningeal metastases and their symptoms is generally more complex and may require therapy with intrathecal administration of GC. In addition, treatment with GC has been shown to be effective and is among the main therapeutic options (along with surgery and radiotherapy) for the treatment of symptoms related to spinal cord compression, especially during spinal cord metastases. Spinal cord metastases occurs in numerous cancers such as breast or lung; in these cases, prompt attention is needed to restore neurological function, relieve pain, and prevent permanent damage [56].

Likewise, GCs appear in clinical guidelines and are currently used systemically in the management and treatment of digestive occlusive symptoms of tumor origin or cause, with the main objective of restoring or improving said occlusive symptoms [57]. This intestinal obstruction can be mechanical or functional, with extrinsic mechanics being representative of a higher percentage of cases.

Said occlusive digestive symptoms may be the result of compression of the digestive tract lumen, by a tumor or primary cancerous mass, by a metastasis (mesenteric or epiploic), by radiation-induced fibrosis secondary to radiotherapy, or by adhesions: abdominal and/or pelvic [58]. Malignant intestinal obstruction is estimated to occur in 10–28% of colorectal cancers and 5–42% of ovarian cancers; in general, it is estimated that it occurs in 2% of all patients in advanced phase of the disease [59].

There is a systematic review on the use of corticosteroids in the management of cancer patients with symptoms of intestinal obstruction, in which several randomized clinical trials with double blind and placebo were analyzed and in which the authors concluded that that there is a tendency towards evidence that GC in a dose range of 6–16 mg of dexamethasone administered intravenously can achieve resolution of intestinal obstruction, presenting an extremely low incidence of side effects in all included studies [60].

At the same time that GCs are used to manage the impact of the symptoms of the disease on the quality of life of cancer patients, they are also used to help control the secondary effects of antitumor treatments, such as chemotherapy, immunotherapy, and radiotherapy, with nausea and vomiting standing out among these symptoms [61].

To control these symptoms, and together with GCs (which are the drugs used for this function for the longest time), 5HT3 or NKl receptor antagonists are associated depending on the emetic response of the chemotherapy protocol. Chemotherapy is known to induce the production of inflammatory mediators, such as eicosanoids, which may explain the efficacy of GC treatment in preventing such nausea and vomiting [62].

In addition to nausea and vomiting, there are certain chemotherapy agents (bleomycin, gemcitabine) or targeted therapies, EGFR (erlotinib, gefitinib), mTOR or MEK inhibitors, or even HER2 monoclonal antibodies (especially trastuzumab and deruxtecan, which are known to cause disease interstitial lung), which can cause respiratory and pulmonary toxicity [63]. Parallel to the interruption of the drug implicated in said toxicity, the use of GC is recommended in the clinical guidelines, since they can help resolve dyspneic symptoms and even improve the pulmonary radiographic presentation [64].

GCs, in addition to being used in the symptoms of cancer patients described above, also have favorable side effects regarding the general condition of cancer patients, due to their orexigenic or appetite-stimulating effect [65] and to having a positive impact on the characteristic asthenia of this type of fragile patients [53]. In general, several studies carried out with cancer patients with advanced and palliative cancer have shown an improvement in the quality of patients receiving treatment with GC, compared to patients receiving placebo [53, 66].

However, and despite the evidence that indicates the benefit of the use of GC in this type of patient, it must also be considered that, like any drug, GC can cause certain clinical side effects in patients (candidiasis oral administration, severe proximal myopathy, insomnia, etc.), which can cause patients to show certain reluctance and difficulties in the use of GCs, even outweighing the possible potential benefits [49]. For all these reasons, even in patients with advanced cancer and palliative patients, GCs should be used with the minimum effective dose and with the shortest possible duration [67].

6 Future Perspectives

Acquired GC resistance emerges over time because the underlying disease process finds mechanisms to evade GC-induced apoptosis [68]. The mechanisms underpinning acquired GC resistance are nevertheless divergent and often cell-type specific, which contributes to the heterogeneity in GC responsiveness observed in patients [69, 70].

Given the extensive number of factors that can contribute to GC resistance, GC resistance is clearly not governed by one single mechanism, but by several mechanisms acting consecutively or alternating to achieve full-blown resistance. Interpatient variability in the underlying mechanisms is also to be expected, as both the GC dose and the duration of GC therapy are often different between patients, in turn affecting the moment of onset and the degree of GC resistance. Frontline technologies such as single cell RNA-sequencing will further help dissecting the gradual emergence and inherent heterogeneity of GC resistance [71].

References

1. Sales-Ca-Campos H, Basso PJ, Alves VB et al (2015) Classical and recent advances in the treatment of inflammatory bowel diseases. Braz J Med Biol Res 48(2):96–107
2. Clark AR, Bervisi MG (2012) Maps and legends: the quest for dissociated ligands of the glucocorticoid receptor. Pharmacol Ther 134:54–67
3. Zhidkova EM, Lylova ES, Savinkova AV et al (2020) A brief overview of the paradoxical role of glucocorticoids in breast cancer. Breast Cancer 14:117822342097466
4. Herr I, Ucur E, Herzer K et al (2003) Glucocorticoid cotreatment induces apoptosis resistance toward cancer therapy in carcinomas. Cancer Res 63:3112–3120
5. Obradovic M, Hamelin B, Manevski NR et al (2019) Glucocorticoids promote breast cancer metastasis. Nature 567(7749):540
6. Pufall M (2015) Glucocorticoids and cancer. Exp Biol Med 872:315–333
7. Wills B, Brahmer J, Naidoo J (2018) Treatment of complications from immune checkpoint inhibition in patients with lung cancer. Curr Treat Options Oncol 19:46
8. Bramer JR, Lacchetti C, Atkins MB et al (2018) Management of immune-related adverse events in patients treated with immune checkpoint inhibitor therapy: American Society of Clinical Oncology clinical practice guideline. J Clin Oncol 36(17):1714–1768

9. Feldmann M, Maini R (2001) Anti-TNF alpha therapy of rheumatoid arthritis: what have we learned? Annu Rev Immunol 19:163–196
10. Gottlieb A (2007) Tumor necrosis factor blockade: mechanism of action. J Investig Dermatol 12(1):1–4
11. Roach D, Bean A, Demangel C, France M, Briscoe H, Britton W (2002) TNF regulates chemokine induction essential for cell recruitment, granuloma formation, and clearance of mycobacterial infection. Immunology 168(9): 4620–4627
12. Pardoll K (2012) The blockade of immune checkpoints in cancer immunotherapy. Nat Rev Cancer 12:252–264
13. Castell-Rodríguez A, Piñón-Zárate G, Herrera-Enríquez M, Jarquín-Yáñez K, Medina-Solares I (2017) Dendritic cells: location, function, and clinical implications. In: Biology of myelomonocytic cells, 3rd edn. Anirban Gosh
14. Arce F, Kochan G, Breckpot K, Stephenson H, Escors D (2012) Selective activation of intracellular signalling pathways in dendritic cells for cancer immunotherapy. Anti Cancer Agents Med Chem 12:29–39
15. Ghiringhelli F, Bruchard M, Apetoh L (2013) Immune effects of 5-fluorouracil: ambivalence matters. Oncoimmunology 2(3):e23139
16. Gato T, Liechtenstein T, Blanco-Luquín I, Zudaire M, Kochan G, Escors D (2015) Dentritic cells in cancer immunotherapy. An Sist Sanit Navar 38(2):279–287
17. Yakimchuk K (2019) Mathematical modeling of immune modulation by glucocorticoids. Biosystems 187:104066
18. Angka L, Khan S, Kilgour M, Xu R, Kennedy M, Auer R (2017) Dysfunctional natural killer cells in the aftermath of cancer surgery. Int J Mol Sci 18(8):1787
19. Chen L, Hasni M, Jondal M, Yakimchuk K (2017) Modification of anti-tumor immunity by tolerogenic dendritic cells. Autoimmunity 50(6):370–376
20. Filipowicz W, Bhattacharyya S, Sonenberg N (2008) Mechanisms of post-transcriptional regulation by microRNAs: are the answers in sight? Nat Rev Genet 9:102–114
21. Robertson S, Diver LA, Alvarez-Madrazo S et al (2017) Regulation of corticosteroidogenic genes by microRNAs. Int J Endocrinol 2017: 1–11
22. Nusrin S, Tong S, Chaturvedi G, Wu R, Giesy J, Kong R (2014) Regulation of CYP11B1 and CYP11B2 steroidogenic genes by hypoxia-inducible miR-10b in H295R cells. Mar Pollut Bull 85:344–351
23. Han Y, Staab-Weijnitz C, Xiong G, Maser E (2013) Identification of microRNAs as a potential novel regulatory mechanism in HSD11B1 expression. J Steroid Biochem Mol Biol 133:129–139
24. de Kloet E, Fitzsimons C, Datson N, Meijer O, Vreugdenhil E (2009) Glucocorticoid signaling and stress-related limbic susceptibility pathway: about receptors, transcription machinery and microRNA. J Brain Res 13(1293): 129–141
25. Moisiadis V, Matthews S (2014) Glucocorticoids and fetal programming part 2: mechanisms. Nat Rev Endocrinol 10(7):403–411
26. Pentkowski N, Bouquin S, Maestas-Olguin C, Villasenor Z, Clark B (2008) Characterization of the vulnerability to repeated stress in Fischer 344 rats: possible involvement of microRNA-mediated down-regulation of the glucocorticoid receptor. Eur J Neurosci 27(9): 2250–2261
27. Vigorito E, Kohlhaas S, Lu D, Leyland R (2013) An ancient regulator of the immune system. Immunol Rev J 253:146–157
28. Kurowska-Stolarska M, Alivernini S, Ballantine L et al (2011) MicroRNA-155 as a proinflammatory regulator in clinical and experimental arthritis. Proc Natl Acad Sci U S A 108:11193–11198
29. Zheng Y, Xiong S, Jiang P et al (2012) Glucocorticoids inhibit lipopolysaccharide-mediated inflammatory response by downregulating microRNA-155. Free Radic Biol Med J 52: 1307–1317
30. Wang Z, Liang Y, Tang H et al (2013) Dexamethasone down-regulates the expression of microRNA-155 in the livers of septic mice. PLoS One 8:e80547
31. Chen Y, Wang G, Liu Z, Wang S, Wang Y (2016) Glucocorticoids regulate the proliferation of T cells via miRNA-155 in septic shock. Exp Ther Med 12:3723–3728
32. Mirzaei S, Zarrabi A, Asnaf S, Hashemi F, Zabolian A (2021) The role of microRNA-338-3p in cancer: growth, invasion, chemoresistance, and mediators. Life Sci 268:119005
33. Guiducci C, Gong M, Xu Z et al (2010) TLR recognition of self-nucleic acids hampers glucocorticoid activity in lupus. Nature 465:937–941
34. Clayton S, Jones S, Kurowska-Stolarska M, Clark A (2018) The role of microRNAs in glucocorticoid action. J Biol Chem 293(6): 1865–1874

35. Shimba A, Ikuta K (2020) Control of immunity by glucocorticoids in health and disease. Semin Immunopathol 42(6):669–680
36. Skor M, Wonder E, Kocherginsky M et al (2013) Glucocorticoid receptor antagonism as a novel therapy for triple-negative breast cancer. Clin Cancer Res 19:6163–6172
37. Moran T, Gray S, Mikosz C, Conzen S (2000) The glucocorticoid receptor mediates a survival signal in human mammary epithelial cells. Cancer Res 60:267–272
38. Greenstein A, Hunt H (2021) Glucocorticoid receptor antagonism promotes apoptosis in solid tumor cells. Oncotarget 12(13): 1243–1255
39. Ramadhani F, Kang S, Kawala R, Chung B, Bai H, Kang B (2021) γ irradiated prednisolone promotes apoptosis of liver cancer cells via activation of intrinsic apoptosis signaling pathway. Mol Med Rep 23:425
40. van den Beuken-van Everdingen MH, Hochstenbach LM, Joosten EA, Tjan-Heijnen VC, Janssen DJ (2016) Update on prevalence of pain in patients with cancer: systematic review and meta-analysis. J Pain Symptom Manag 51(6):1070–1090
41. Foley KM (2011) How well is cancer pain treated? Palliat Med 25(5):398–401
42. Fallon M, Hanks G, Cherny N (2006) Principles of control of cancer pain. BMJ 332:1022–1024
43. Haywood A, Good P, Khan S, Leupp A, Jenkins-Marsh S, Rickett K, Hardy JR (2015) Corticosteroids for the management of cancer-related pain in adults. Cochrane Database Syst Rev 2015:CD010756
44. Chouahnia K, Luu M, Baba-Hamed N, Des Guetz G (2009) Bone metastases pain in the elderly. Rev Med Suisse 5:1126, 1128, 1130
45. Chow E, Meyer RM, Ding K, Nabid A, Chabot P, Wong P, Ahmed S, Kuk J, Dar AR, Mahmud A et al (2015) Dexamethasone in the prophylaxis of radiation-induced pain flare after palliative radiotherapy for bone metastases: a double-blind, randomized placebo-controlled, phase 3 trial. Lancet Oncol 16:1463–1472
46. Barghi K, Edmonds KP, Ajayi TA, Atayee RS (2018) Prescribing trends of palliative care team's use of dexamethasone for cancer-related pain. J Pain Palliat Care Pharmacother 32(1): 37–43
47. Bordag N, Klie S, Jurchott K, Vierheller J, Schiewe H, Albrecht V, Tonn JC, Schwartz C, Schichor C, Selbig J (2015) Glucocorticoid (dexamethasone)-induced metabolome changes in healthy males suggest prediction of response and side effects. Sci Rep 5:15954
48. Moricke A, Zimmermann M, Valsecchi MG, Stanulla M, Biondi A, Mann G, Locatelli F, Cazzaniga G, Niggli F, Aricò M (2016) Dexamethasone vs prednisone in induction treatment of pediatric ALL: results of the randomized trial AIEOP-BFM ALL 2000. Blood 127(17):2101–2112
49. Kalfeist L, Galland L, Ledys F, Ghiringhelli F, Limagne E, Ladoire S (2022) Impact of glucocorticoid use in oncology in the immunotherapy era. Cell 11:770
50. Zimmerman M, Drings P (1984) Guidelines for therapy of pain in cancer patients. In: Pain in the cancer patient. Recent results in cancer research, vol 89. Springer, Berlin/Heidelberg, pp 1–12
51. Lin RJ, Adelman RD, Mehta SS (2012) Dyspnea in palliative care: expanding the role of corticosteroids. J Palliat Med 15:834–837
52. Hui D, Kilgore K, Frisbee-Hume S, Park M, Tsao A, Delgado Guay M, Lu C, William W, Pisters K, Eapen G et al (2016) Dexamethasone for dyspnea in cancer patients: a pilot double-blind, randomized, controlled trial. J Pain Symptom Manage 52:8–16.e1
53. Paulsen O, Klepstad P, Rosland JH, Aass N, Albert E, Fayers P, Kaasa S (2014) Efficacy of methylprednisolone on pain, fatigue, and appetite loss in patients with advanced cancer using opioids: a randomized, placebo-controlled, double-blind trial. J Clin Oncol 32(29): 3221–3228
54. Ryken TC, McDermott M, Robinson PD, Ammirati M, Andrews DW, Asher AL, Burri SH, Cobbs CS, Gaspar LE, Kondziolka D et al (2010) The role of steroids in the management of brain metastases: a systematic review and evidence-based clinical practice guideline. J Neuro-Oncol 96:103–114
55. Roth P, Happold C, Weller M (2015) Corticosteroid use in neuro-oncology: an update. Neurooncol Pract 2:6–12
56. Sodji Q, Kaminski J, Willey C, Kim N, Mourad W, Vender J, Dasher B (2017) Management of metastatic spinal cord compression. South Med J 110:586–593
57. Laval G, Marcelin-Benazech B, Guirimand F, Chauvenet L, Copel L, Durand A, Francois E, Gabolde M, Mariani P, Rebischung C et al (2014) Recommendations for bowel obstruction with peritoneal carcinomatosis. J Pain Symptom Manag 48:75–91
58. Ferguson HJ, Ferguson CI, Speakman J, Ismail T (2015) Management of intestinal obstruction in advanced malignancy. Ann Med Surg 4(3):264–270

59. Tuca A, Guell E, Martinez-Losada E, Codorniu N (2012) Malignant bowel obstruction in advanced cancer patients: epidemiology, management, and factors influencing spontaneous resolution. Cancer Manag Res 4:159
60. Feuer DJ, Broadley KE (2000) Corticosteroids for the resolution of malignant bowel obstruction in advanced gynecological and gastrointestinal cancer. Cochrane Database Syst Rev 2000: CD001219
61. Grunberg SM (2007) Antiemetic activity of corticosteroids in patients receiving cancer chemotherapy: dosing, efficacy, and tolerability analysis. Ann Oncol 18:233–240
62. Gupta K, Walton R, Kataria SP (2021) Chemotherapy-induced nausea and vomiting: pathogenesis, recommendations, and new trends. Cancer Treat Res Common 26:100278
63. Tarantino P, Modi S, Tolaney SM, Cortes J, Hamilton EP, Kim S-B, Toi M, Andre F, Curigliano G (2021) Interstitial lung disease induced by anti-ERBB2 antibody-drug conjugates: a review. JAMA Oncol 7:1873–1881
64. Vahid B, Marik PE (2008) Pulmonary complications of novel antineoplastic agents for solid tumors. Chest 133:528–538
65. Miller S, McNutt L, McCann MA, McCorry N (2014) Use of corticosteroids for anorexia in palliative medicine: a systematic review. J Palliat Med 17:482–485
66. Yennurajalingam S, Bruera E (2014) Role of corticosteroids for fatigue in advanced incurable cancer: is it a “wonder drug” or “deal with the devil”. Curr Opin Support Palliat Care 8: 346–351
67. Sturdza A, Millar BA, Bana N, Laperriere N, Pond G, Wong RKS, Bezjak A (2008) The use and toxicity of steroids in the management of patients with brain metastases. Support Care Cancer 16:1041–1048
68. Harding T, Baughn L, Kumar S et al (2019) The future of myeloma precision medicine: integrating the compendium of known drug resistance mechanisms with emerging tumor profiling technologies. Leukemia 33:283–883
69. Ramamoorthy S, Cidlowski J (2016) Corticosteroids: mechanisms of action in health and disease. Rheum Dis Clin N Am 42:15–31
70. Rasche L, Kortüm K, Raab M, Weinhold N (2019) The impact of tumor heterogeneity on diagnostics and novel therapeutic strategies in multiple myeloma. Int J Mol Sci 20:1–14
71. Clarisse D, Offner F, De Boscher K (2020) Latest perspectives on glucocorticoid-induced apoptosis and resistance in lymphoid malignancies. Biochim Biophys Acta Rev Cancer 1874(2):188430

Index

A

B

C

D

E

Carlos Barreiro and José-Luis Barredo (eds.), *Microbial Steroids: Methods and Protocols*, Methods in Molecular Biology, vol. 2704, https://doi.org/10.1007/978-1-0716-3385-4,

MIX
Papier aus verantwortungsvollen Quellen
Paper from responsible sources
FSC® C105338

If you have any concerns about our products,
you can contact us on
ProductSafety@springernature.com

In case Publisher is established outside the EU,
the EU authorized representative is:
Springer Nature Customer Service Center GmbH
Europaplatz 3, 69115 Heidelberg, Germany

Printed by Libri Plureos GmbH
in Hamburg, Germany